# LIFEScience

## Student Edition

purposeful design®
p u b l i c a t i o n s

Colorado Springs, Colorado

# Development Team

| | |
|---|---|
| *Vice President for Purposeful Design Publications* | Steven Babbitt |
| *Directors for Textbook Development* | Don Hulin<br>Lisa Wood |
| *Editorial Team* | Lindsey Duncan<br>Janice Giles<br>Julie Holmquist<br>Macki Jones<br>Adelle Moxness |
| *Design Team* | Claire Coleman<br>Steve Learned |
| *Art Illustrations* | Aline Heiser |

*Purposeful Design Publications is grateful to Christian Schools International for the contributions they made to the original content of the Purposeful Design Life Science course.*

# LIFEScience

### Student Edition

## Purposeful Design

Purposeful Design Publications is the publishing division of the Association of Christian Schools International (ACSI) and is committed to the ministry of Christian school education, to enable Christian educators and schools worldwide to effectively prepare students for life. As the publisher of textbooks, trade books, and other educational resources within ACSI, Purposeful Design Publications strives to produce biblically sound materials that reflect Christian scholarship and stewardship and that address the identified needs of Christian schools around the world.

References to books, computer software, and other ancillary resources in this series are not endorsements by ACSI. These materials were selected to provide teachers with additional resources appropriate to the concepts being taught and to promote student understanding and enjoyment.

Purposeful Design Publications
*A Division of ACSI*
PO Box 65130 • Colorado Springs, CO • 80962-5130
Customer Service Department: 800/367-0798
Website: www.purposefuldesign.com

# Table of Contents

Life

UNIT
1

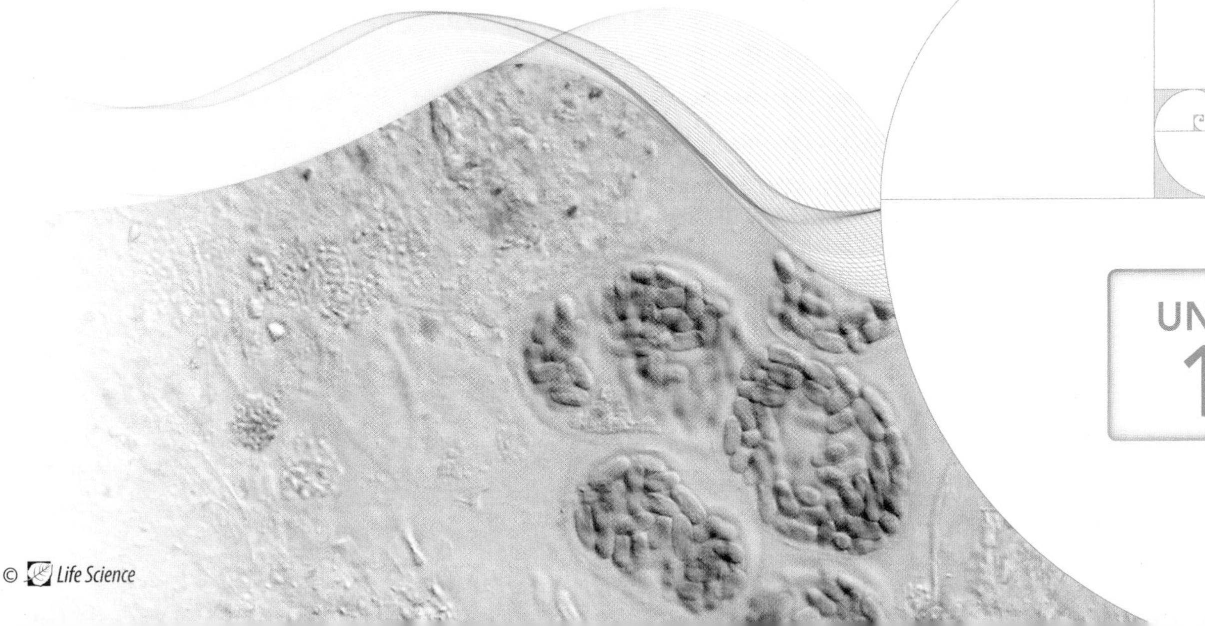

**Viruses, Bacteria, Archaea, Fungi, and Protists**

Plants

UNIT
3

UNIT
4

**Animals**

**Human Body**

UNIT
5

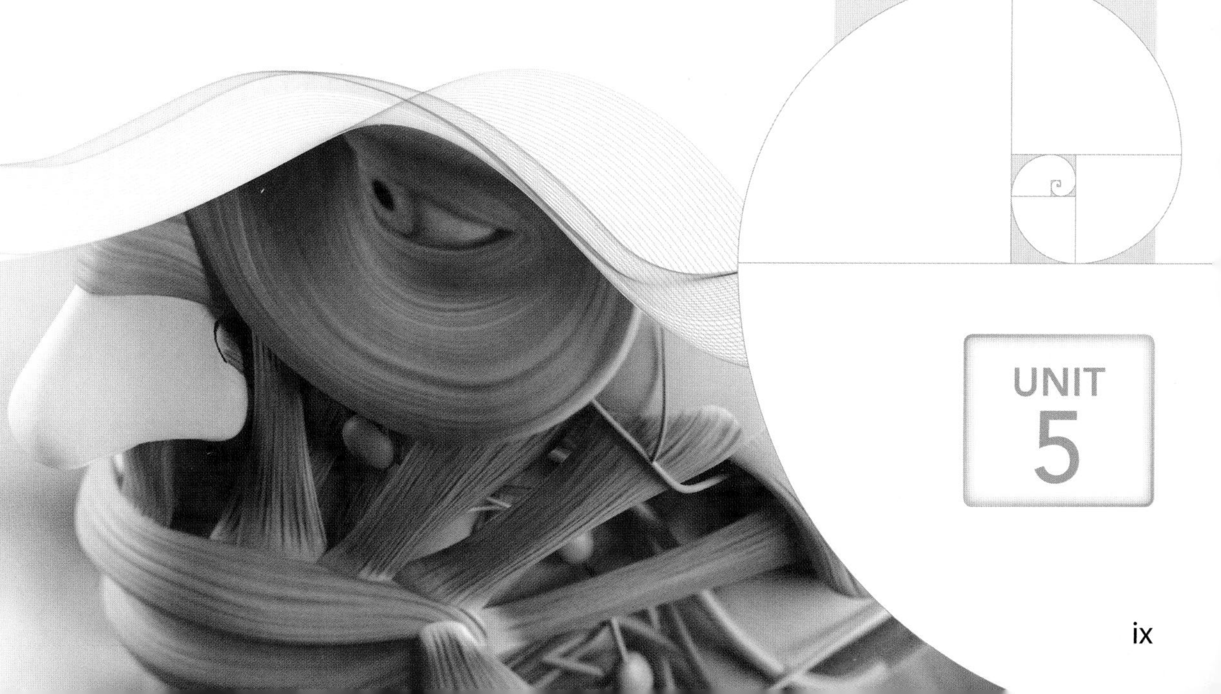

# UNIT
## 6

**Genetics and Heredtiy**

## Chapter 1: *Genetics*

## Chapter 2: *Application of Genetics*

Ecology

UNIT
7

# The Scientific Method

## Background

The word *science* comes from the Latin word *scire*, which means "to know." A scientist is someone who does science—a person who wants to understand the natural world and all of its complexities.

Even after scientific experiments, many scientific hypotheses are disputed within the scientific community; all are subject to revisions and changes as new data and experimental evidence are collected. Theories and laws, however, are supported by a substantial number of experiments and observations. Ideally, scientists are open to constructive criticism about their hypotheses. When scientists communicate in scientific journals or at conventions, they work toward finding new evidence and creating solutions to resolve conflicting interpretations and viewpoints.

The scientific method is an orderly, systematic approach to solving a problem or answering a question. Although different scientific endeavors require different approaches and steps, the following terminology and steps are often used. You need to become familiar with the vocabulary of science as well as realize that scientists understand that the scientific method is not a rigid set of steps that is always followed. Various orders are possible, and sometimes steps are eliminated or repeated.

## Vocabulary

**control**—the sample in an experiment in which the variables are kept at a base level

**hypothesis**—a prediction of what you think will happen and which can be tested to see if it is true

**inference**—an educated guess based on observation

**observation**—something noticed through the senses

**scientific law**—a generalization based on observations that describe the ways an object behaves under specific conditions

**scientific method**—the series of steps that scientists follow when they investigate problems or try to answer questions

**theory**—an explanation of the scientific laws

**variable**—a changeable factor that could influence an experiment's outcome

## Scientific Method

**1. Identify or define the problem.**

2. **Make a hypothesis**. Not all predictions are hypotheses; they must be measurable. For example, "If the rate of fermentation is related to temperature, then increasing the temperature will increase gas production" is a hypothesis. On the other hand, "If yeast is heated, then more gas will be produced" is not a hypothesis because it offers no proposition to test. It does not show a relationship or suggest variables. A hypothesis may also be a question ("Does temperature affect the fermentation of yeast?") or a conditional statement ("Temperature may affect the fermentation of yeast").

3. **Experiment, controlling the variables**. Scientists repeat and refine the experiment based on the findings. They will not allow a hypothesis to remain untested; it must be tested and shown valid.

4. **Make observations and record the results**. Most scientists accurately record their observations and measurements in a journal or on a computer.

5. **Make inferences and conclusions**. A conclusion is a statement about what a scientist has learned and whether or not the scientist's hypothesis was supported. Scientists often learn as much from an incorrect hypothesis as from a correct one. You may feel pressure to have your experiment show that your "guess" was correct, but you need to understand that sometimes scientists arrive at completely unexpected conclusions and that an incorrect hypothesis is never considered a failure. For example, sometimes an unexpected conclusion raises new questions that will lead to another experiment and conclusion. The purpose of science is not to prove what is already known; the purpose is to increase the understanding of God's world!

6. **Apply the findings**. Scientists add their conclusions to their understanding of the natural and technical universe. To do so with wisdom and stewardship benefits all creation.

# Measurements

## Background

The metric system is a universal system of measurements that serves as a standard for scientific research throughout the world and in many countries for everyday use. It is helpful to become more familiar with the metric system and have practice estimating measurements. Throughout this course, the metric system is used predominantly.

Since 1899, international conferences have been held to standardize the metric system. In 1960, the 11th International Conference of Weights and Measures substantially changed the system, renaming it the International System of Units (abbreviated SI).

Using the SI system of measurement has several advantages. It is based on standards that have been recognized by the international science community for use in the sciences and commerce, making the system universally adopted. All of the unit conversions used in the metric system are based on the number 10, and only decimal multiples of the basic units of length, volume, and mass are employed. The metric system is also used by all of the major countries of the world except the United States.

Although students in the United States are more familiar with the Fahrenheit scale, the Celsius (Centigrade) scale is much more common around the world. Scientists use the Celsius scale or the Kelvin scale, which is an absolute scale. The Fahrenheit scale assigns a temperature of 32° for the freezing point of water and 212° for the boiling point of water. The Celsius scale standardized the temperature scale by assigning a value of 0° for the freezing point of water and 100° for the boiling point of water. The size of the divisions used in the Fahrenheit and Celsius scales are not equal. The formula used to convert Celsius to Fahrenheit is $F = \frac{9}{5}C + 32$. The formula for converting Fahrenheit to Celsius is $C = \frac{5}{9}(F - 32)$.

One way to record these measurements is on graphs. A graph is a pictorial representation of statistical data or of relationships between variables. While graphs can serve a predictive function because they show general tendencies in the quantitative behavior of data, as approximations they are sometimes inaccurate and misleading. Most graphs include two axes. The horizontal axis represents the independent variables, and the vertical axis represents the dependent variables. In line graphs (the most common type), the horizontal axis often represents time. Bar graphs can be used to depict the relationship between two nontemporal numerical values. Such information can also be expressed in a circular graph (pie graph) that illustrates the part-to-whole relationship. The size of each sector is directly proportional to the percentage of the whole it represents; this graph is often used to emphasize proportions.

Scientific measurements require specific words to be used in descriptions so the language of science is understood worldwide. You need to have a working knowledge of the following terms:

### Vocabulary

**Celsius scale**—a temperature scale in which 0° represents the freezing point of water and 100° represents the boiling point of water

**gram**—the standard unit for mass in the metric system

**liter**—the standard unit for the volume of liquid in the metric system

**meter**—the standard unit for length in the metric system

**metric system**—a universal system of measurement based on the number 10 used by scientists around the world

# Laboratory Safety

## Background

Each lab has specific safety needs, and you must focus on the teacher's instructions. Take only your notebook, student text, and a pen or pencil to the laboratory area. Leave all other items at your desk. Read and review your teacher's procedures to be certain that you understand the instructions.

## Safety Rules

Some standard safety rules include the following:

- Wash your hands before and after any hands-on activity or experiment. Wear protective gear appropriate to the activity. Some activities require goggles or gloves. Sanitize safety goggles after each use.
- Report any spills or breakage to your teacher immediately.
- Store equipment and materials in a safe location and make sure that you return them to their designated places after use. Always test and check the materials prior to use.
- Do not eat or handle supplies unless instructed to do so.
- Be aware of allergies that you may have to science activity materials and take any necessary precaution.
- Carry only one jar at a time.
- Never lean on aquarium glass.
- Never prop a meterstick over your shoulder.
- Carry microscopes with two hands.
- Never eat or drink during a science activity.
- Remove loose or bulky clothing to reduce the chance of spreading fire or knocking over equipment.
- Keep equipment (especially breakable or expensive items such as glassware or microscopes) away from the edge of work surfaces.
- Wear safety goggles, especially for activities involving potentially harmful chemicals, glassware, or heat. (When your teacher is wearing goggles, you must wear goggles.)
- Use plastic equipment if possible. However, if glassware is used and does break, back away from the accident and do not touch the shattered glass. Designate someone to retrieve a dustpan and broom and alert the teacher.
- Know basic chemical safety. Many chemicals are labeled with three hazards—health, flammability, and reactivity—rated on a scale of 0 to 4. A 0 rating indicates that the substance is not a threat for this hazard, and a 4 means it is a substantial threat.
- Be aware of two common safety symbols—flames (signifying flammability) and a test tube being poured on a hand (signifying a corrosive substance). These symbols are

primarily for storage purposes, but they also warn the user of these two dangers. Different chemicals have different disposal guidelines. Follow your teacher's directions regarding how to properly dispose of the chemicals and materials you use.

### Safety Procedures

- Be a good example. Follow the safety rules at all times.
- Inspect the lab area to look for possible hazards before you start an experiment.
- Inspect your equipment.
- Report hazards to your teacher immediately.
- Know the location of your lab's fire bucket, fire blanket, fire extinguisher, and safety shower, if available.

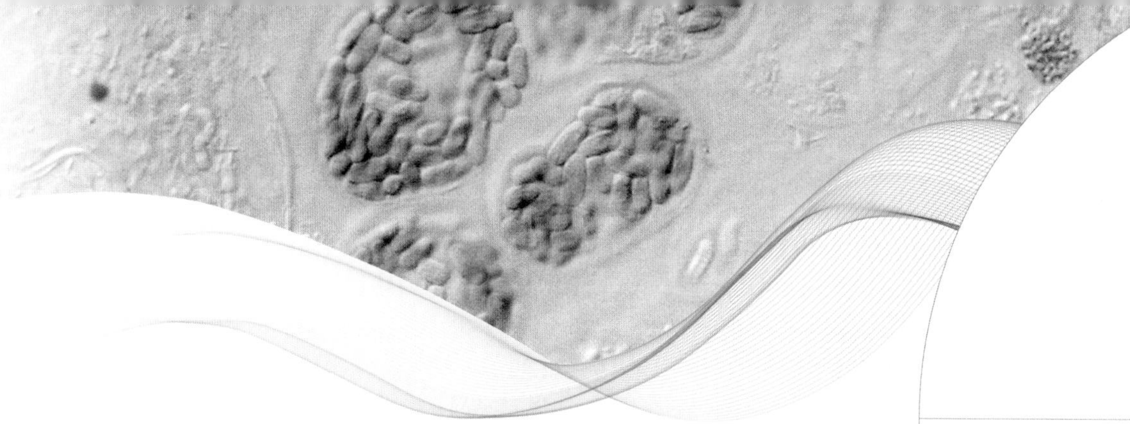

Chapter 1: *Introduction to Life Science*
Chapter 2: *Cells*
Chapter 3: *Taxonomy*

*Vocabulary*

archaea
asexual
  reproduction
bacteria
binary fission
binomial
  nomenclature
botanist
cell theory
cellular respiration
chromosome
diffusion
DNA
endoplasmic
  reticulum

eukaryote
fungi
gene
Golgi apparatus
homeostasis
lichen
life science
lipid
lysosome
meiosis
metabolism
mitochondrion
mitosis
nucleic acid
nucleotide

organic compound
osmosis
prokaryote
protist
replication
ribosome
sexual reproduction
species
taxonomy
technology
vacuole
zoologist

Life

*Key Ideas*

- Systems, order, and
  organization
- Evidence, models, and
  explanation
- Change, constancy, and
  measurement
- Abilities necessary to do
  scientific inquiry
- Understandings about
  science and technology

- Structure and function in
  living systems
- Reproduction and heredity
- Personal health
- Science as a human
  endeavor
- Nature of science
- History of science

SCRIPTURE

[Solomon] spoke about
plant life, from the cedar of
Lebanon to the hyssop that
grows out of walls. He also
spoke about animals and
birds, reptiles and fish. From
all nations people came to
listen to Solomon's wisdom,
sent by all the kings of the
world, who had heard of his
wisdom.

1 Kings 4:33–34

### OBJECTIVES

- Defend why it is important for Christians to pursue scientific inquiries.
- Explain how technology has both enriched and harmed God's creation.

### VOCABULARY

- **life science** the study of living things
- **technology** the application of science

God has blessed Earth with a vast variety of life. Living things, called *organisms*, are sometimes too small to be seen—thousands can fit inside the period at the end of this sentence. One drop of water can be home to an array of living things, each with its own characteristics. Whales, Earth's largest creatures, can have a mass up to 180,000 kg, but they feed on plankton, some of Earth's tiniest creatures. The rain forests that stretch across the equator teem with beautiful parrots, 10 m pythons, orchids that mimic bees and spiders, moths with a 22 cm wingspan, and trees with huge trunks. Living things thrive in every corner of the world. Even in some of the coldest and hottest places, amazing creatures such as polar bears, penguins, chameleons, and bacteria are found.

The human body is also one of God's astounding creations. It is made of trillions of cells and dozens of chemicals, and it can function continuously for many years. No toy, appliance, or car can do that!

We interact with living things every day, but we often do not pay much attention to them. At other times, though, we are excited by living things—a nest with squirming baby robins in a tree outside the window, dolphins or whales leaping from the water, a crocus poking through the snow in early spring, or a flutter of butterflies

The humpback whale can consume 1,360 kg of food per day.

shimmering by. Is it enough for Christians to just enjoy the beauty of living things? Why should Christians study living things? Why is studying life science important?

Some Christians are suspicious of science. They believe that science tries to do away with God or that people will depend on scientific solutions to their problems instead of on God. Other Christians do not trust certain aspects of science because some scientific theories, such as that of evolution, disagree with a literal interpretation of Genesis. But Christians should embrace science. Science is the systematic study of God's creation using methods based on observation and experimentation. In fact, God calls many Christians to be scientists. Science helps Christians better understand God and His world by observing, experimenting, and finding relationships among the things of the cosmos. Scientists strive to understand more than what is already known about this complex and amazing world, a gift from God.

**Life science**, the study of living things, can help people better understand and appreciate God and His creation. Learning about the wonderful pattern of genetic information stored in a DNA molecule or the intricate connections among coral reef creatures can also deepen a Christian's reverence and awe for God and can help glorify Him.

Another important reason to study life science is so that people can apply the knowledge they gain about living things to everyday life. The application of science is called **technology**. For example, studying the interactions between fungi and bacteria led to the discovery of penicillin, the first antibiotic. In this case, applying the knowledge gained from studying living things has saved many lives. In North America the polio vaccine, penicillin, and other antibiotics are taken for

There are about 35 species of spider orchids that are found in southeastern North America, the West Indies, and parts of Central and South America.

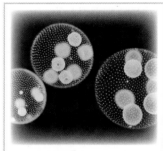

*Volvox aureus* is a type of phytoplankton, also known as *microalgae*.

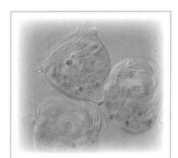

*Campanella umbellaria* is a unicellular organism that exhibits the creative and unique hand of God.

## 🔷 BIBLE CONNECTION

Although the Bible is not a science textbook, it tells us about God's involvement with the natural world. God created the natural world, and He sustains all natural processes (Genesis 1:1; Job 38:3–39:30). God sees His creation as good (Genesis 1:4, 10, 12, 18, 21, 25, 31). God wanted variety and bounty in His creation (Genesis 1:21–22), and He cares about all of His creatures and the earth (Genesis 6:19–20; Psalm 65:9–13). God established His covenant with Noah and every living thing on the Earth (Genesis 9:8–11). And one day, God will set free all of His creation (Romans 8:19–21).

Studying the genetics of plants has led to greater crop yields to feed more people.

granted, but in many parts of the world people still die from diseases that could be prevented by these medicines. Studying the genetics of plants has led to greater crop yields to feed more people. Phosphates were widely used in laundry detergents, but studying their harmful effects on aquatic life has led to phosphate-free detergents that do not kill water plants and fish. Other technologies that have sprung from the study of living things include everything from the formulation of shampoos, aspirin, artificial flavorings, fertilizers, and processed foods, to responsible fishing methods. Technology enriches peoples' lives and helps them care for God's world.

The purpose of technology is to use science to solve problems, but some technologies do not meet this goal. Some technologies damage God's world. Certain chemicals used in manufactured products break down the ozone layer, the region of the atmosphere that protects Earth from the sun's harmful rays. Some technologies that involve burning fossil fuels pollute the atmosphere, which some scientists believe leads to global warming. Drilling for oil can seriously damage the environment and its living things. Other technologies are abused. For

example, nuclear technology can be used to heat homes or to destroy life. Biological technology can cure diseases or deliberately infect people with diseases.

Science is a powerful tool that, like any other tool, can be used wisely or unwisely. The root of the world's problems is human depravity, or corruption, so science cannot solve all world problems. Science and technology themselves are not good or evil. The people who apply science and technology determine whether their results are helpful or harmful. People must develop and use science and technology to God's glory.

Studying living things can help Christians learn how to be wise stewards of God's marvelous gift of creation. By protecting the environment and using technology wisely, Christians can fulfill God's desire for His children to care for His world as it waits to be set free.

## CHALLENGE

**Make the World a Better Place**

Think of a problem that you could help solve through your study of science. What kind of technology could you develop to help solve this problem? Study the problem, and apply your knowledge to suggest a solution.

## LESSON REVIEW

1. What would you say to a person who told you that Christians should not be too interested in science?
2. What are some ways in which technology has benefited the world?
3. What are some harmful effects or products of technology? What can be done to solve some of these problems?
4. What determines whether science is used for good or bad purposes?

NATIONAL WILDLIFE REFUGE

Ever since God created Adam and Eve, people have asked questions about the natural world: How are living things related to each other? How are other forms of life different from humans? What is life? What is death? These are hard questions. Biologists still have not agreed on a definition for the word *life*. Most living things have many things in common, but scientists cannot agree on whether some things—such as viruses, for example—are even alive. People study life science to find answers for some of these challenging questions and to better understand God, the Creator.

The living things that God created are amazing! They come in a vast array of sizes—from a microscopic one-celled, or unicellular, bacterium to a gigantic whale. And they are everywhere! Even the air you breathe is inhabited by living things, from spores to bacteria to airborne spiders. God created each living thing and provided for its needs!

When you see a frog leap out of a mud puddle, you do not bother to wonder whether or not the mud "made" the frog. You know that the frog came from other frogs. But most early scientists believed that organisms such as worms, beetles, frogs, and salamanders came from dust or mud because that is where

The fact that frogs come from frogs may seem obvious today, but many early scientists thought that such creatures were products of their environments.

they saw these creatures. They thought that rodents came from moist grains and that plant lice formed in dew drops. The theory that living things can develop from nonliving matter is called *spontaneous generation*.

In the 17th century, Francesco Redi conducted a famous experiment that disproved this theory. He put decaying meat in jars. He kept some of the jars open and covered the others with a fine mesh. Maggots grew on the meat in the open jars because flies were able to lay their eggs on the meat. The meat in the mesh-covered jars produced no maggots. This experiment proved that maggots do not come from decaying meat but from eggs laid on the meat by flies.

By the 19th century, most scientists rejected the idea that living things could develop by spontaneous generation. Surprisingly, though, improvements in microscopes caused some scientists to revisit that theory. Biologists noticed that when they placed decaying substances, such as spoiled food, in closed containers and then warmed them, they could see "live beasts" under the microscope lens. They concluded that very simple organisms could develop spontaneously from water and other substances.

In the 1860s, the Academy of Sciences in Paris offered a reward to anyone who could settle the matter. Louis Pasteur won the prize in 1864 for his experiment demonstrating that microorganisms are not spontaneously generated. Instead they are carried through the air and onto other organisms or substances where they begin to grow. Pasteur put broth in glassware with swan-shaped necks and boiled the broth

### Redi's Experiment

Flies could enter the open jars to lay eggs on the meat. The netting kept the flies out of the jar, so no maggots appeared on the meat. Maggots did grow on the mesh.

## BIOGRAPHY

Louis Pasteur (1822–1895) was a French chemist whose experiments with bacteria contradicted the theory of spontaneous generation. Pasteur's experiments led to the theory that germs infect people. His work on distilling wine, vinegar, and beer led to the development of pasteurization—the heat-treatment process that kills harmful microorganisms in foods and beverages such as milk.

Pasteur also developed a solution for the control of silkworm disease, and he studied chicken cholera. In addition, he developed a vaccine against rabies in 1885. In 1888, Pasteur founded the Pasteur Institute in Paris to study infectious diseases. Pasteur continued to work on rabies, to teach, and to research contagious diseases until his death.

## Pasteur's Experiment

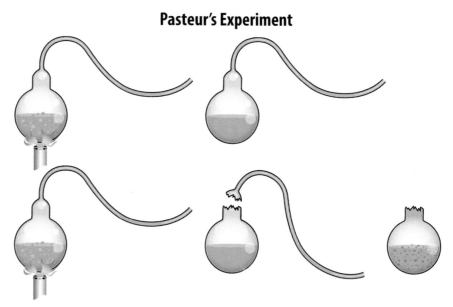

Microorganisms in the air could not travel through the long, curved tube to the broth. When the tube was broken, microorganisms could easily reach the broth.

to kill any live microorganisms. He then waited to see if microorganisms would spontaneously generate. But nothing happened because the long, curved necks of the glassware allowed air and oxygen into the glassware but prevented microorganisms from entering. When Pasteur broke the curved necks of the glassware, the broth soon swarmed with microorganisms, proving that such organisms do not spontaneously generate. "Life is a germ, and a germ is life," Pasteur proclaimed. "Never will the doctrine of spontaneous generation recover from the mortal blow of this simple experiment."

Despite Pasteur's prediction, some 20th century scientists returned to the idea of spontaneous generation. In 1922, Russian scientist Aleksandr I. Oparin hypothesized that life could have spontaneously arisen during early earth conditions using heat energy from volcanoes, electrical energy from lightning, certain elements, and water vapor. Most biologists ignored Oparin because Pasteur's experiment demonstrated that spontaneous generation could not occur. In the 1950s, American scientist Stanley Miller tested Oparin's hypothesis. He mixed hydrogen, methane, and ammonia with water vapor and passed electricity through the mixture. The experiment yielded amino acids and other substances found in living things. He

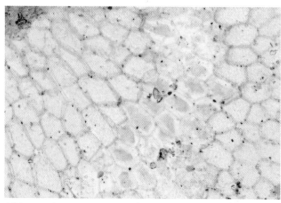

All living things are made of cells, the building blocks of life.

produced the chemicals and organic molecules needed to support life, but no living things developed. Neither Oparin's theory nor Miller's experiment could produce life.

Although biologists sometimes have trouble drawing the line between living and nonliving things, they do know that all living things have certain similarities. Certain characteristics distinguish between living and nonliving things; a nonliving object may have one or more of these characteristics, but only a living thing has all of the following characteristics:

All living things reproduce, and they reproduce after their own kind.

**All living things are made of cells, the building blocks of life.** Cells not only build living things, but they also perform the life functions that living things need to do in order to thrive.

**All living things reproduce, and they reproduce after their own kind.** Gorillas always reproduce gorillas, hollyhocks always reproduce hollyhocks, and bacteria always reproduce bacteria.

All living things reproduce, but they do this in two different ways. Most plants and animals reproduce sexually. **Sexual reproduction** involves two parents and the offspring have characteristics from both parents. Plants and animals reproduce sexually in different ways. For example, when some species of fish mate, they gather into large schools. The female fish release millions of eggs into the water, and then the males release sperm over the eggs. An earthworm has both male and female reproductive organs but must still join with another earthworm to reproduce. Flowering plants such as roses, maple trees, and poison ivy have male and female parts on their flowers. When their flowers are pollinated, sperm and egg cells unite and grow into seeds.

All living things grow and develop.

Unlike sexual reproduction, **asexual reproduction** involves only one parent and the offspring is identical to the parent. For example, bacteria reproduce by dividing into two parts, and most yeast reproduce by forming buds that break off.

**All living things grow and develop.** Growing and developing is not limited to getting bigger. It also often involves becoming more complex. For example, a newly conceived child is made of only a couple of cells, which divide and develop into a baby that is ready to be born. After the child is born, it continues to grow and develop throughout life. Bacteria grow in size by adding to their cell membrane and cytoplasm. Trees germinate from seeds, which eventually sprout stems that grow many meters tall and develop branches, leaves, and fruit. Many insects start as eggs, hatch into larvae, and form cocoons during their pupa stage, and emerge as an adult with a very different form.

**All living things respond to their environments.** Anything that affects an organism's activity is called *a stimulus*. The organism's response to a stimulus can be an action, a movement, or a behavior change. Maple leaves turn red in response to changing temperatures and shorter daylight hours. Cats jump at the sound of a dog barking. Your mouth waters when you smell chocolate cake baking. The response to a stimulus is designed to help an organism survive, such as deer running at the sight of a wolf, plant roots growing down in response to gravity, and plant stems growing up in response to light.

**All living things use energy.** God created living things so that they could obtain energy through chemical activities. These activities include such things as making food, breaking down food, moving materials in and out of cells, and building cells. Some of these activities combine simple substances into more complex substances, and some of these activities break the complex substances down into simpler substances that the organism can

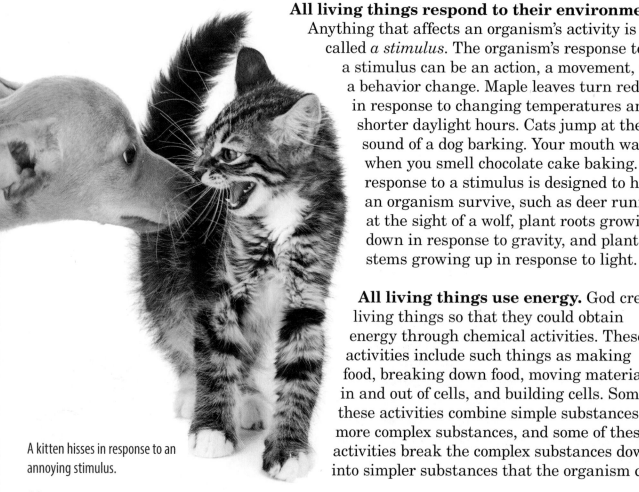

A kitten hisses in response to an annoying stimulus.

use. The sum of the chemical activities taking place inside a living cell or organism is called **metabolism**.

Before metabolism can begin, all living things must take in food or produce their own food. Most animals obtain food for metabolism by eating it, but green plants and some other organisms, such as algae and certain bacteria, produce their own food. They use water, carbon dioxide, and energy from the sun to make their own food in a process called *photosynthesis*. Many one-celled organisms simply engulf their food. Most fungi secrete enzymes to break down foods into small molecules that they can absorb. Other organisms, such as sponges, eat by filtering food from the water.

Once an organism has eaten food, metabolism can begin. The first step in metabolism is digestion. Digestion breaks down the food into simpler substances. The organism uses some of these substances to build more complex substances needed for growth. It uses others to store energy needed for its activities.

Metabolism also includes respiration, the process in which living things use gases to produce energy. The respiration of most living things includes oxygen. Animals combine the energy from oxygen with the energy from food. Humans, for example, get energy from the combination of digested food and oxygen.

 **TRY THIS**

**Exhale!**
Carbon dioxide is a by-product of respiration. You give off carbon dioxide when you breathe. The pH indicator phenol red turns yellow in the presence of carbon dioxide.

Put 100 mL of water in a 150 mL flask. Add four drops of phenol red indicator to the flask. Place a drinking straw into the flask. Wrap a paper towel around the opening of the flask. Inhale through your nose and exhale through the straw. Record how long it takes for the phenol red solution to turn yellow. Run in place for three minutes and then repeat the process. Describe any change.

Sweating is an example of homeostasis.

Digestion and respiration produce by-products that living organisms do not need. These unneeded products must be released. Humans release carbon dioxide when they breathe out. Plants release oxygen and water through photosynthesis. The process of getting rid of unneeded materials is called *excretion*. These materials, once excreted, are used by other organisms. People breathe in the oxygen that the plants release, and the plants use the carbon dioxide that people release. This oxygen cycle is one of the recycling plans God built into creation.

To carry out these chemical activities, an organism must maintain a stable internal environment. The ability to do this is called **homeostasis.** Living things display this ability in many ways. For example, a person's body temperature is about 37°C at all times. When you get cold, your muscles twitch to generate heat, which is what makes you shiver. When you get too hot, you sweat to cool off. A dog pants when it is hot because it cannot sweat. Another example of homeostasis is your body's ability to maintain a stable amount of sugar in your blood even when you have not eaten for a while. This ability is important because low blood sugar can sometimes cause dizziness, confusion, or even a coma.

You have just learned that living things have certain characteristics in common. Even though all plants, animals,

## TRY THIS

**Homeostasis**

Take your temperature. Then run in place (or do some other form of exercise) until you break out in a sweat. Take your temperature again. Is it higher, or is it the same? Why?

## Cellular Respiration

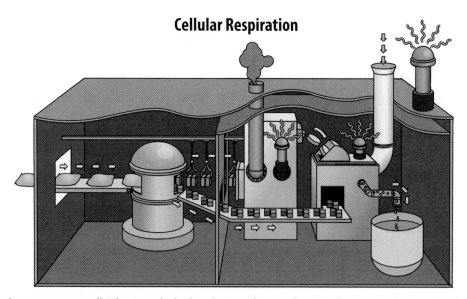

As sugar enters a cell, it begins to be broken down in the cytoplasm. As the sugar and oxygen reach the mitochondria, they are broken down further and they release carbon dioxide, water, and energy.

## HISTORY

**Spontaneous Generation**

- **384–322 BC:** Aristotle and early philosophers accept that organisms like worms, beetles, frogs, and salamanders arise spontaneously from dust or mud.
- **1600s:** Francesco Redi contradicts the spontaneous generation of large organisms with his decaying meat experiment.
- **1800s:** Few scientists accept the spontaneous generation of large organisms, but many scientists accept that microorganisms can develop through spontaneous generation.
- **1860s:** The Academy of Sciences in Paris offers a reward for an experiment that can settle the spontaneous generation debate.
- **1864:** Louis Pasteur's experiment using jars with swan-shaped necks shows that microorganisms are carried through the air and do not arise spontaneously from broth or other media.
- **1922:** Russian scientist Aleksandr I. Oparin hypothesizes that life could spontaneously arise during early Earth conditions.
- **1950s:** Stanley Miller recreates the climatic conditions of early Earth, and in his experiment, simple amino acids and organic molecules are formed from hydrogen, water vapor, methane, and ammonia.
- **2000:** Scientists have never made life or DNA (the genetic origins of life) in the lab; however, there is widespread acceptance in the scientific community that the "primordial soup" present in early Earth spontaneously gave rise to living cells.

and other organisms have these general things in common, the specific processes displayed by organisms are very different. For example, an oak tree responds to the stimulus of a moth chewing on its leaves by producing bad-tasting chemicals in its other leaves. A moth is not a stimulus to you, however. On the other hand, your mouth might water when you smell brownies baking, but a good smell is not a stimulus to an oak tree. Trees receive their energy from the sun, but we receive ours by eating plants and animals. Fish move by swimming, but we move by walking. Each living thing has its own unique, God-given design.

## LESSON REVIEW

**1.** Give three examples of evidence used in the past to support spontaneous generation.

**2.** What are five characteristics of living things?

**3.** How can reacting to a stimulus help an organism?

**4.** How does a dog maintain homeostasis?

All living things need water.

All organisms have certain needs in common—food, water, energy, gases, and space—and God provided for these needs in a wide variety of ways.

All living things need food. Food provides the energy and raw materials needed for metabolism. The foods that organisms eat are as different as the organisms themselves. Termites thrive on wood, lions eat zebras, frogs eat flies, giant whales filter microscopic plankton from seawater, amoebas feed on bacteria, earthworms ingest soil to consume microorganisms, and plants make their own food through photosynthesis. Different living things need different chemicals and energy, so they obtain foods that contain the chemicals and energy they need. Imagine what would happen to the food supply if all things—fungi, bacteria, fish, humans—ate the same food! God provided a balance in creation by creating all organisms with different food needs.

All living things need water. People can probably live for about three weeks without food, but could survive only a few days without water. Water sustains humans and other living

All living things need food.

things because of its many extraordinary properties. Water dissolves more substances than any other liquid. Blood, which is mostly water, dissolves minerals and nutrients and carries them throughout the body. Sap, which is mostly water, dissolves sugars and nutrients and carries them throughout a tree. Without water, metabolism could not happen. Water is also easily absorbed by plants and animals. Frogs, for example, absorb water through their skin.

All living things need energy. Living things produce energy through metabolism, but how do they use this energy? Organisms use energy in countless ways—a mushroom uses some of its energy to grow, a bird uses some of its energy to build a nest, and a tree uses some of its energy to produce leaves.

The sun is the source of energy for living things. Plants use energy directly from the sun to make their food. Animals do not use energy from the sun in this way, but they do eat plants, gaining the energy from the sun secondhand. Other animals get this energy when they eat plant-eating animals. By eating food like a granola bar, an apple, or a sandwich, people gain energy from the sun in the same way.

All living things need certain gases. Some species of bacteria live on gases such as ammonia, sulfur dioxide, or methane that are poisonous to most other organisms. But most animals and microorganisms need oxygen to survive. Organisms that live on land get this oxygen directly from the air. Organisms that

## TRY THIS

**Needs of a Living Thing**
Choose an organism and observe it in its natural habitat. How are its needs met? How much space does it need?

 **FYI**

**Water-Saving Rat**
All living things need water. But what about animals whose habitat is a desert that receives just 12.5 cm of precipitation (or less) a year and almost all the precipitation comes in the wintertime?

The kangaroo rat is adapted to this type of situation. Its home is a U-shaped burrow or tunnel underground. It comes out only at night, when it is much cooler, to eat. It uses its two front feet to stuff seeds or grass into its mouth. The kangaroo rat produces the water it needs from the food that it ingests and the oxygen that it breathes. And the kangaroo rat has special kidneys that help it dispose of waste without much water. God provided for even the very smallest members in creation!

All living things breathe.

live under water either come up to the surface for oxygen or use the oxygen dissolved in the water.

When organisms use oxygen during respiration, they exhale carbon dioxide, the by-product of respiration, back into air or water. Plants use this carbon dioxide for photosynthesis, and then they release oxygen, the by-product of photosynthesis, back into the air. This process is a recycling plan that continues to provide Earth with oxygen and carbon dioxide.

All living things need space. The place in which an organism lives is its habitat. Have you ever wondered what chirping birds are saying to each other? Often they are saying, "Go away! Back off!" Birds and other creatures defend their territory because all living things need space. Any given amount of space offers only a certain amount of food, water, and shelter, so only a certain number of living things can flourish in that space. Squirrels chase each other, coyotes howl, male peacocks flaunt their feathers—these are just some of the ways that organisms defend their space.

All organisms need enough space to thrive, but the amount of space that a group of organisms needs differs greatly. Many

All living things need space. Some creatures even fight each other for it!

microorganisms can live in a drop of water the size of a grain of sand, but a pair of jaguars requires 400 hectares of healthy rain forest to supply them with enough food. Plants need space too. When you walk in the woods, you will notice that mature trees grow a certain distance apart. Each tree needs its own space for sunlight and water. Many small plants grow only in clearings. They cannot compete with the large trees or their shadows for the space they need for sunlight and water.

People are using more and more natural resources for food and living space. If people are to take good care of God's marvelous creation, it is important to understand that all God's creatures need habitats, food, water, gases, and space to be fruitful and multiply. Otherwise humans will upset the carefully balanced systems that God intended to be a part of His world.

## LESSON REVIEW

**1.** What are the basic needs of all living things? Choose an organism and describe how these basic needs are met.
**2.** What do you think would happen if all organisms' basic needs were met in the same way?
**3.** Why is water so important for life?

All living things are made of cells, but what are cells made of? God used basic substances to build and sustain every part of creation, both living and nonliving. These basic substances are called *elements*. An element is a pure substance that cannot be broken down into simpler substances by physical or chemical means. For example, water can be broken down into hydrogen and oxygen, and the elements hydrogen and oxygen are in their simplest forms. Today scientists recognize over 100 elements. Ninety-eight of these elements occur naturally.

When elements chemically join together, they make a compound. A compound is a pure substance consisting of atoms of two or more elements that are chemically combined in a fixed proportion. Water is a compound composed of oxygen and hydrogen. A molecule—a particle consisting of two or more atoms chemically tied together—is the smallest part of a compound.

Millions of substances in the world are mostly compounds. These compounds can be divided into two categories: inorganic compounds and organic compounds. A compound that contains carbon is called an **organic compound**. The word *organic* refers to something that is or was alive. These compounds are

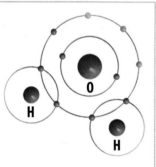

Water is a compound of two elements. Two hydrogen atoms bond to one oxygen atom to form a molecule of water.

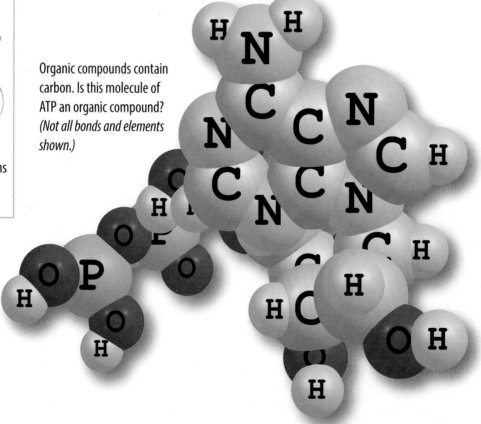

Organic compounds contain carbon. Is this molecule of ATP an organic compound? *(Not all bonds and elements shown.)*

referred to as *organic compounds* because they were once thought to be found only in living things. Inorganic compounds do not contain carbon.

Living things contain over three million organic compounds. Living things are not only made of organic compounds but they also depend on organic compounds to live because most food is made of organic compounds. Organic compounds that are common to living things are carbohydrates, lipids, proteins, and nucleic acids.

Carbohydrates are organic compounds made from one or more sugar molecules. Sugars are made from carbon, hydrogen, and oxygen. Carbohydrates provide and store energy. Ideally, people should get their energy from foods with carbohydrates. There are two types of carbohydrates: simple and complex. Simple carbohydrates are made of one or two sugar molecules linked together. Table sugar and the sugar in fruit are examples of simple carbohydrates. Starches are complex carbohydrates made of many sugar molecules linked together. Pasta, bread, rice, and potatoes are high in starch. Dietary fiber is also a complex carbohydrate. Most vegetables are a good source of dietary fiber.

Candy, soft drinks, and other high-sugar foods are considered junk food because even though they are high in carbohydrates

## TRY THIS

**Design a Meal**
Research the nutritional needs of someone your age and how those needs can be met. Plan a breakfast, lunch, and supper that would meet those needs. Do your usual meals meet those needs?

Carbohydrates

---

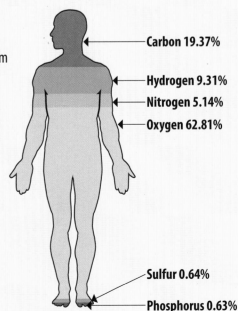

## FYI

**Major Elements**
Most organic compounds contain a combination of six major elements: carbon, hydrogen, nitrogen, oxygen, phosphorus, and sulfur. (The acronym CHNOPS can help you remember these six elements.)

| Element | Human | Alfalfa | Bacterium |
|---|---|---|---|
| Carbon | 19.37% | 11.34% | 12.14% |
| Hydrogen | 9.31% | 8.72% | 9.94% |
| Nitrogen | 5.14% | 0.83% | 3.04% |
| Oxygen | 62.81% | 77.90% | 73.68% |
| Sulfur | 0.64% | 0.10% | 0.32% |
| Phosphorus | 0.63% | 0.71% | 0.60% |
| **Total:** | **97.90%** | **99.60%** | **99.72%** |

- Carbon 19.37%
- Hydrogen 9.31%
- Nitrogen 5.14%
- Oxygen 62.81%
- Sulfur 0.64%
- Phosphorus 0.63%

How much (by mass) of you is carbon, hydrogen, nitrogen, oxygen, phosphorus, and sulfur?

## TRY THIS

**Starchy Food**
Gather some food samples. Use an eyedropper to drop iodine on each sample. Because iodine turns black in the presence of starch, this will indicate which food samples contain starch.

Lipids

(sugar), they provide little or no other nutrients. On the other hand, a potato, which is high in starch, provides not only carbohydrates but also protein, vitamins, iron, and minerals.

The cells in an organism break carbohydrates down into glucose, which is a simple sugar. Glucose provides the energy that the organism needs. Glucose that is not used immediately is stored as starch.

**Lipids** are organic compounds that are fats and oils. They provide and store energy and make up most of the membrane of cells. Most fats are solid at room temperature; most oils are liquid at room temperature. Although the energy gained from carbohydrates is used right away, the body often uses energy from fats or oils as a reserve supply. Like carbohydrates, lipids are made of carbon, hydrogen, and oxygen, but the proportions are different than in carbohydrates. Many people eat more lipids than necessary and as a result are often overweight.

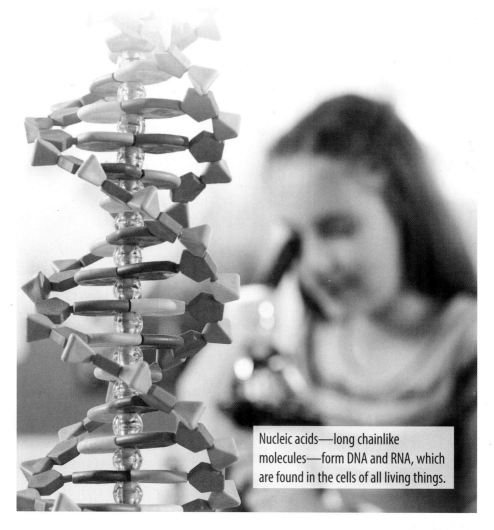

Nucleic acids—long chainlike molecules—form DNA and RNA, which are found in the cells of all living things.

Proteins are organic compounds made of amino acids. Proteins make up many structures of living organisms such as muscles, fingernails, antlers, and feathers. They help build and repair cells and tissues. Proteins also provide some energy, and some proteins carry important materials through the body. For example, hemoglobin, a protein in red blood cells, carries oxygen. Other proteins, such as the yolk of bird or reptile eggs, store food. Foods that are high in protein include fish, meats, eggs, beans, peas, and nuts.

Proteins

Enzymes are special proteins that speed up chemical reactions in an organism or a cell. Enzymes are another important part of God's design to sustain life. For example, without the work of enzymes, metabolism would take place so slowly that humans could not live.

**Nucleic acids** are organic compounds that contain genetic information. DNA and RNA are two important nucleic acids. DNA (deoxyribonucleic acid) contains the information needed to build protein and carries the genetic information about an organism. RNA (ribonucleic acid) reads the genetic messages in the DNA so that proteins can be built from the genetic information. These two nucleic acids are sometimes compared to blueprints.

## LESSON REVIEW
**1.** What are organic compounds?
**2.** What are organisms made of?
**3.** List and describe the four main organic compounds found in living things.

God created the earth like no other place in the solar system, and He blessed the earth with the conditions necessary to support life. He positioned the earth at just the right distance from the sun to create temperatures that are not too hot and not too cold for life. He provided water on the earth. He created Earth with an atmosphere that contained the gases necessary for living things and that allowed the right amount of sunlight. God arranged this complex set of conditions so that organisms could thrive.

God filled the earth with millions of species of living things and despite their differences, they have a common foundation: all are made of cells. Many organisms, such as bacteria, are unicellular. Other organisms, such as beetles, are multicellular. In a multicellular organism, cells divide and organize themselves into tissues, organs and organ systems.

Modern evolutionary scientists use simple tree diagrams to show how they believe life evolved. All animals shown on various branches of a tree are related to one common ancestor shown on the tree's base. This theory suggests that the ancestor is a parent to several different groups of organisms. For example, mammals and birds are shown as evolving from reptiles. There is little scientific evidence to support this theory.

Cells testify to God's creative and sustaining work. As technology for viewing the microscopic world improves, scientists observe that cells are complex systems in which hundreds of structures perform millions of chemical activities each minute. Most cells house tiny bits of genetic material that contain enough information to create a new, complete organism. Cells are not merely boxes of organic molecules or bricks that build more complicated life. Rather, they are like miniature worlds that undergo intricate life processes.

God gave living things a spectacular variety of forms, behaviors, and survival needs that is almost unimaginable. Microscopic organisms live in a drop of water. Some bacteria live in

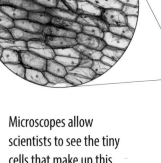

Microscopes allow scientists to see the tiny cells that make up this onion epidermis. Stains are often used to make the cells more distinguishable.

sulfur dioxide and methane, which are poisonous to people. These same bacteria would die if they were exposed to the oxygen you breathe. Redwood trees tower hundreds of meters above the ground. Slime molds cling to the bottom of rotting logs. Aquatic worms thrive far beneath the surface of the sea, surviving tremendous water pressure while they wait for nutrients to drift down to them. Polar bears roam the frozen wastelands. What could these living things possibly have in common? All of them are made of cells.

For most of history, people did not have the technology to see tiny individual cells. Dutch lens makers Hans and Zacharias Jansen invented the microscope sometime between 1590 and 1610. In 1665, English scientist Robert Hooke became the first person to see cells when he used his microscope to look at thin slices of tree cork. He named the empty compartments *cells* because they reminded him of the tiny rooms where monks lived.

In the 1670s, Antonie van Leeuwenhoek, a Dutch microscope builder who was blessed with amazing eyesight and great skill at grinding lenses, discovered thousands of microorganisms living in a single drop of water. He called these organisms *animalcules* because he thought they were animals, but they were actually unicellular organisms. "This was for me, among all the marvels that I have discovered in nature, the most marvelous of all," he wrote. Van Leeuwenhoek was also the first to observe

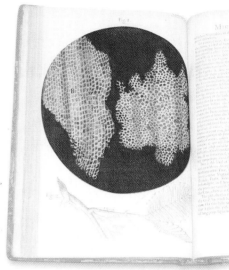

Hooke's drawings of tree cork cell structure

Leeuwenhoek's microscope

**FYI**

**Life Under the Lens**
Galileo Galilei, an Italian astronomer, was one of the first people to look at living things under a lens. Early in the 17th century he looked at an insect through an adapted telescope and saw the amazing geometric patterns of an insect's eyes.

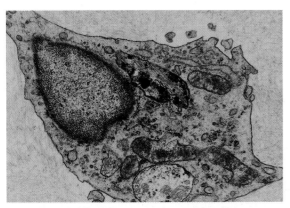

Neuroglia cells like this one provide insulation between nerve cells and help maintain homeostasis.

the even smaller structures within these tiny creatures. He thought that these structures were organs, such as the heart and lungs.

For nearly 200 years after the microscope was invented, scientists learned little more about cells. Lens designs improved in the 1820s, so biologists could see tiny cells better. Scottish botanist Robert Brown observed a structure in plant cells he called *the nucleus*. In the 1830s, scientists made many new discoveries about cells. Their work defined **the cell theory**, which is a foundation of modern biology. German botanist Matthias Schleiden viewed different plant parts and found that plants were made of cells too. In 1839, German biologist Theodor Schwann discovered that animal tissues were also made of cells with nuclei. He hypothesized that the nucleus was the center of all cell reproduction. Schwann proposed that all organisms are formed from one or more cells and that the cell is the basic unit of all life. A decade later, German scientist Rudolf Virchow experimented with growing and reproducing cells. He concluded that all cells arise from other cells.

## LESSON REVIEW

**1.** What is a cell?
**2.** What is the cell theory?
**3.** How does the account of Creation compare to the cell theory?
**4.** Name the contributions that three scientists made to the knowledge of cells.
**5.** Name the 11 parts of a compound microscope.

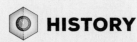 **HISTORY**

**The Cell Theory Timeline**
• **1665:** Robert Hooke first saw cells in slices of cork and gave them their name.
• **1674–76:** Antonie van Leeuwenhoek discovered microorganisms living in a drop of water.
• **1820s:** Lens designs improved dramatically.
• **1831:** Robert Brown first observed the nucleus in a plant cell.
• **1839:** Theodor Schwann discovered cells in animal tissues.
• **1840s:** Theodor Schwann and Matthias Schleiden proposed that the nucleus was the center of cell reproduction.
• **1840s:** Theodor Schwann proposed that all organisms were made of one or more cells, and he declared the cell the basic unit of all life.
• **1850s:** Rudolf Virchow concluded that all cells develop only from existing cells.

You probably think that all cells are very small, but a single nerve cell in a giraffe's leg can be 3 m long. A chicken egg is a single cell that you can easily see. But most cells are microscopic, 5–20 micrometers in diameter. An average-sized cell is only one ten-billionth of a gram. The human body has about 60 trillion of these tiny cells. There are about 1,000 times more cells in your body than there are stars in the Milky Way galaxy!

Individual cells have many differences. Plant cells are generally large and rectangular, but animal cells tend to be small and oval. Most cells have a structure called *the nucleus*, but bacteria cells lack this structure. Most cells are part of a larger organism, but some cells make up an organism all by themselves—some of these cells can even move on their own! Despite their differences, cells have similarities. Cells contain 75%–85% water; the other 15%–25% is mostly fats, sugars, and proteins. These ingredients carry on billions of chemical reactions each minute. Even the smallest cells have structures that provide protection and support, repair cell parts, store and release energy, transport materials, get rid of wastes, and reproduce.

If you were to explore a plant cell, you would first have to remove the cell wall, a thick outer layer of cellulose that gives plant cells added support and strength. Cellulose is a strong organic material. For example, the stringy part of celery is cellulose. Even though the cell wall is stiff and strong, it allows important substances such as oxygen, water, carbon dioxide, and minerals to travel in and out of the cell.

**Egg White**
88% Water
11% Protein

**Egg Yolk**
49% Water
33% Fat
17% Protein
2% Carbohydrates and Minerals

Animal cells do not have cell walls. The outer layer of an animal cell is the cell membrane, a flexible, thin protective layer around a cell that helps control what goes in and out of the cell. Plant cells have cell membranes too, which are located just inside the cell wall. A plant's cell membrane supports and protects the cell, even though the membrane is not rigid like the plant cell wall. Everything

## OBJECTIVES

- State characteristics common to all cells.
- Observe and describe the differences between plant and animal cells.
- Identify the major parts of a cell and explain the function of each.

## VOCABULARY

- **chromosome** a structure that contains genetic information and directs cell growth
- **endoplasmic reticulum** an organelle that functions as the cell's transportation system
- **Golgi apparatus** an organelle that receives, packages, and disperses materials
- **lysosome** an organelle that breaks down food particles and old cell organelles
- **mitochondrion** an organelle that produces energy for the cell
- **ribosome** an organelle that produces proteins
- **vacuole** an organelle that stores materials

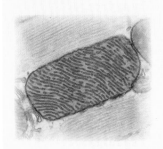
the cell needs enters through the cell membrane, and the cell's waste products exit through the cell membrane.

A watery fluid called *cytoplasm* surrounds everything inside the cell. Cytoplasm moves constantly and is filled with tiny structures called *organelles*, structures inside a cell that perform a particular function for the cell. All cells except bacteria have a nucleus. The nucleus is the cell's control center. It stores all the cell's information and controls the cell's functions. The nucleus is the largest and often the most visible organelle.

The thin membrane around the nucleus is called *the nuclear membrane*, or *the nuclear envelope*. Like the cell membrane, it lets materials into and out of the nucleus. The nucleus is filled with a fluid in which structures called **chromosomes** may be visible. The chromosomes are structures that direct cell growth and reproduction and hold the genetic information that is passed on to new cells. For example, because of the information in their chromosomes, white blood cells always reproduce more white blood cells rather than any other type of cells.

A **mitochondrion** is an organelle that produces energy for the cell. Mitochondria metabolize sugars into water and carbon dioxide. As they metabolize these sugars, they release

## Plant Cell

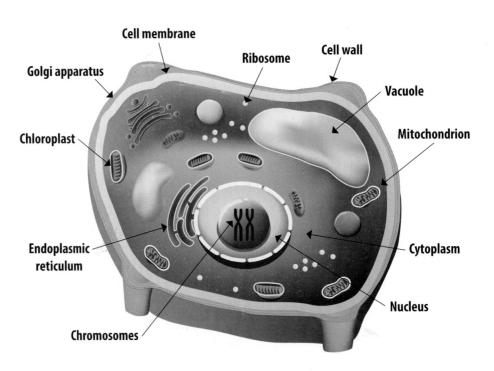

energy that the cell can use to do work. Active cells have more mitochondria than less active cells. For example, one liver cell may have 1,000 mitochondria while a skin cell has only a few.

The **endoplasmic reticulum** (ER) is an organelle that functions as the cell's transportation system. The endoplasmic reticulum is a system of tubes, flattened sacs, and channels within the cytoplasm. The ER transports proteins and other substances throughout the cell. This organelle comes in two forms: rough ER, which is studded with organelles called *ribosomes*, and smooth ER, which has no ribosomes.

**Ribosomes** are tiny organelles that produce protein for the cell. They float throughout the cell, but they are often attached to the ER. Once the ribosomes have made their proteins, they can deposit them into the passages of the ER, which then carry the proteins throughout the cell. The proteins can also be transported by the **Golgi apparatus**, an organelle that modifies and transports proteins and lipids. The Golgi apparatus (also called *the Golgi body* or *the Golgi complex*) looks a lot like the ER, but it is located closer to the cell membrane.

**Lysosomes** are organelles that recycle worn-out cell organelles and break down food particles. They then pass these materials

## TRY THIS

**School on a Cellular Level**
Imagine that your school building is a cell. Think about the parts of a cell and their functions and compare them to the parts of your school building.

## Animal Cell

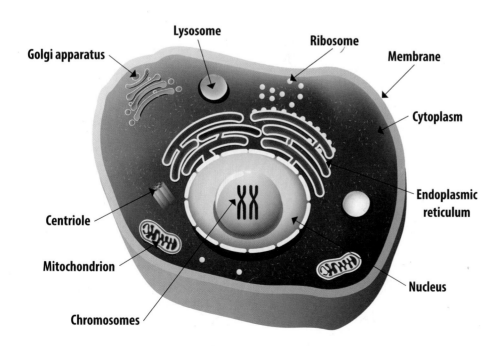

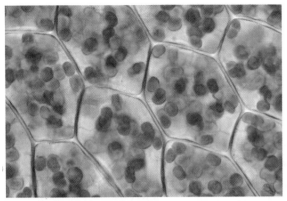

Chloroplasts in the cells of a *Plagiomnium affine*

on to the mitochondria. Lysosomes also destroy invaders such as bacteria. Lysosomes are more common in animal cells than in plant cells.

**Vacuoles** are organelles that store water, food, and other materials that the cell needs. They also store waste products. When you water a plant, the vacuoles fill up with water—that is why plants "plump up" after you water them.

Chloroplasts, which are found only in plant cells, are metabolic centers that contain the green pigment *chlorophyll*. The chloroplasts capture sunlight, which the plant uses to make its own food in a process called *photosynthesis*.

## LESSON REVIEW

**1.** What do cells teach you about God's power?
**2.** What characteristics are common to all cells?
**3.** List the organelles that are common to both animal and plant cells. Explain the function of each organelle.
**4.** How are plant cells different from animal cells? How do these different cell structures help plants thrive?

### Human Cells

**Columnar epithelial cells**

**Red blood cells**

**Smooth muscle cells**

**Nerve cell**

**Bone cell**

**Sperm cell**

**Ovum cell**

Cells come in many sizes and shapes. Each type of cell is perfectly designed for its function.

Early in the 20th century, scientists concluded that a cell's chromosomes contained material essential for a cell to reproduce. Since this material was found in the nucleus and it was acidic, scientists called it *nucleic acid*—specifically, deoxyribonucleic acid, or **DNA**. DNA is the molecule within a cell that carries the genetic information of an organism. A small segment of DNA that carries hereditary information is called a **gene**.

A DNA molecule is shaped like a twisted ladder. This shape is called *a double helix*. Long chains of sugar and phosphate molecules form the sides or supports of the ladder. The genetic information is stored in the nucleotides. A **nucleotide** is the combination of a nitrogen base, sugar, and a phosphate molecule that form the basic unit of the DNA chain. One nucleotide is attached to the next to make a long chain. DNA has four different nucleotide bases: adenine, thymine, guanine, and cytosine. Two of these bases are paired together to make each rung of the DNA ladder, or double helix.

Some DNA molecules are several million nucleotides long! Genes, which are very small sections of the DNA molecule, are anywhere from 75 to hundreds of thousands of nucleotides long.

DNA plays a very important role in cell division. Before a cell can divide to produce two new cells, the DNA molecules must be copied exactly. The process in which DNA molecules make exact duplicates is called **replication**. In replication, the DNA double helix unwinds. The nucleotide pairs on each rung let go of each other, and the DNA separates into two strands of nucleotides. This separating is called *unzipping*. An enzyme

### OBJECTIVES

- Describe the structure and components of a DNA molecule.
- Describe DNA replication.

### VOCABULARY

- **DNA** the molecule within a cell that carries the genetic information
- **gene** a small segment of DNA that carries hereditary information
- **nucleotide** the basic structural unit of DNA
- **replication** the process in which molecules make exact duplicates

**Adenine**          **Thymine**

**Guanine**          **Cytosine**

## Replication

1. The double helix unwinds and splits lengthwise, separating the nucleotide bases.

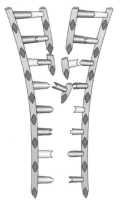

2. Free nucleotide bases match up with the exposed DNA bases. Only matching bases pair up.

3. Two new DNA molecules are built, each a duplicate of the original.

then selects a free complementary nucleotide to make a new DNA strand. Only adenine fits across from thymine, and only guanine fits across from cytosine. In this way, a new complementary strand grows opposite the old strand. This process takes place on each of the old strands until finally two DNA molecules are present where there had previously been one DNA molecule.

## LESSON REVIEW

**1.** What is the importance of DNA?
**2.** Describe the structure and components of a DNA molecule.
**3.** How does a DNA molecule copy itself?
**4.** What might happen if a DNA molecule copied incorrect

 **CAREER**

### Forensic Scientists

Do you like using clues to solve puzzles? If so, then you might like to be a forensic scientist. Amazing clues can be gleaned from the evidence at a crime scene. Analyzing and interpreting those clues require the expertise of many different branches of science. For example, some forensic scientists use chemistry to identify drugs or poisons found in a body by analyzing

samples. Others study burn patterns and conduct the chemical analysis of a fire scene to decide whether a flame accelerant was used or whether arson was involved.

Forensic scientists use physics to analyze ballistic data to identify what kind of gun was used in a crime. No two guns fire identically, so individual guns leave identifying marks on bullets. In fact, every tool, such as a hammer or a pair of pliers, also leaves characteristic marks on surfaces, which are analyzed by forensic scientists working in crime investigation.

Comparing DNA from blood, hair, or other body tissues to those of suspects can also help forensic scientists identify probable criminals. Fingerprints are important evidence but DNA "fingerprints" are even more valuable. The hair and fibers left at a crime scene or found on a person can connect a victim to a suspect. By analyzing pollen or soil on clothing, a forensic scientist can determine where a person has been. Forensic scientists use numerous other details to solve crimes. Their careful examination of handwriting or of teeth marks can pinpoint a suspect. Mud or blood spatters at a crime scene can indicate the direction of a blow, a blow's angle and force, and the type of instrument used. Forensic scientists even study the types of insects found on a body, which indicate how long the person has been dead and perhaps even where the person was before he or she died.

## Crick and Watson

Francis Crick (1916–2004) was a British biophysicist. He wanted to determine the 3-D structures of molecules in living organisms. He joined the Medical Research Council Unit at Cavendish Laboratories, where he met American biologist James Watson in 1951.

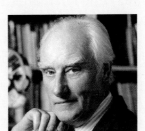

James Watson (born 1928) enrolled at the University of Chicago when he was 15 years old and graduated when he was 19. His PhD research on viruses helped him prove that DNA affects hereditary traits. When Watson went to Cavendish Laboratories, it was already known that DNA played an important part in heredity. Watson and Crick wanted to find out about the structure of DNA.

Prompted by Rosalind Franklin's work, Watson and Crick developed the double-helix model of DNA, which also shows how DNA can duplicate itself. They determined that the strands of DNA could separate and that each strand served as a pattern for the formation of an identical new strand. This model unlocked the secret of how genes and chromosomes could be copied, and it led to the discovery of how DNA replicates. The double-helix model was one of the most important discoveries of 20th century biology.

## Rosalind Franklin

Rosalind Franklin (1920–1958) was a British scientist who made important contributions to the discovery of the molecular structure of DNA. When Franklin was 15, she knew that she wanted to be a scientist. She graduated from Cambridge University in 1941 and started work on her doctorate. In 1950, Franklin joined a team of scientists who were studying living cells. She extracted finer DNA fibers than had been done before and arranged them in parallel bundles, discovering important keys to DNA's structure. Maurice Wilkins, who worked with Franklin, shared the information of Franklin's breakthroughs with James Watson and Francis Crick. Watson and Crick used this knowledge along with their own work to propose the structure of DNA in 1953.

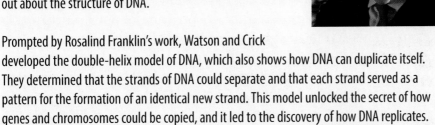

Franklin died of cancer at the age of 37. Before she died, Franklin turned her attention to viruses. Her research group's findings laid the foundation for structural virology. In 1962, Wilkins, Watson, and Crick shared a Nobel Prize and acknowledged Franklin's contribution to their discovery. Had Franklin still been living, she would have also shared in the prize.

Later, Crick was a professor at the Salk Institute for Biological Studies in San Diego, California, and Watson taught at Harvard from and helped direct the Human Genome Project.

### The Scene of the Crime

Although much of human DNA is identical, the molecules of DNA are so long that within a person's DNA there are about three million places where a single nucleotide pair can differ. The DNA of family members is very similar, but the DNA of unrelated people is very different.

Scientists can use a sequence of a person's DNA for identification. For example, a hair or skin cells found at a crime scene can be used to identify a criminal. A match between DNA from a hair or skin cells and the DNA of a suspect is evidence that the suspect was present at the crime scene. A jury can use DNA evidence to decide whether a person is innocent or guilty of a crime.

## OBJECTIVES

- Explain the relationship between energy, ATP, and cellular respiration.
- Describe the purpose of enzymes.

## VOCABULARY

- **cellular respiration** the breaking down of food molecules by cells into usable energy

It is not hard to tell when you have run out of energy—you feel tired and hungry. Your body uses these feelings to tell you that your cells need energy to carry out the chemical activities needed to live, reproduce, and grow. Energy is the ability to do work. It is the ability to make things happen and to cause processes to occur. A cell is the smallest living unit that uses energy for life.

God designed the sun as the source of energy for life on Earth. Plants capture light energy and change it into food through photosynthesis. In this process plants use energy to change carbon dioxide and water into glucose. Plants then use some of the glucose as food and store some of it as starch. Photosynthesis also produces oxygen.

Cells must be able to convert energy into a form that they can use. The conversion of energy generally happens through **cellular respiration**—the breaking down of food molecules by cells into usable energy. Cellular respiration is not breathing respiration, which is the way your body takes in oxygen and expels carbon dioxide. The two processes are related, however, because breathing supplies an organism's cells with the oxygen needed for cellular respiration.

In most cells, cellular respiration happens in the mitochondria. During cellular respiration, glucose is metabolized into water and carbon dioxide, and energy is released for cellular work. Some of the energy is stored in a molecule called *ATP*. ATP (adenosine triphosphate) is the molecule that carries small packets of energy for a cell's activities. Most of the energy is released in the form of heat. In many organisms (humans included), this heat helps maintain body temperature.

Cells use energy to move materials within their cell membranes through metabolic pathways. A metabolic pathway is a series of chemical reactions that break down or make materials that the cell needs. Metabolic pathways usually contain about a dozen different steps. Plant cells, for example, have metabolic pathways that use energy for building cell walls. Red blood cells have metabolic pathways that use energy to manufacture hemoglobin molecules.

A healthy diet gives your cells energy and keeps you feeling good.

Many of the reactions that take place in a cell involve enzymes, which are special proteins that speed up chemical reactions in an organism or a cell. Enzymes are required to start the many chemical reactions that happen inside each cell. Without enzymes, the reactions inside the cell would not happen fast enough for the cell and the organism to live. Enzymes speed up the metabolic pathways so that the cell flourishes.

Cellular respiration (breaking down sugar to make ATP) requires oxygen. Sometimes cells cannot get enough oxygen

## Cellular Respiration

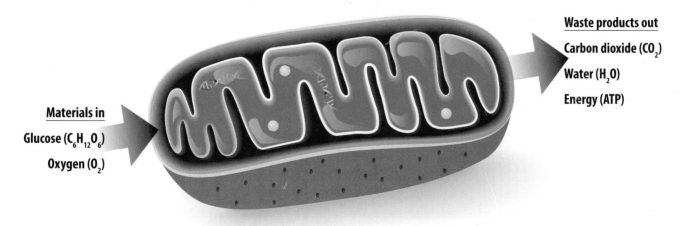

Materials in
Glucose ($C_6H_{12}O_6$)
Oxygen ($O_2$)

Waste products out
Carbon dioxide ($CO_2$)
Water ($H_2O$)
Energy (ATP)

needed for sufficient cellular respiration to occur. This lack of oxygen causes the production of ATP to diminish. When this happens, cells use fermentation. Fermentation produces a small amount of ATP and other products by partially breaking down glucose without using oxygen. In muscles, fermentation produces lactic acid, which contributes to muscle fatigue. The burning sensation that sometimes occurs after exercise is caused by lactic acid. Some bacteria and yeast also use fermentation for energy. However, alcohol, not lactic acid, is a product of yeast fermentation.

## LESSON REVIEW

1. What is cellular respiration?
2. What is a metabolic pathway?
3. Would you expect to find more mitochondria in liver cells, which are very active, or in teeth cells, which are less active? Why?
4. What is the relationship between energy, ATP, and cellular respiration?
5. What is the purpose of enzymes?

Bromelain, an enzyme found in pineapples, is known for its many health benefits. Bromelain is a natural anti-inflammatory that is often used to treat muscle injuries. It also acts as a digestive aid, helps with sinus inflammation, and helps thin blood.

Cells use both passive and active transport to move materials around. Passive transport requires no energy on a cell's part, just as it requires no energy from you to coast down a hill on your bike. Active transport requires a cell to expend energy, just as you use energy to pedal up a hill on your bike. Cells are small because they rely on passive transport to bring materials they need to survive across the cell membrane.

One method of passive transport is diffusion. **Diffusion** is the movement of particles from an area of higher concentration to an area of lower concentration. Diffusion happens because particles are always moving and bumping into each other, causing them to spread out. Picture a group of people crowded into the corner of a room, bumping into each other; given the chance, they would probably spread out into the room's empty spaces. They would move from an area of higher concentration (where there are a lot of people) to an area of lower concentration (where there are fewer people).

Diffusion is the primary way that substances move around inside a cell. Although diffusion is a slow method for transporting

### OBJECTIVES

- Describe the processes of diffusion and osmosis.
- Summarize how cells depend on diffusion and osmosis.
- Determine why cells must be small in order to sustain life.

### VOCABULARY

- **diffusion** the movement of particles from an area of higher concentration to an area of lower concentration
- **osmosis** the diffusion of water through a membrane

## Diffusion

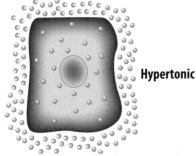
Hypertonic Isotonic

Particles from the highly concentrated area outside the cell (hypertonic) move to the less concentrated area inside the cell until the concentration is equal (isotonic).

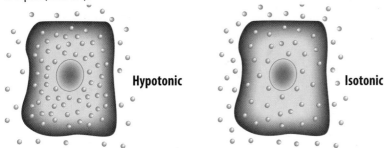
Hypotonic Isotonic

Particles from the highly concentrated area inside the cell move to the less concentrated area outside the cell (hypotonic) until the concentration is equal (isotonic).

As tea steeps in water, tea molecules move from the tea leaves—the area of greater concentration—to the water—the area of lower concentration.

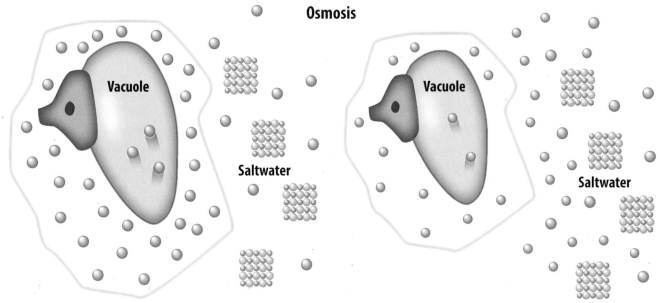

**Osmosis**

Vacuole

Saltwater

Vacuole

Saltwater

The higher concentration of water moves out of the cell to mix with the saltwater.

The cell begins to shrink as the water continues to pass through the membrane.

Turgor pressure is a result of osmosis. When plant cells have plenty of water, turgor pressure keeps the cells plump, or turgid, and the plant stands upright. When plant cells do not receive enough water, the turgor pressure drops, cells become limp, and the plant wilts.

substances over large distances, it is an efficient transport system inside a tiny area like a cell. Cells assist the diffusion process by forming streams in their cytoplasm, which form currents throughout the cell. The most important substances that move by diffusion in a cell are oxygen, water, and carbon dioxide.

**Osmosis** is the diffusion of water through a membrane. Cell membranes are selectively permeable membranes—membranes that allow some substances to pass through them while blocking other substances. For example, water will move across a membrane from an area that contains more water (and less salt) to an area that contains less water (and more salt). How could this affect a cell? Imagine putting a cell into saltwater. The concentration of water outside the cell is lower than the concentration of water inside the cell; the water will travel from the inside of the cell to the outside, and the cell will shrink. The opposite will occur if you put a cell into freshwater—the water will travel from outside the cell to the inside, and the cell will expand. If the cell becomes too full of water, it bursts. Vacuoles contract to expel the water if the cell gets too full.

Osmosis could present many difficulties for living cells, but God designed them to overcome these problems. For example, osmosis could cause microorganisms living in the ocean to dry out because of the salt content of the water, but God designed these organisms to have the same concentration of salt in their bodies as the ocean water, so they maintain homeostasis.

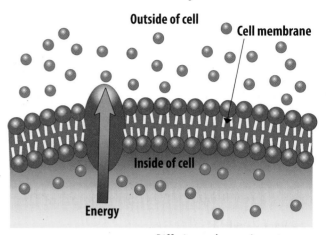

**Passive Transport**

Outside of cell

Cell membrane

Inside of cell

**Active Transport**

Outside of cell

Cell membrane

Inside of cell

Energy

Diffusion and osmosis are types of passive transport because they do not require energy. Active transport requires energy to move molecules across a cell's membrane.

Plant cells are designed to allow water in. The movement of water into a plant cell causes an internal pressure called *turgor pressure*. Turgor pressure causes the cell to expand. When a plant does not receive enough water, the turgor pressure drops, water is lost from the vacuoles, and the cell walls of the plant fold in on themselves. This drop in turgor pressure makes the plant wilt.

Not all substances that move across a cell membrane or within a cell do so through diffusion or osmosis. Cells also move materials using active transport systems. Unlike passive transport systems, active transport systems use up energy. Proteins are the active transporters within a cell membrane. For example, some proteins push small molecules across the cell membrane in the opposite direction from diffusion. Vesicles, saclike structures that form on organelles and cell membranes, are another type of active transporter. Vesicles move their contents out of the cell or into the cytoplasm. Another type of active transport system is the sodium-potassium pump, which uses proteins to move sodium and potassium across the cell membrane.

## LESSON REVIEW

**1.** What is passive transport, and what are two types of passive transport?
**2.** What is the difference between diffusion and osmosis?
**3.** If you put a raisin in a glass of water, what will happen to the raisin? Why?
**4.** How do cells depend on diffusion and osmosis?
**5.** Why do cells have to be small to sustain life?

## TRY THIS

**Turgor Pressure**
Use a microscope to examine the cells of an elodea leaf. Look at the vacuoles, walls, and chloroplasts. Then drop saltwater on the cells. What happens? Why?

### OBJECTIVES

• Name and summarize the phases of mitosis.
• Distinguish between mitosis and meiosis.
• Explain how cells without a nucleus reproduce.

### VOCABULARY

• **binary fission** the cell division process in which one cell splits into two identical cells, each having a complete set of DNA
• **meiosis** the cell division process that forms sex cells that contain half the usual number of chromosomes
• **mitosis** the cell division process that forms new cells with an identical copy of the parent's chromosomes

## TRY THIS

**Multiplying Cells**
Given the right conditions, some bacterial cells can reproduce as often as every 20 minutes! At that rate, how many bacteria would there be in 24 hours?

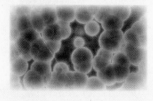

Living things grow. You have grown since you were a small child. Maybe you have even grown in the past few weeks. When you plant a tree or a garden, you are eager to watch the new living things grow. But how do living things grow? You might think that their cells get bigger, but they do not. A baby's cells are the same size as yours, but you have many more cells than a baby does. Living things grow by adding cells.

Cells are limited in size because of their surface area. Remember that materials enter and leave a cell through the cell membrane. If a cell grew too big, the cell membrane would not be able to handle all of the materials that would have to flow through it. Food and important gases could not enter the cell quickly, and wastes could not leave the cell fast enough. The cell would die.

God created organisms to thrive through cell reproduction. Cells produce new cells by dividing. During cell division, each cell (called *a parent cell*) divides into two new cells (called *daughter cells*). The daughter cells are identical to the parent cell. Organisms grow as more and more cells are produced.

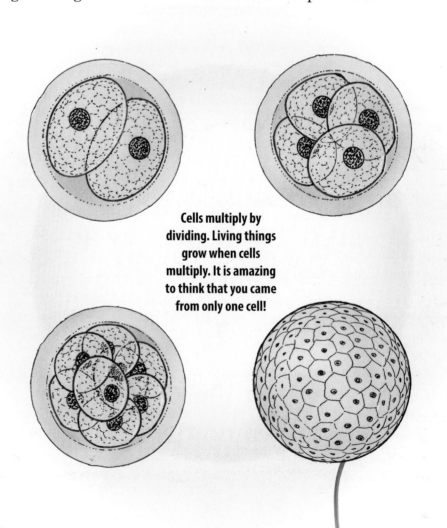

Cells multiply by dividing. Living things grow when cells multiply. It is amazing to think that you came from only one cell!

Each new cell usually receives a complete set of chromosomes, which carries the blueprints, or DNA, used to make the cell membrane and organelles in the new cell. Without a complete set of chromosomes, a cell cannot function or reproduce. But how can daughter cells each receive a complete set of chromosomes if they both come from the same cell? God designed cell reproduction to follow a series of steps that ensure that each daughter cell is identical to its parent cell.

The series of reproductive steps is different for a cell without a nucleus, like bacteria, and a cell with a nucleus, such as plant or animal cells. Bacteria reproduce by **binary fission**—cell division in which one cell splits into two identical daughter cells, each having a complete set of DNA. In about 30 minutes, a bacterium will duplicate its single DNA strand and distribute each strand to two new cells.

Cells that have nuclei use a more elaborate method of cell division to reproduce. Each of these cells has at least two DNA strands. The pair of strands each have a DNA pattern similar to the other. Human body cells usually contain 23 pairs of DNA strands; these strands may also be called *chromosomes*. Some cells contain hundreds of DNA strands. Each length of DNA needs to be duplicated, and this happens through mitosis.

**Mitosis** is the cell division process in which each daughter cell receives an identical copy of the parent cell's chromosomes. New daughter cells are produced by mitosis. During this process, the parent cell's chromosomes are duplicated and the cell divides to form two new daughter cells. Mitosis normally ensures that each new cell will receive the correct number of chromosomes.

Cell division has three stages: interphase, mitosis, and cytokinesis. During the interphase stage, which is before mitosis begins, the cell's nucleus produces a copy of each chromosome. Each chromosome now consists of two sister chromatids, or identical chromosome copies. Mitosis has four phases of its

## FYI

**Mitosis**
Mitosis results in two new cells, each with the same number of chromosomes as the original cell.

Interphase

Prophase

Metaphase

Anaphase

Telophase

Cytokinesis

 **TRY THIS**

**Mitosis**
Use string to represent a nuclear membrane. Twist two pipe cleaners together to represent two chromosomes. Twist two more pipe cleaners together to represent duplicated chromosomes. Manipulate the "chromosomes" through the various phases of mitosis.

## TRY THIS

**Dividing Cells**

Soak the bottom of an onion bulb in water for three days or until roots develop. Prepare a microscope slide with the roots and methylene blue. Carefully crush the root tip with the end of a glass rod and observe the cells through a microscope. Study the slide carefully until you notice a cell dividing. Observe the process. Record your observations.

own. During the prophase phase, the chromatids condense into rodlike structures and the nuclear membrane disappears. During metaphase, the chromatids line up at the equator, or center, of the cell. During anaphase, the sister chromatids move apart to opposite sides of the cell. Finally, during telophase, the nuclei reorganize at the opposite sides of the splitting cell, and a nuclear membrane forms around the chromosomes. Cytokinesis is the final stage of cell division. In animal cells the cell membrane constricts at the equator of the cell until the cytoplasm is divided in half and two new daughter cells are formed. In plant cells the division of the cytoplasm begins in the center of the cell, moving out to the cell membrane; cell formation follows this. Each new cell contains the same number and kind of chromosomes as the parent cell.

**Meiosis**, a process similar to mitosis, is the cell division process that forms sex cells. The new cells have half the usual number of chromosomes. During meiosis, the cell undergoes a double division in the chromosome number because the chromatids divide twice. The first part of meiosis (Meiosis I) is just like mitosis; the chromosomes are duplicated as chromatids, but the chromatids stay together during the first separation. During the second part of meiosis (Meiosis II), the original duplicated chromatids separate, leaving one half the original number of chromosomes. Normal body cells have paired sets of

## Meiosis I

| Parent Cell | Prophase I | Metaphase I | Anaphase I | Telophase I |
|---|---|---|---|---|

**Meiosis**

Meiosis results in four gametes (sperm cells or egg cells), each with half the number of chromosomes found in other cells in that organism.

chromosomes; meiosis produces cells with one member of each paired set. These cells are called *gametes*, which are sex cells. Each sex cell, a sperm cell or an egg cell, contains half the number of chromosomes found in non-gamete cells. When an egg cell and a sperm cell join together, a new cell called *a zygote* is formed. The zygote has the full number of chromosomes. God designed this process to ensure that new organisms end up with the correct number of chromosomes.

## LESSON REVIEW

1. Compare the way a bacterium reproduces to the way a skin cell reproduces.
2. Explain the difference between mitosis and meiosis.
3. What are the phases of mitosis?
4. What would happen if meiosis did not halve the number of chromosomes before gametes are made?
5. Why are organisms that reproduce by fission more exact copies of their parents than organisms that reproduce using meiosis?

## TRY THIS

**Reproducing Cells**
Your fingernails and toenails are actually the outer layer of your skin. The cells in your nails are dead, but nails grow as new cells are added to the base of the nail. Nail growth is evidence that the cells in your body are making new cells. Design an experiment to test how quickly the cells in your nails are replenished. Do all of your nails grow at the same rate? Why? Do you think that the rate of nail growth could be a sign of good health or poor health? Why?

## Meiosis II

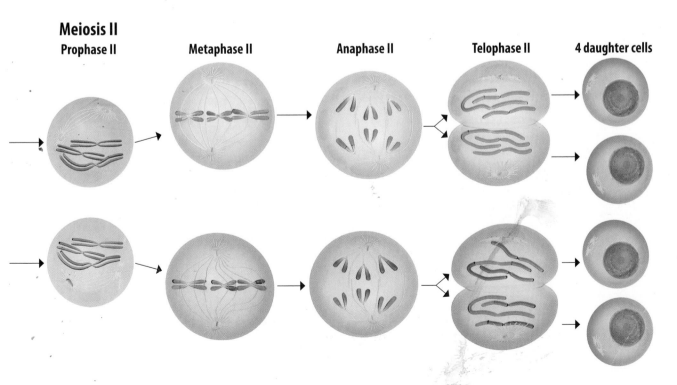

Prophase II · Metaphase II · Anaphase II · Telophase II · 4 daughter cells

### OBJECTIVES

- Summarize the history of taxonomy.
- Explain the use of binomial nomenclature.
- Describe how the science of taxonomy supports God's unlimited power and creativity.

### VOCABULARY

- **binomial nomenclature** the two-part scientific naming system
- **taxonomy** the scientific classification of organisms

Imagine putting on scuba diving equipment and plunging into the warm waters of the Caribbean Sea. You swim 20 m beneath the water's surface to explore the breathtaking world of the coral reef. Schools of fish streak past, flashing in the sunlight. You dive deeper to a maze of ravines and hills on the ocean floor that swarm with life. Colorful corals spread out across the ocean landscape. White and purple flowers are scattered across the coral, but they disappear into long, thin tubes when you touch them. Red and green worms bristling with hairs crawl across coral that resemble black and white feathers.

Fish swim all around you. You recognize the torpedo shape of a 2 m long barracuda as it glides past. Schools of vibrant blue fish huddle around yellow pillars of coral. Large, multicolored fish with parrotlike beaks grind away at the coral. Some fish seem to have wings, others are shaped like long sticks, and a few look like colorful rocks or speckled pancakes resting quietly on the ocean floor.

Before diving into the sea, you expected to see ordinary, gray fish, so you are amazed at the variety of organisms thriving on the coral reef. You now want to know more about these strange creatures you have observed. Where do you start?

The books that divers use to help identify all the beautiful creatures on the reef rely on the science of taxonomy. **Taxonomy** is the scientific classification of organisms. Understanding taxonomy makes it easier for you to learn that the colorful flowers on the reef are actually worms and the speckled pancake-shaped fish are peacock flounders. Learning about taxonomy can help you better appreciate the wide variety of marvelous creatures that God placed on Earth.

"The earth is the Lord's, and everything in it, the world, and all who live in it; for He founded it on the seas and established it on the waters" (Psalm 24:1–2).

**Ladder of Life**

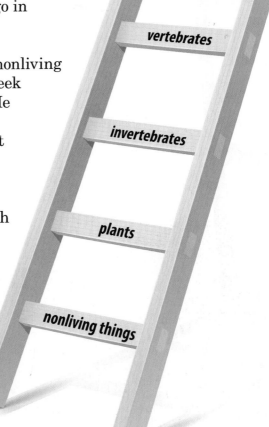

If you walked past a display of every organism on Earth, how many could you identify? Three hundred? One thousand? Scientists categorize the millions of organisms that live on Earth into separate groups. These groups are called *species*. Scientists so far have identified about 1.4 million species of living things, but biologists believe that about 10 million different species of organisms live on Earth. What incredible variety! Millions of organisms have yet to be discovered and identified, so taxonomists (scientists who study taxonomy) have a long way to go in identifying all of Earth's creatures.

The first scientist to organize living and nonliving things into categories was Aristotle, a Greek philosopher who lived from 384–322 BC. He called his classification system the *Scala Naturae*—"the ladder of life." Aristotle put nonliving things at the lowest rung of his ladder. He placed plants on the second rung and invertebrates (animals with no backbone) on the third rung. On the fourth rung he put vertebrates (animals with a backbone), and he put mammals on the top rung.

In the 14th century, philosophers restructured Aristotle's ladder of life into what was called *the Great Chain of Being*. The Great Chain of Being started with Earth's lowest life-forms, moved up to humans, and ended with spiritual beings

Spiritual beings

Humans

Animals

Plants

Nonliving beings

such as angels. They believed that once they had discovered, named, and described all the links in the Great Chain of Being, they would know the meaning of life.

In the 1700s, the Swedish naturalist Carolus Linnaeus revised this system. He created a workable system for classifying plants and animals. Linnaeus knew that God created an orderly world so its members could be classified into groups. He also believed that studying creation could teach people more about God's wisdom.

Linnaeus set up two large categories to classify all living things: one for plants and one for animals. These two categories were kingdom Plantae and kingdom Animalia. He also devised a naming method for organisms called **binomial nomenclature**, which means "consisting of two names." His system gave every known organism a unique two-part Latin name, and scientists still use his naming system today. For example, scientists use the Latin word *Acer* as the name for all the maple trees in the world. *Acer saccharum*, *Acer nigrum*, and *Acer rubrum* are the names for three different species of maple trees. *Acer saccharum* is the name for the sugar maple, *Acer nigrum* is the name for the black maple, and *Acer rubrum* is the name for the red maple. Their unique two-part names

distinguish these different species of maple trees from all the other millions of species of living things on the earth. Yet the two-part names also show that all maples are similar to each other.

With the invention and development of the microscope, scientists began to see microorganisms that were neither plants nor animals, so they formed a new kingdom, Protista, for microorganisms. Scientists continued to find great differences between various kinds of microorganisms. Scientists also discovered major distinctions between plants and mushrooms. In 1969, taxonomist Robert H. Whittaker proposed a five-kingdom system to separate all of these various types of organisms into their own kingdoms: animals, plants, fungi, bacteria, and protists. As a result of new discoveries made through technological advancements, taxonomists organized organisms into three domains—Bacteria, Archaea, and Eukarya. The Eukarya domain, the largest of the three domains, is divided into four kingdoms—Protista, Fungi, Plantae, and Animalia.

Taxonomy is a useful science that helps scientists discuss, gather, and organize information about the millions of types of organisms. This information helps them find ways to control pests, grow better crops, and fight diseases caused by microorganisms. Taxonomy is a tool that helps people understand how different organisms relate to one another in an ecosystem. And taxonomy gives Christians a glimpse of the power and creativity God used to breathe life into His world.

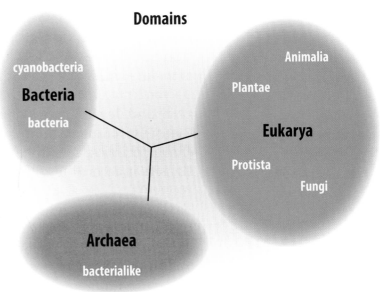

**Domains**

## LESSON REVIEW
**1.** What is taxonomy?
**2.** Name three scientists who influenced taxonomy.
**3.** What is binomial nomenclature? How is it used?
**4.** Why should a Christian study taxonomy?
**5.** How did technology change taxonomy?

*Nerodia fasciata* — banded water snake

*Nerodia cyclopion* — green water snake

*Nerodia sipedon* — northern water snake

Aristotle included nonliving things in his classification system, but today, biologists classify only living things. So the first thing that scientists do when they want to classify something is to determine whether or not it is alive or was once alive. As you know, all living things are made of cells, and they grow and develop, reproduce after their own kind, respond to their environment, and use energy. If something has these characteristics, scientists categorize it as a living thing and classify it according to taxonomy.

Taxonomists put living things into categories with other similar types of organisms. The eight major categories that taxonomists use are domain, kingdom, phylum (or division to classify plants and fungi), class, order, family, genus, and species. The two-part name used in binomial nomenclature comes from the genus and species name. For example, *Canis lupus* is the scientific name for the gray wolf, also known as the timber wolf. *Canis* is the genus, and *lupus* is the species.

The largest category, which includes the greatest variety of organisms, is the domain. Within domains the next largest

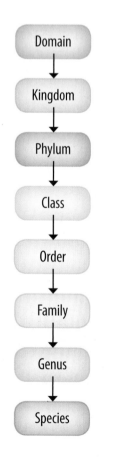

Domain
↓
Kingdom
↓
Phylum
↓
Class
↓
Order
↓
Family
↓
Genus
↓
Species

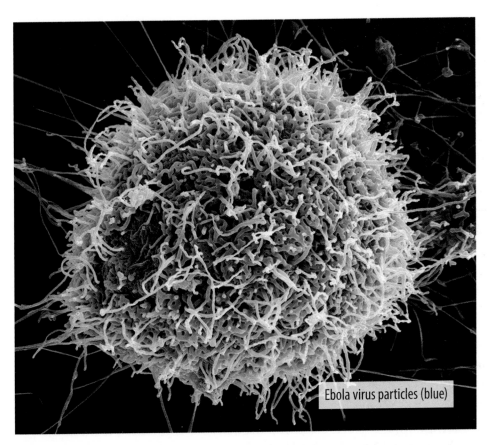
Ebola virus particles (blue)

categories are called *kingdoms*. Each kingdom is made of different phyla (plural for phylum). Each phylum is made up of classes. Classes contain orders. Orders are made up of families, families are made up of genera (plural for genus), and genera are made from species.

The word *species* has a precise scientific definition. A **species** is a group of organisms that can mate with one another and produce fertile offspring. *Fertile* means that the offspring can grow up to produce their own offspring. A female horse and a male donkey can mate and produce a mule; a male horse and a female donkey will produce a hinny. A male tiger and a lioness can mate and produce a tiglon; a female tiger and a male lion will produce a liger. But mules and hinnies are sterile, which means they cannot reproduce, so horses and donkeys are considered different species. Because most tiglons and ligers are sterile, and because they do not exist in the wild, tigers and lions are also considered separate species.

Taxonomists use the basic characteristics of organisms to classify them. They usually put organisms in a group with other organisms that have similar internal and external features. Sometimes organisms are grouped according to their role in the

Kingdom Animalia

Kingdom Plantae

Kingdom Fungi

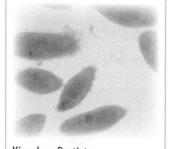

Kingdom Protista

ecosystem. For example, scientists decided to make a category of all mammals that eat meat, so they placed all cats, dogs, bears, raccoons, weasels, mongooses, and hyenas under order Carnivora.

Currently, scientists use three domains—Bacteria, Archaea, and Eukarya. The Bacteria domain has one kingdom—Bacteria; the Archaea domain also has one kingdom—Archaea. The Eukarya domain has four kingdoms: Animalia, Plantae, Protista, and Fungi.

- Kingdom Animalia contains all the animals, both vertebrates and invertebrates.

- Kingdom Plantae includes all green plants.

- Kingdom Fungi contains all **fungi** (plural of fungus), nongreen organisms that reproduce from spores and absorb their food. Examples of fungi include mushrooms, lichen, and molds.

- Kingdom Protista contains organisms that have a nucleus that are not animals, plants, fungi, or bacteria. These organisms are usually unicellular.

Some things that seem to be alive do not fit into any of these kingdoms. Viruses such as influenza, polio, Ebola, and HIV act like living organisms but are not considered living things by scientists. Viruses can reproduce and grow but only when they are inside a host cell. Viruses are made of proteins rather than cells. Some scientists consider viruses to be unusual life-forms, but others disagree. Because of this debate, taxonomists have not classified viruses in the kingdom classification system. Do you think that taxonomists should create a new kingdom just for viruses?

## LESSON REVIEW

1. What are the eight major categories that taxonomists use to classify living things?
2. What are the three domains?
3. What are the kingdoms of the Eukarya domain?
4. Why might taxonomists want to recognize a new kingdom?
5. What problems might you encounter in classifying an organism?

Most people think of plants and animals when they think about living things. But God designed other organisms for His world. Many of these organisms are classified under the Bacteria and Archaea domains and the Fungi kingdom. These organisms may be smaller and less noticeable than plants and animals, but they play a very important part in God's world.

Domains Bacteria and Archaea separate bacteria from bacterialike organisms. The bacteria that cause human disease, the bacteria that live inside your body to keep you healthy, and the bacteria that enrich the soil belong to domain Bacteria. Cyanobacteria, commonly known as *blue-green algae*, are also members of this domain.

**Bacteria** are unicellular organisms without a nucleus, which means they are **prokaryotes**. Bacteriologists, scientists who study bacteria, have identified approximately 10,000 species of bacteria that live in the human body alone, but scientists believe that most species of bacteria have yet to be named or identified.

Bacteria live almost everywhere and in great numbers. A spoonful of soil can contain billions of bacteria. A small scraping from the film on your gums can contain 1 billion bacteria. In fact, the number of bacteria in your mouth right now may be greater than the number of people who have ever lived!

## OBJECTIVES

- Compare and contrast bacteria and archaea.
- Recognize the important role that all organisms are given in God's world.

## VOCABULARY

- **archaea** a prokaryotic unicellular organism belonging to domain Archaea
- **bacteria** a prokaryotic unicellular organism belonging to domain Bacteria
- **prokaryote** a unicellular microorganism that lacks a distinct nucleus

*Neisseria meningitidis* is a eubacteria responsible for causing meningitis, which affects the brain and spinal column in humans.

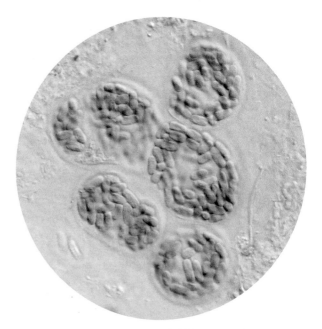

Cyanobacteria are sometimes called *blue-green algae*.

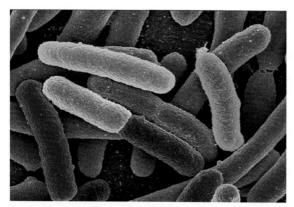

*E. coli* bacteria is commonly found in the human intestinal tract. Only a few varieties of *E. coli* are harmful to humans.

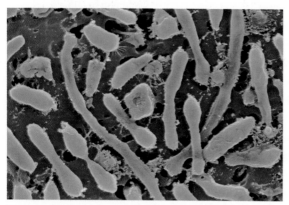

This archaea lives in water that has a high saline concentration—that of 2.5 times normal seawater.

### TRY THIS

**Growing Bacteria**
Pour 300 mL of milk into two beakers. Place one beaker in the refrigerator and the other on a windowsill. Examine the beakers daily for a week. Can you see bacteria? How do you know they are there?

You may think of bacteria as harmful or disgusting because some kinds of bacteria can cause cavities or make you sick. However, most bacteria are extremely useful. They decompose dead materials by recycling nitrogen, phosphorus, and other valuable nutrients back into an ecosystem. Like green plants, some bacteria can use photosynthesis to produce food and oxygen. Other bacteria are a main food source for many microorganisms. Bacteria also enrich soil so that plants can grow. Bacteria in the human digestive tract help digest food and provide beneficial substances that are used by the body. Although some types of *Escherichia coli* (*E. coli*) can be harmful, other varieties are beneficial. *E. coli* in the large intestine makes Vitamin K, which is absorbed by the body and used to clot blood.

Domain Archaea contains **archaea**. These organisms used to be considered a different kind of bacteria, but now they are classified as separate organisms. Archaea are unicellular and lack a nucleus like bacteria, but their structure is very different. Most archaea live in extreme environments such as hot springs, very salty water, the tops of high mountains, petroleum deposits underground, and mud that is lacking oxygen. There are approximately 500 known species of archaea, although like bacteria, scientists believe there are many species yet to be discovered. Along with bacteria, archaea can be both helpful and harmful to ecosystems.

### LESSON REVIEW

1. What kind of attitude do you think most people have toward the organisms in the two domains mentioned in this lesson? Why do you think this is?
2. How do bacteria help ecosystems?
3. Compare bacteria and archaea.
4. What might cause scientists to study the bacteria and archaea domains in greater depth?

Domain Eukarya contains all the **eukaryotes**—organisms composed of cells that have a visible nucleus. This domain includes protists, fungi, plants, and animals. A **protist** is an organism that has a nucleus, but it is not an animal, a plant, or a fungus. It also is not a bacterium because bacteria do not have nuclei. Because most of these organisms are not major food sources for humans and they do not cause diseases in people, scientists have not studied them as much as they have the organisms in the other kingdoms. Protists are still an important part of creation, however. For example, some of these organisms generate much of the oxygen that is in the atmosphere. Certain protists live in the digestive systems of cattle, helping them digest the plant material they eat; without them, the cattle could not live. Other protists live in the guts of termites; without them, termites could not digest wood.

Most organisms in kingdom Protista contain only one cell and cannot be seen without a microscope, but the giant kelp, which grows off the coasts of North and South America, can grow up to 30 m long. Green algae and seaweed are usually classified under kingdom Protista, even though some scientists still consider these organisms plants. They are

Slime molds are classified as protists.

## OBJECTIVES

- Distinguish between eukaryotes and prokaryotes.
- Describe the four kingdoms in domain Eukarya.

## VOCABULARY

- **botanist** a scientist who studies plants
- **eukaryote** an organism composed of one or more cells containing a visible nucleus
- **lichen** an organism formed by a symbiotic relationship between a fungus and an alga or a cyanobacterium
- **protist** a eukaryotic organism that cannot be classified as a fungi, a plant, or an animal
- **zoologist** a scientist who studies animals

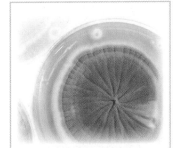

Penicillin is made from the *Penicillium* fungi—*Penicillium chrysogenum.*

Fungi grow where a tree's bark has been cut or damaged.

not generally classified as plants because they do not reproduce through seeds or spores like plants.

Kingdom Fungi contains nongreen organisms that reproduce through spores and absorb their food. Common types of fungi are molds, yeast, and mushrooms. These organisms were once classified as plants, but since they do not have chlorophyll, which is necessary for photosynthesis, and they must rely on other organisms for food, they were placed into a separate kingdom.

Fungi are important for many different reasons. Fungi keep trees healthy by transporting important nutrients from the soil to tree roots, and the trees supply moisture and food to the fungi. People eat fungi such as mushrooms and use mold and yeast to make cheese and bread. Fungi are also used to make antibiotics such as penicillin. **Lichen** is an organism formed by a symbiotic relationship between a fungus and an alga or a cyanobacterium. It plays an important role in many ecosystems by giving off chemicals that wear down rocks into soil so that plants can grow. Lichen is very sensitive to pollution and its presence often indicates the health of an ecosystem.

In Genesis 1:11–12, it says, "Then God said, 'Let the land produce vegetation: seed-bearing plants and trees on the land that bear fruit with seed in it, according to their various kinds.' And it was so. The land produced vegetation: plants bearing seed according to their kinds and trees bearing fruit with seed in it according to their kinds. And God saw that it was good." The plants that sprang up at God's command and those that continue to spring up today show the imagination of the Creator. Think of the wide variety of plants that you see every day; they are only a small sample of the plants that God placed on Earth.

Kingdom Plantae includes multicellular organisms that produce their own food through photosynthesis. Plants reproduce through seeds or spores. Scientists who study plants are called **botanists**. Botanists have named over 300,000 species of plants, but they believe many species remain undiscovered, especially in the tropical rain forests. Botanists use the word *division* instead of the word *phylum* in the classification system. They have created 10 divisions for classifying plants. The most common plants belong to the following divisions:

**Divisions Bryophyta and Hepaticophyta.** Mosses and liverworts grow around the world. These plants cover damp areas like green carpets. Although most plants draw water up through tissue in their roots and stems, these plants absorb water and minerals through their entire body surfaces. They reproduce with spores rather than seeds.

**Division Filicinophyta.** Ferns also reproduce with spores, but they have internal tissues that conduct water and minerals. Most ferns grow in tropical rain forests. Very few species are found in cold, dry areas.

**Division Coniferophyta.** Almost all of these cone-bearing plants are evergreens with needlelike leaves—spruces, pines, and firs, for example. In places where most of the landscape turns brown during the winter, these trees add bright spots of green.

**Division Angiospermophyta.** Plants that use flowers for reproduction make up this division. This classification includes not only plants with obvious flowers, such as daisies, roses, and apple trees, but also plants whose flowers are not as obvious, such as corn, oak trees, and grasses.

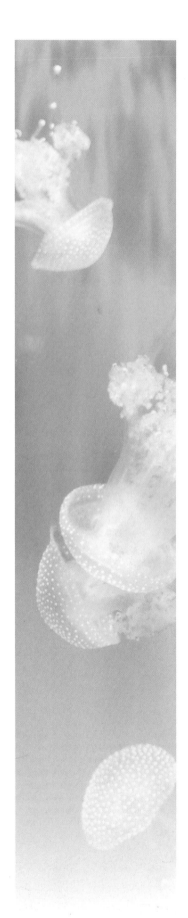

**Division Ginkgophyta.** Only one species, the ginkgo tree, makes up this division. The ginkgo is an unusual tree. It does not use flowers for reproduction, but unlike other nonflowering trees, it is not evergreen. It has broad leaves that drop every year.

At Creation, God commanded the waters to teem with creatures, the sky to be filled with birds, and the land to produce all living creatures that move across the ground. The amazing variety of organisms classified under kingdom Animalia shows God's creativity. God created animals with different shapes, behaviors, and sizes. Animals are multicellular organisms that have to obtain their food to survive.

**Zoologists**, scientists who study animals, have divided animals into at least 34 phyla. It may be surprising that most of these phyla are types of aquatic worms. The animals that people are most familiar with are not necessarily the ones that are most plentiful on Earth. The most common animals fall into nine of the currently identified phyla. Notice the wide variety of creatures that belong to each one.

**Phylum Porifera.** Most sponges live in the ocean, but a few species live in freshwater. Sponges show God's creativity; they range in size from 1 cm to 2 m tall.

**Phylum Cnidaria.** This phylum contains many kinds of animals, but cnidarians all have something in common: they have a circular design and stinging tentacles arranged around their mouths. Sea anemones, jellyfish, and coral are in this phylum.

**Phylum Platyhelminthes.** Flatworms belong to this phylum. They range in color from white to bright green. The planarian is a flatworm that is about 3 mm long. Tapeworms, which can live inside human intestines, are the largest species of flatworm—they can grow to be longer than 25 m.

**Phylum Rotifera.** If you look at a sample of pond water under a microscope, you will see many of the species of rotifers—microscopic animals that rotate. Most of the rotifers seen are females. In fact, the males of some rotifer species have never been seen!

**Phylum Mollusca.** Mollusks have soft bodies. There are more than twice as many mollusk species as vertebrate (animals with a backbone) species. Giant squid are mollusks that grow over 13 m long and live deep in the ocean. Slugs, snails, and octopi also belong to this phylum.

**Phylum Annelida.** Segmented worms, including earthworms and leeches, belong to this phylum. The smallest species of segmented worms is 0.5 mm; one of the largest, the giant Australian earthworm *Megascolides australis*, is 3 m long.

**Phylum Echinodermata.** Organisms from this phylum are all marine animals and they have five radiating body parts. Many echinoderms can regrow lost body parts. Starfish, sand dollars, and sea urchins are echinoderms.

**Phylum Arthropoda.** All the animals in this phylum have segmented bodies and they have exoskeletons, which is a hard covering on the outside of the body. Insects, spiders, and crabs are arthropods. This phylum is the largest of the phyla. Zoologists do not even know the total number of species of arthropods living on Earth. The insects in this phylum have a valuable role in the world because they pollinate flowering plants and crops, making it possible for these plants to reproduce.

**Phylum Chordata.** Vertebrates have a backbone or a notochord, which is a simple backbone. You are probably the most familiar with this phylum because you are part of it. Phylum Chordata contains five classes:

Class Pisces, which includes about 30,000 species of fish.

Class Amphibia, which includes approximately 7,000 species of amphibians like salamanders, newts, toads, and frogs.

Class Reptilia, which includes about 10,000 species of reptiles like lizards, snakes, turtles, and crocodiles.

Class Aves, which includes about 10,000 species of birds.

Class Mammalia, which includes over 5,400 species of mammals, more than 1,000 of which are bats. Humans are a member of this class as well.

## LESSON REVIEW

1. How are eukaryotes different from prokaryotes?
2. Comment on the variety of protists, fungi, plants, and animals that God placed on the earth.
3. Describe the characteristics of each kingdom in domain Eukarya.
4. There are over 900,000 known insect species on Earth as compared to the approximate 5,400 known mammal species. What does this tell you about God's creativity?

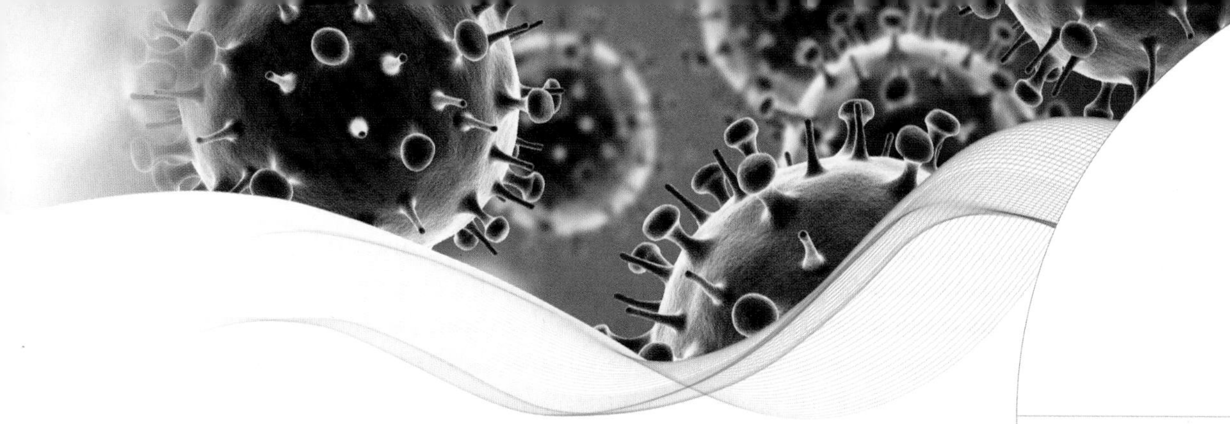

## Chapter 1: *Viruses*
## Chapter 2: *Bacteria and Archaea*
## Chapter 3: *Fungi*
## Chapter 4: *Protists*

*Vocabulary*

| | | |
|---|---|---|
| AIDS | heterotroph | plankton |
| air bladder | HIV | preservative |
| antibiotic | holdfast | protozoa |
| autotroph | hyphae | pseudopod |
| bacteriophage | influenza | rabies |
| blade | malaria | retrovirus |
| budding | mold | saprophyte |
| cilia | mushroom | spore |
| conjugation | mutualism | sporozoan |
| decomposer | mycelium | symbiosis |
| electron microscope | nitrogen-fixing | unicellular |
| endospore |   bacteria | viral infection |
| extremophile | pasteurization | virus |
| flagellum | pathogen | yeast |

**Viruses, Bacteria, Archaea, Fungi, and Protists**

*Key Ideas*

- Systems, order, and organization
- Evidence, models, and explanation
- Change, constancy, and measurement
- Form and function
- Abilities necessary to do scientific inquiry
- Understandings about science and technology
- Structure and function in living systems
- Diversity and adaptations of organisms
- Science as a human endeavor
- Risks and benefits
- Science and technology in society
- Science as a human endeavor
- Nature of science
- History of science

SCRIPTURE

For in Him all things were created: things in heaven and on earth, visible and invisible ... all things have been created through Him and for Him.

Colossians 1:16

Have you ever had a virus? The answer must be yes because even if you have never had influenza or the chicken pox, you cannot have escaped the common cold. When people say they have a virus, they mean that they have an illness caused by a virus. Some of these illnesses are minor; others are serious. Today viruses can cause deadly diseases such as AIDS and hepatitis. Even though viruses are extremely small, they play a destructive role in the world.

What comes to mind when you hear the word *virus*? You probably think of feeling sick or of being vaccinated for measles or mumps, two illnesses caused by viruses. But what exactly are viruses? What are these things that can cause so much misery?

A **virus** is a tiny particle that contains nucleic acid—either DNA or RNA—encased in protein. Nucleic acids are organic compounds that contain genetic information and the information necessary for an organism to make the protein it needs. Some viruses are also surrounded by a membrane. Given all the trouble they cause, you might be surprised to learn how small viruses are. They are much smaller than the tiniest cell—only 17–300 nanometers. (A nanometer is one-billionth of a meter.) Tens of thousands of the largest viruses could line up in a centimeter.

## TRY THIS

**Virus Model**
Construct a simple virus model. Twist two pipe cleaners to represent DNA or use single pipe cleaners to represent RNA strands of nucleic acid. Use plastic connecting bricks or other building materials to surround the virus and construction paper to surround the genetic material. Compare your virus to electron micrographs of actual viruses. In relation to the approximate size ratio of a virus to a cell, how large a cell would your virus infect?

Because viruses are about 300 times smaller than a cell, for centuries people knew them only by the diseases that they caused. But when the electron microscope was invented in 1931, microbiologists were able to look at viruses. An **electron microscope** uses a beam of electrons instead of light to detect objects, which allows objects as small as 0.5 nanometers in diameter to be seen. The electron microscope allowed scientists to see that just as each disease caused by a virus is different, so the size

**Virus Shapes**

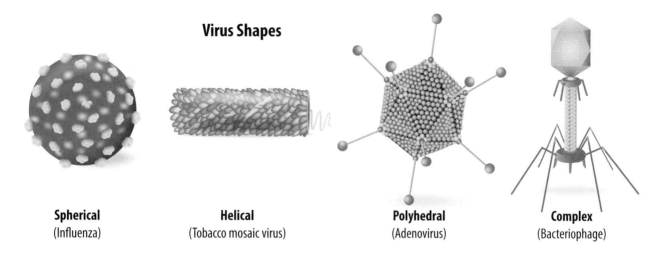

**Spherical**
(Influenza)

**Helical**
(Tobacco mosaic virus)

**Polyhedral**
(Adenovirus)

**Complex**
(Bacteriophage)

and shape of each virus is unique. In fact, a virus can be sphere-shaped, rod-shaped, spiral-shaped, or thread-shaped. Some viruses even look like tiny spaceships.

Despite everything that scientists know about viruses, they do not know the answer to one basic question: Are viruses living things? Viruses can reproduce and mutate, but only by infecting a living cell. They contain proteins that are found only in living things. Viruses also have genetic material—the RNA or DNA that a cell needs to reproduce itself. In these ways viruses are like living things.

However, viruses are not made of cells, which are the building blocks of life. Viruses cannot mutate or reproduce outside of host cells. They can survive in an inactive form for years outside of host cells. They have to take control of cells to reproduce, which means they do not develop like other organisms do. A virus's genetic material resembles its host cell's genetic material more than it resembles the genetic material of other viruses. In these ways, viruses are not like living things.

## LESSON REVIEW

**1.** How did the invention of the electron microscope help scientists learn more about viruses?

**2.** How are viruses like living things?

**3.** How are viruses not like living things?

**4.** Describe the basic characteristics of viruses.

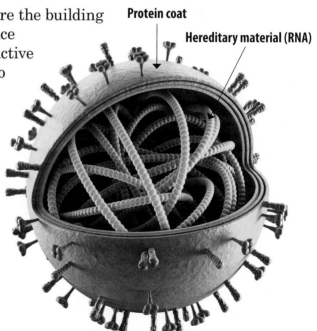

Protein coat

Hereditary material (RNA)

The protein spikes that extend from the influenza virus mutate easily. That's why each flu season involves a different strain of influenza. People must build a new immunity with each mutation.

## 2.1.2 *Virus Reproduction*

Occasionally, a new computer virus appears. A computer virus is a piece of software designed to enter a computer system and "infect" its files. Computer viruses damage or even destroy the computer's data. Once they are in a computer, computer viruses reproduce themselves. Then they enter as many files as possible. Sometimes a person passes on a computer virus to another computer without even knowing it through an e-mail or a USB.

Viruses that infect organisms act a lot like computer viruses. One researcher said that viruses are genetic programs that carry one simple message: Reproduce me! Viruses cannot reproduce unless they are in host cells. They are very specialized. Each virus can work only in certain types of cells, and it can infect only certain species of bacteria, plants, or animals.

A **viral infection** is the penetration of a virus or its nucleic acid into a host cell. The infection can begin in several ways. Sometimes it begins when a virus injects its genetic material into a cell. A virus that infects bacteria is called a **bacteriophage**. The bacteriophage moves into a bacterial cell in a step-by-step invasion. First, it attaches its protein coat to the surface of the bacterial cell. Then the virus injects its nucleic acid into the cell. The protein coat is left outside the cell. The nucleic acid instructs the cell to produce new protein coats and nucleic acids, which then form new viruses. This process continues until the bacteria cell bursts, releasing hundreds of new viruses.

### Bacteriophage Reproduction

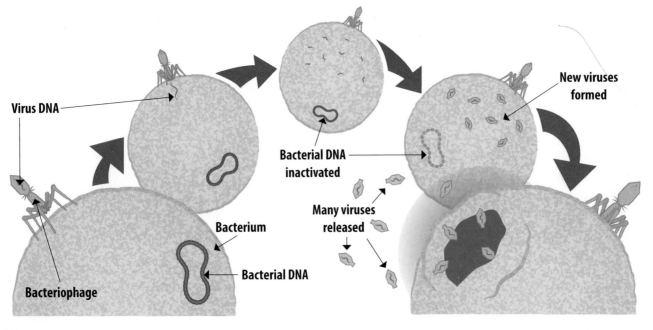

Virus DNA

New viruses formed

Bacterial DNA inactivated

Bacterium

Many viruses released

Bacterial DNA

Bacteriophage

The viruses that infect animals and plants work a little differently from bacteriophages. Infection in animal cells begins when an entire virus is engulfed by the host cell. Infection in plant cells begins when a virus enters a wound in the plant and penetrates the plant's cells. Once inside the plant or animal cell, the virus takes control of the cell's activities. Some viruses make toxins that destroy a host cell's genetic material. Then they convert the cell into a virus-making factory. Other viruses use the host cell's genetic material to produce more virus particles.

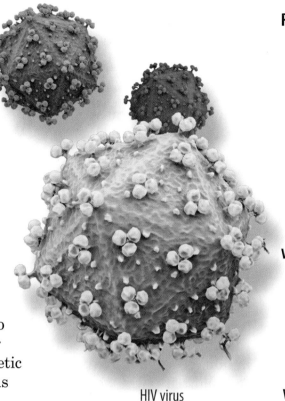

HIV virus

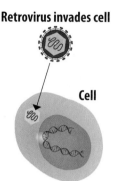

## Retrovirus Reproduction

**Retrovirus invades cell**

**Cell**

**Virus RNA changes to viral DNA**

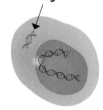

**Viral DNA is inserted into host cell genetic material**

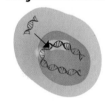

**Duplicates of viral RNA**

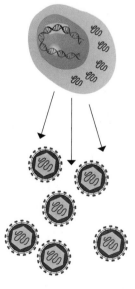

**New retroviruses**

Still another type of viral infection is caused by retroviruses. A **retrovirus** is a virus that has its genetic material in the form of RNA. After invading a host cell, the virus uses a special enzyme to change its single-strand RNA into a double-strand of viral DNA. Then the viral DNA is "retrofitted," or custom fitted, into the chromosomes of the host. This "retro" DNA is added to the host's genetic material, hiding among the host's DNA strands. The viral DNA may not do anything for a time. At some point, however, the DNA makes viral protein and RNA, which are assembled as new retroviruses in the host cell. Retroviruses are stealth viruses that may not show any sign of their presence for months or even years. HIV, the virus that causes AIDS, is a retrovirus.

The process of infection varies from virus to virus, but the basic pattern is the same. The virus gets itself or its genetic material into the host cell. Then the host cell makes more virus particles, and the virus particles leave the host cell to infect more cells.

## LESSON REVIEW
**1.** What is the basic way that viruses reproduce?
**2.** Name three ways a virus might infect a cell.

## OBJECTIVES

- Name several diseases that viruses can cause.
- Describe how vaccines work.

## VOCABULARY

- **AIDS** a condition that destroys the body's immune system
- **HIV** the virus that causes AIDS
- **influenza** a contagious viral infection that causes muscle aches, inflammation of the respiratory system, fever, and chills
- **rabies** a potentially fatal viral disease transmitted by the bite of an infected animal

Tobacco leaf with TMV

Viruses have a great impact on living things. They can invade all forms of life, including humans, animals, plants, and bacteria. Some scientists believe that almost every living thing on Earth has a viral infection associated with it.

Plant viruses are quite common. A well-known plant virus is the tobacco mosaic virus, or TMV. If the virus is rubbed onto the leaves of a tobacco plant, TMV forms in those spots. The rubbing damages the cells, enabling the virus to enter.

Animal viruses are widespread and varied. Viruses from the papillomavirus family cause warts in humans, rabbits, and dogs. Some of these viruses can cause cancer in small rodents such as mice and hamsters. The influenza virus also affects animals. Birds and mammals are easily affected by strains of the influenza virus. In 1997, chickens in Hong Kong became sick with avian influenza. Poultry farmers had to kill 1.5 million of the infected birds. **Rabies** is another viral disease that affects mammals. Rabies is a potentially fatal viral disease that is transmitted by the bite of an infected animal. The carrier can be infected for weeks without showing visible signs of the disease.

Viruses cause many human diseases. The most familiar to you is probably the common cold. A cold can be caused by any one of hundreds of different strains of the cold virus, which is why people are not immune after catching one cold. The cold virus enters a person's body through the nose or mouth. After a two- to five-day incubation period, the person exhibits cold symptoms. These symptoms usually last four to six days. Other viral

 **BIOGRAPHY**

Carlos Finlay (1833–1915) was a Cuban physician and scientist who identified the mosquito as the carrier of the yellow fever virus. Yellow fever is an often-fatal disease that involves fever, chills, rapid heartbeat, headache, back pains, vomiting, jaundice, and internal bleeding.

In 1881, Finlay suggested that if mosquito populations could be controlled, yellow fever would not spread so quickly. Unfortunately, few people believed him; instead, he was called *Mosquito Man*. Not until 1900 was a study initiated that supported his hypothesis. Finlay was appointed chief sanitation officer in Cuba in 1902, and after his death the Cuban government created the Finlay Institute for Investigations in Tropical Medicine to honor his work.

diseases include chicken pox, measles, mumps, hepatitis, and influenza (flu). Worldwide, viral infections cause up to 11 million deaths a year, not counting severe influenza epidemics.

Two well-known viruses are influenza and Human Immunodeficiency Virus, or HIV. **Influenza** is a contagious viral infection that causes muscle aches, inflammation of the respiratory system, fever, and chills. There are several strains of the influenza virus. Some strains have been devastating. The Spanish influenza virus caused a tragic worldwide epidemic in 1918–1919 that killed 50 million people. Other strains that caused severe outbreaks include the Asian flu in 1957, the Hong Kong flu in 1968, and the Russian flu in 1977.

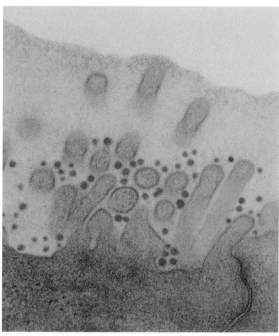

This electron micrograph shows rhinoviruses, which cause the common cold, infecting epithelial cells in the nose.

Influenza is dangerous. It can be fatal for the elderly or for people with weak immune systems. Some strains have killed healthy people just as quickly as people in the high-risk categories. Researchers and government officials worry about another worldwide epidemic, or pandemic, of a dangerous strain. In 2009, the H1N1 influenza virus was declared a pandemic. It caused great concern and was the cause of most of the human influenza that year.

One of the most deadly infectious viruses is **HIV** (Human Immunodeficiency Virus). HIV is transmitted from an infected person to another person through bodily fluids. It is the virus that causes **AIDS** (Acquired Immunodeficiency Syndrome). AIDS is not a specific disease. It is a condition that destroys the body's immune system, making the victim vulnerable to many different diseases. People with AIDS become unable to fight off diseases such as pneumonia and cancer. One of the first symptoms of AIDS is fever and swollen lymph glands. People with AIDS also lose a lot of weight and feel tired.

Viruses spread through animal or human populations in many different ways. Some viruses, such as measles, smallpox, and influenza, are spread from person to person by droplets in the air from a sneeze or a cough. The hepatitis A virus spreads through infected food or drink, hepatitis B through bodily fluids, and hepatitis C through infected blood products. Some viral infections, such as yellow fever, are spread by mosquitoes.

### FYI

**Virus Epidemic**
In 1793, yellow fever wiped out 10% of Philadelphia's population. At that time, Philadelphia was the largest city in the United States and the country's capital.

Although many viruses are harmful, some are used for human benefit. Scientists have found ways to make vaccines from weakened or "killed" viruses. This process depends on the way that live viruses function. Live viruses have two effects—they make people feel sick and they trigger an immune response. This response strengthens until the viruses become fewer in number, the symptoms disappear, and the person recovers. The immune response is caused by proteins called *antigens*. The antigens trigger the antibodies that fight infection. Memory cells are also produced, and these cells remain in the bloodstream, often for a lifetime. The memory cells respond quickly if the same infection ever reenters the body. Memory cells prevent infection if a person is exposed to the same virus at a later time. The weakened or inactive viruses that scientists use for vaccines do not cause illness; they just cause the body to produce the necessary antibodies to ward off the infection. So if you have received a measles vaccination, you will not get measles if you are exposed to it later. The influenza virus mutates so quickly that each year new strains arise, requiring a new, different vaccine.

 **BIOGRAPHY**

Jonas Salk (1914–1995) was born in New York City, where he graduated from New York University School of Medicine in 1939. He spent years researching epidemics, what caused them, and how to control them. Salk made many important contributions to the understanding and control of influenza and other infectious diseases. He is most famous for developing the first successful polio vaccine. The polio virus enters the body through the mouth, invades the bloodstream, and may be carried to the central nervous system, where it affects nerve cells in the spinal cord and brain, causing muscle paralysis. Many polio victims recover

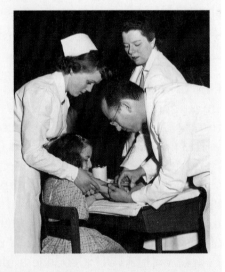

fully, but polio can be fatal or cause permanent disabilities. In 1953, Salk announced that he had developed a trial vaccine for polio. The following year 1,830,000 children took part in a nationwide field trial to test the safety and effectiveness of the vaccine. Salk's vaccine and a second polio vaccine developed by Albert Sabin are being used in an international effort to make polio extinct throughout the world. Salk received many honors, including a Congressional gold medal for his achievement in the field of medicine. In 1960, Salk established the Salk Institute for Biological Studies in La Jolla, California. He dedicated the rest of his life to improving human health through his medical research, lectures, and writing.

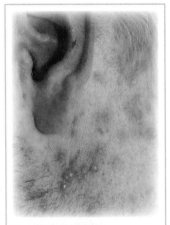

The chicken pox virus causes blistery sores all over the body.

A recent medical breakthrough is the chicken pox vaccine. During the 1900s, an estimated 4 million children under the age of 15 had chicken pox. Those who did become sick then carried the dormant virus strain all their lives. Sometimes the virus would reappear in the form of shingles later in life. Scientists in Japan created a chicken pox vaccine in 1981. The vaccine was made available in the United States in 1995. Since introduction of the vaccine, the occurrence of chicken pox has declined 97%. Scientists use vaccines to make the world a healthier place.

Some viruses have been used to kill bacteria, insects, and other pests that cause diseases. Scientists are looking for ways to treat hereditary diseases using viruses. They insert specific hereditary material into viruses to program them to treat certain diseases such as cystic fibrosis. In addition, scientists have infected some plants with viruses that carry hereditary material in order to improve the plants.

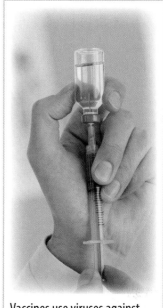

Vaccines use viruses against themselves, allowing the person receiving the vaccine to escape that illness in the future.

## LESSON REVIEW

**1.** What do viruses infect?
**2.** What are some ways that viral infections spread or are transmitted?
**3.** Name four diseases that are caused by viruses.
**4.** Describe how vaccines work.

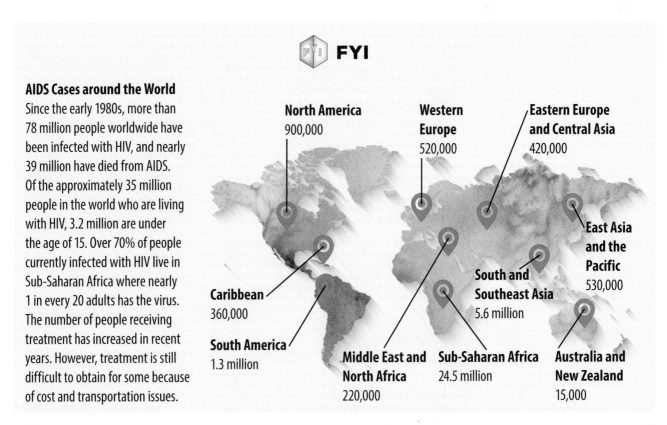

**FYI**

**AIDS Cases around the World**
Since the early 1980s, more than 78 million people worldwide have been infected with HIV, and nearly 39 million have died from AIDS. Of the approximately 35 million people in the world who are living with HIV, 3.2 million are under the age of 15. Over 70% of people currently infected with HIV live in Sub-Saharan Africa where nearly 1 in every 20 adults has the virus. The number of people receiving treatment has increased in recent years. However, treatment is still difficult to obtain for some because of cost and transportation issues.

**North America**
900,000

**Western Europe**
520,000

**Eastern Europe and Central Asia**
420,000

**East Asia and the Pacific**
530,000

**South and Southeast Asia**
5.6 million

**Caribbean**
360,000

**South America**
1.3 million

**Middle East and North Africa**
220,000

**Sub-Saharan Africa**
24.5 million

**Australia and New Zealand**
15,000

### OBJECTIVES

• Describe the structure of bacteria.

### VOCABULARY

• **flagellum** a thin, whiplike structure that helps an organism move through liquid

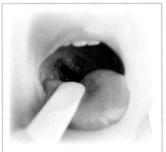

The creatures in your mouth right now outnumber all of the people living on Earth!

### FYI

**Bacteria by the Number**
Bacteria are everywhere. If each bacterium on Earth were the size of a penny, and you stacked all of these bacteria end to end, the stack would be at least 1 trillion light-years high, which is 9.46 trillion km long!

Imagine a group of creatures that live on land, in freshwater, in saltwater, and in the air. These creatures are so small that most of them are invisible without a powerful microscope. Although they are small, they are tough; in fact, they can live almost anywhere. Some live in boiling acid, some live in sulfur gases rising from deep ocean trenches, and some spend much of their lives in the air or on the highest mountaintops. Others survive being frozen for years and then thawed. Still others can even live inside nuclear reactors! One billion of them can live in a small drop of saliva. These tiny creatures keep the entire world alive—but they can also cause diseases. What are these amazing creatures? Bacteria.

Bacteria are prokaryotes—unicellular organisms without a nucleus. Although bacteriologists and microbiologists, the scientists who study bacteria, have identified thousands of species of bacteria, they believe that most species are as yet undiscovered. Scientists used to think that bacteria were very simple organisms, but they now know that bacteria are very complex. Most bacteria are between 0.3 and 2 micrometers in diameter. (One micrometer is equal to one millionth of a meter.) Bacteria can be seen only through a microscope. But even though bacteria are so small, God crafted each one with a distinct structure.

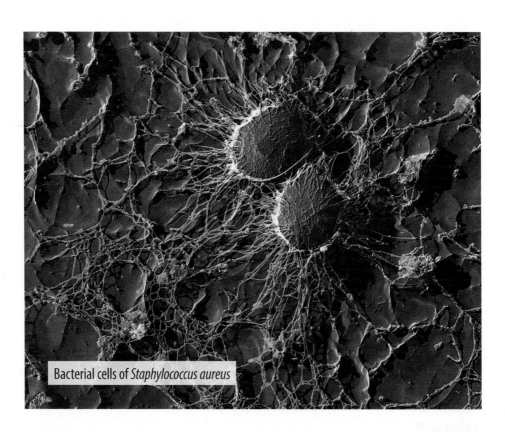

Bacterial cells of *Staphylococcus aureus*

Each bacterium is enclosed in a tough protective cell wall. The cell wall determines a bacterium's shape and makes it possible for a bacterium to live in a wide variety of environments. Some bacteria have a slimy layer outside the cell wall. This layer, called *the capsule*, protects the cell from destructive chemicals. Inside the cell wall is a cell membrane that controls the substances moving in and out of the cell. The cell membrane allows small food particles to pass through but keeps large particles out. Within the membrane is a jellylike substance called *cytoplasm*. The cytoplasm contains enzymes, which help break down food and build cell parts. Most cells, such as plant or animal cells, contain hereditary material in a nucleus surrounded by a nuclear membrane. Because bacteria do not have nuclei, their hereditary material moves freely around the cytoplasm. Bacteria lack most organelles, including the Golgi apparatus, endoplasmic reticulum, lysosomes, and mitochondria. They do, however, have ribosomes to create new proteins.

The cell wall and cell membrane of this cyanobacterium (magnification 5,335×) can be easily identified in this transmission electron micrograph.

Some bacteria have a **flagellum**, a thin, whiplike structure that helps an organism move through liquids. Other bacteria move by wriggling. Many bacteria cannot move on their own. They

## Structure of Bacteria

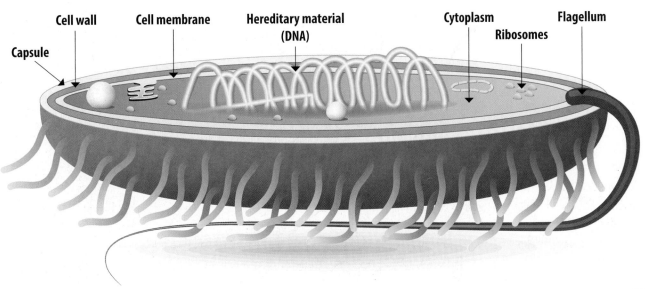

Capsule

Cell wall

Cell membrane

Hereditary material (DNA)

Cytoplasm

Ribosomes

Flagellum

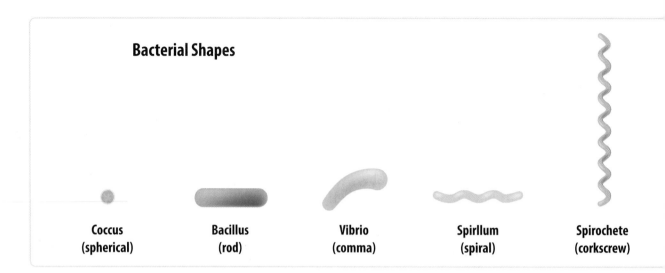

**Bacterial Shapes**

| Coccus (spherical) | Bacillus (rod) | Vibrio (comma) | Spirllum (spiral) | Spirochete (corkscrew) |

are carried by wind, water, clothing, animals, people, and other objects that they happen to land on.

Because bacteria are so small, it is easy to think that they all look alike, but they do not. Bacteria come in many different colors and shapes. Bacteria are often grouped according to shape. Cocci are round; bacilli are rod-shaped, vibrios look like bent rods and spirilla and spirochetes are spiral-shaped.

Some bacteria live alone, whereas others are attached to one another. The names of bacteria linked together in pairs or groups have descriptive prefixes. *Diplo* means "pair," *staphlo* means "cluster," and *strepto* means "chain." The name of a bacteria can tell you a lot. What are streptococci? They are a type of round bacteria that are linked in chains.

## FYI

**Bacteria Giants!**

In April 1999, scientists broke the news that not all bacteria are microscopic. Dr. Heide Schulz discovered *Thiomargarita namibiensis*, bacteria that can be up to 0.75 mm in diameter (about the size of the period at the end of this sentence). If an ordinary bacterium were the size of a bee, these newly discovered bacteria would be the size of a blue whale.

 **HISTORY**

**First Glimpse**

"The fourth of June in the morning I saw a great abundance of living creatures; and looking again in the afternoon of the same day, I found great plenty of them in one drop of water. . . . They looked to my eye, through the Microscope, as common as sand doth to the naked eye." —from the writing of Antonie van Leeuwenhoek

In 1676, Antonie van Leeuwenhoek wrote this description of bacteria. He was the first person to see these microscopic wonders. Since Van Leeuwenhoek's time, scientists have filled volumes with information about bacteria.

## Bacterial Groups

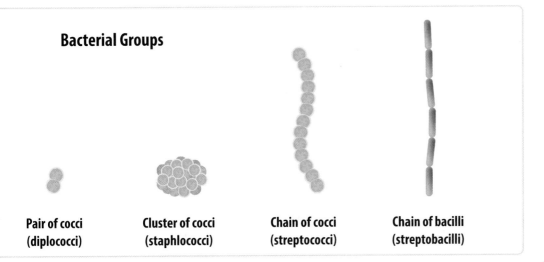

**Pair of cocci (diplococci)**

**Cluster of cocci (staphlococci)**

**Chain of cocci (streptococci)**

**Chain of bacilli (streptobacilli)**

## LESSON REVIEW

**1.** What are bacteria?

**2.** Describe the structure of bacteria.

**3.** Describe the various shapes that bacteria can have.

**4.** Where do bacteria thrive?

## TRY THIS

**Bacteria on Beans**

Place several lima beans in a wide-mouthed jar. Cover them with water and let them soak for three or four days. Then put the lid on the jar and keep it in a warm, dark place for three or four more days. Using an eyedropper, remove a small sample of water from the surface. Place the sample on a microscope slide. Place a cover slip over the sample. Using the microscope, observe the bacteria.

 **CAREER**

### Microbiologists

A microbiologist is a scientist who studies organisms and other infectious agents (such as viruses) that are usually only visible with a microscope. Microbiologists study how these microorganisms interact with people and the rest of the world, and they help find cures for diseases caused by microorganisms.

Some microbiologists are highly specialized. Biochemists are microbiologists who study how microorganisms use energy, use nutrients, and reproduce. Mycologists study molds, yeast, and other fungi. They determine how fungi infect other living things and what role fungi play in the world. Parasitologists study microorganisms that are parasites. Virologists study viruses 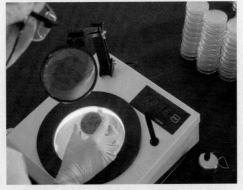 and the diseases they cause. Bacteriologists research bacteria, develop antibiotics to fight bacteria-caused diseases, and study how bacteria help the environment.

Even though microorganisms are so small that you often forget they exist, they are a big part of the world. Just ask a microbiologist!

## OBJECTIVES

- Describe two methods of bacterial reproduction.
- Identify bacteria according to how they obtain food.

## VOCABULARY

- **autotroph** an organism that makes its own food
- **budding** a type of asexual reproduction in which a small outgrowth of the parent develops into an independent organism
- **decomposer** an organism that breaks down dead plant and animal material to return it to the soil
- **endospore** a protective capsule that some bacteria form
- **heterotroph** an organism that obtains food from an outside source

You wake up one morning feeling miserable. None of your friends are sick; everyone else in your family is healthy. How could you be sick? Of course, you have no way of knowing that your finger picked up a couple of bacteria from an elevator button yesterday and a few seconds later, with a quick rub, you planted them in your eye. You know that bacteria can survive harsh conditions, so it should not surprise you to know that they could live on an elevator button, but how can a couple of tiny bacteria make you feel so awful—and within only a few hours?

Bacteria can grow to full size and reproduce with amazing speed. This explains why you can quickly become sick by being exposed to only a few bacteria. Bacteria reproduce only when the conditions are right: when food is plentiful and the environment is favorable. But under the right conditions, many bacteria grow and are ready to reproduce after only 20 minutes. With that rate of reproduction, it may seem that bacteria would take over the world, but that will not happen because bacteria reproduce only under certain conditions.

All bacteria reproduce asexually. Asexual reproduction is reproduction that involves only one parent and in which the offspring is identical to the parent. The organism reproduces by

Bacteria can be found on all kinds of surfaces. Washing your hands after touching objects that other people or animals have touched is the best way to prevent illness.

itself, usually by splitting into two cells. Different species do this in different ways. Most bacterial species reproduce through binary fission. Binary fission is cell division in which one cell splits into two identical daughter cells, each having a complete set of DNA. Some bacterial species reproduce through **budding**, asexual reproduction in which a small outgrowth of the parent develops into an independent organism.

Bacteria need the right conditions not only for reproducing but also for living. Just as God created bacteria with different colors and shapes, God designed different methods to maintain them. Aerobic bacteria need oxygen to live, but anaerobic bacteria do not. In fact, some anaerobic bacteria die in the presence of oxygen. For example, fermenting bacteria do not use oxygen. Fermentation is the process by which sugars and starches are converted into commercially important products such as cheeses and yogurts.

When some bacteria find themselves in conditions that might harm them, they take action. These bacteria put a copy of their DNA in an **endospore**, a protective capsule that protects it from extreme heat or other harsh conditions. Endospores may resemble brightly colored spheres, tiny trees, or dark-colored cysts. Endospores allow some bacteria to remain dormant, or in a resting state, for decades. When conditions become favorable, the bacteria once again become active.

Some bacteria are **autotrophs**, organisms that make their own food from simple substances. Many autotrophic bacteria use sunlight and water to produce energy through photosynthesis. Cyanobacteria (once called *blue-green algae* and considered a part of the plant kingdom) are aerobic bacteria that perform photosynthesis and are an important source of oxygen for the earth. Other bacteria use hydrogen sulfide instead of water during photosynthesis. Instead of making oxygen as green plants do, they make sulfur. Have you ever smelled rotting food? The sulfur-like smell comes from these anaerobic bacteria.

Some bacteria are **heterotrophs**, organisms that must obtain their food from an outside source.

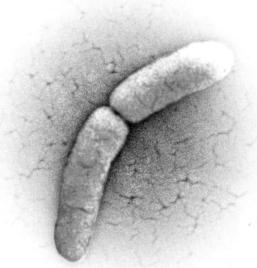

This *Salmonella typhimurium* bacterium (scanning electron micrograph 15,000× magnification) is undergoing binary fission.

## TRY THIS

**Stop the Bacteria!**
Pour milk into two jars, and screw on the lids. Place one jar in a refrigerator; place the second jar in a warm place. Examine the milk in each jar daily for seven days. Record what happens each day.

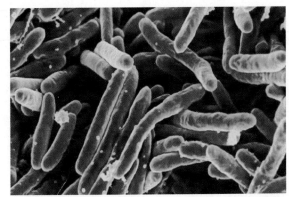

These *Mycobacterium tuberculosis* bacteria reproduce by budding.

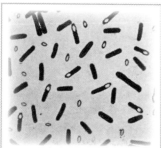

*Clostridium botulinum* can be found in soil and grow best in low-oxygen conditions. The bacteria form endospores, which allow them to survive in a dormant state until conditions can support their growth.

**Binary Fission**

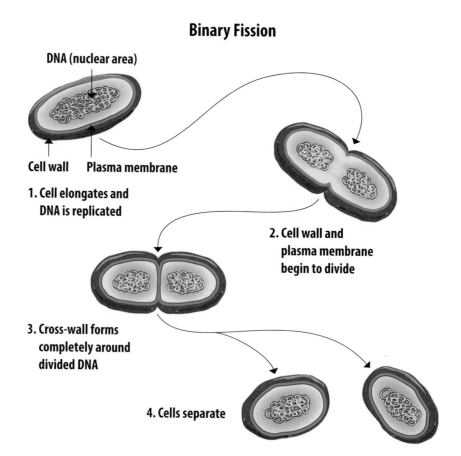

DNA (nuclear area)

Cell wall    Plasma membrane

**1. Cell elongates and DNA is replicated**

**2. Cell wall and plasma membrane begin to divide**

**3. Cross-wall forms completely around divided DNA**

**4. Cells separate**

Some of these bacteria feed on other living things, causing infections in plants, animals, and people. As you might have guessed, the bacteria that make you sick are heterotrophs. Other bacteria are **decomposers**, organisms that get their energy by breaking down the remains of dead organisms or animal wastes. They decompose, or break down, dead matter into simpler substances. They help keep dead organisms (both plants and animals) from piling up, and at the same time they return important materials to Earth's water and soil.

## LESSON REVIEW

**1.** Describe two ways that bacteria can reproduce.
**2.** What are endospores?
**3.** Describe three ways that bacteria can get food.

Some marine bacteria degrade petroleum, such as this oil-oxidizing bacteria.

Take a moment to think about the goodness of creation. What comes to mind? You probably thought of your family, friends, favorite animals, trees, or foods. Or maybe you thought about mountains and lakes or the moon, the stars, or the sun. You probably did not think of bacteria. If you believe the hype, all bacteria are disgusting. It is true that some bacteria are harmful, but if all the bacteria on Earth were killed, all other life on the planet would be in big trouble. Most bacteria are helpful. Bacteriologists are learning more about the role of bacteria in the environment. Knowing that God created bacteria to help creation—including mankind—can help people better appreciate bacteria.

Bacteria are a crucial part of the food and energy relationships in the world. Decomposers, including bacteria, break down dead material to form simpler substances and return them to the soil. Then autotrophs such as green plants use these substances for life processes. These plants are food for some protists and animals, which are in turn eaten by larger animals. When these organisms die, decomposers break them down to begin the cycle again. So bacteria play an important role in the food chain. Bacteria also interact with the environment in ways that help maintain the balance of the environment. For example, bacteria populate bare rock left by volcanoes, helping to break it down into soil so that plants, fungi, protists, and animals can live there. Eventually the bare rock is covered with life.

You may think of plants and algae as the main organisms that use photosynthesis, but cyanobacteria also use energy from the

## OBJECTIVES

- Identify ways that bacteria are instrumental in the world's food chain.
- Discuss the importance of bacteria in maintaining a well-balanced environment.

## VOCABULARY

- **nitrogen-fixing bacteria** a type of bacteria that turns nitrogen into nitrogen compounds
- **symbiosis** a relationship in which one organism lives on, near, or inside another organism and at least one organism benefits

Bacterial mats cover these boulders at East Diamante volcano near the Mariana Islands.

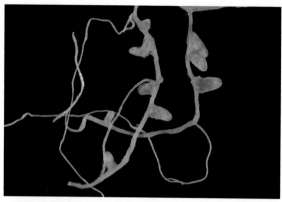

Nodules form when bacteria infect a growing root. The plant supplies carbon compounds to the bacteria, and the bacteria convert nitrogen from the air into compounds the plant can use.

sun to perform photosynthesis, which produces oxygen. These bacteria help maintain the balance of oxygen in the atmosphere for life to thrive.

God created many relationships between bacteria and other organisms. One type of relationship is **symbiosis**—a relationship in which one organism lives on, near, or inside another organism and at least one of the organisms benefits. For example, bacteria help cows, goats, elephants, people, and other plant-eating creatures digest plant materials. Millions of bacteria live in a termite's intestines. Without these bacteria, termites, which are decomposers, could not digest wood. Another example of symbiosis is the relationship between some bacteria and plants. **Nitrogen-fixing bacteria** turn nitrogen gas into nitrogen compounds that plants need to grow. Without these bacteria, most plants could not live. For growth, plants need nitrogen compounds such as ammonia and urea. Plants use these nitrogen compounds to make proteins. Without nitrogen-fixing bacteria to make nitrogen compounds, plants could not make proteins, which would affect the food chain.

Bacteria help clean the environment. Some bacteria are decomposers. These bacteria can be used to treat sewage or decompose garbage. Some bacteria can even break down oil spills, chemicals, and certain kinds of plastic. Scientists are changing the genetic material of bacteria to make them even more useful for these purposes. But some scientists worry that these new strains of bacteria will interact with natural bacteria. Scientists fear that this interaction may produce new bacteria species that may harm the environment or cause disease.

 **FYI**

**Super Bacteria!**
Bioremediation is a natural way of cleaning up the environment. Plants, bacterial decomposers, and enzymes are used to remove contaminants and restore balance to ecosystems.

*Alcanivorax*, a rod-shaped bacterium, is currently the world's most important oil-eating organism. These microbes use oxygen to gain energy. They are also halophilic, meaning they live in salty environments, such as salty water, which makes them ideal for cleaning up oil spills in the ocean.

Many fuels can be traced back to bacteria. Decomposing bacteria can break down garbage, such as plant material, manure, and sewage, to produce methane, a natural gas used for cooking and heating. Bacteria also helped create fossil fuels; heat and pressure inside the earth changed organisms, including bacteria, into petroleum. Petroleum is used for heating oil, gasoline, and many other useful substances.

Bacteria even have many industrial uses. Certain bacteria are used to extract valuable metals from rock, such as copper or gold. Bacteria are also used to color food and cosmetics, tenderize meat, process fabric and paper, and remove stains from clothing or furniture. However, harmful bacteria can interfere with industrial processes and damage products. Bacteria can damage leather and fabric, ruin paper pulp, break down asphalt (damaging paved surfaces), infest the water supply, and turn fruit juice into vinegar.

## LESSON REVIEW
1. How are bacteria part of the food and energy relationships in the world?
2. Give one example of symbiosis involving bacteria.
3. How do bacteria help clean up the environment?

### FYI

**Useful Bacteria**
Researchers have found that certain bacteria expand and contract when exposed to heat. Scientists are now combining technology with these bacteria to create new sportswear. The clothing adapts to a person's heat. The technology is not limited to sportswear. There are ideas for lamp shades that move in response to the heat of a light bulb and tea bags that signal when they are perfectly steeped. Bacteria and its uses are not limited to the field of science. Many areas of interest benefit from utilizing bacteria.

© Life Science

## 2.2.4  Bacteria and Humans

### OBJECTIVES

- Identify several foods made with the help of bacteria.
- Explain how bacteria help keep people healthy.

### VOCABULARY

- **antibiotic** a drug used to kill harmful bacteria

The bad reputation of bacteria is undeserved. Although some bacteria cause diseases, the benefits bacteria provide are far greater than the harm they cause. Bacteria help humans in a variety of ways.

Even though certain kinds of bacteria in and on food can make people sick, bacteria are responsible for much of the food that humans eat. For example, nitrogen-fixing bacteria help fertilize crops. Bacteria are also involved in the production of foods such as vinegar, cheese, buttermilk, sour cream, soy sauce, and pickles. The tangy taste of yogurt and the sour taste of sourdough bread come from bacteria. A few people even eat cyanobacteria (once called *blue-green algae*). These bacteria, which grow in water as large masses of strands, are nutritious and easy to digest, although it might not seem appetizing to make a lunch of bacteria!

Bacteria help keep people healthy. The bacteria in your intestines help digest food and make Vitamins K and $B_{12}$. Other good bacteria in your body help crowd out disease-causing bacteria.

Scientists are always looking for more ways to use bacteria to help humans. Scientists use genetic engineering—the manipulation of the genetic information in organisms to alter

## FYI

**Overusing a Good Thing**

More than 100 antibiotics and thousands of antibacterial products are available today. Antibiotic use has risen sharply in recent decades—the United States alone produced 1 million kg of antibiotics in 1954 and over 16 million kg in 2013. The overuse of these drugs and antibacterial products allow other bacterial species known as *superbugs* to flourish. Superbugs develop when the genetic material of some bacteria make them resistant to antibiotics. These few bacteria survive and reproduce and their offspring inherit the same genetic resistance. No members of the new bacteria community are affected by the antibiotic.

Consumers can help keep antibiotics effective by taking such medicines only as directed, by not asking the doctor for an antibiotic when it is not needed, and by washing their hands with standard soap rather than antibacterial soap.

Bacteria help make cheese delicious!

The green tablets are spirulina, a cyanobacterium known for its health benefits. The green powder and drink are chlorella, a green algae also known for its health benefits.

the hereditary traits of the organism—to alter bacteria for many uses. These uses include medicines, adhesives, and foods. Not all genetically modified food is beneficial. Scientists have found that some genetically modified foods lead to certain diseases.

Many bacteria have been genetically engineered to make large quantities of antibiotics. **Antibiotics** are drugs that are used to kill harmful bacteria. Scientists have also created genetically engineered bacteria to produce other medicines such as insulin. People who have diabetes take insulin because their bodies do not produce it. The body needs insulin to properly use sugars and carbohydrates.

## LESSON REVIEW

**1.** Name three foods that are produced using bacteria.

**2.** How do bacteria help keep humans healthy?

 **FYI**

**DNA Fingerprinting**

God created all DNA (genetic material) with the same chemical structure, but the order of the base pairs in DNA is different for each person. The sequences of DNA are a way to identify a person or to match tissue samples, such as hair and blood, to a person. DNA is often used to identify people's remains or to link a person to a crime scene.

Using someone's DNA for identification purposes is called *DNA fingerprinting*. DNA fingerprinting requires the enzyme DNA polymerase. Because the process of DNA fingerprinting requires high temperatures to break down the bonds in the DNA, the enzyme must be able to withstand high temperatures. The solution? Scientists use *Thermococcus litoralis*, bacteria that thrive in very high temperatures, as the source of DNA polymerase.

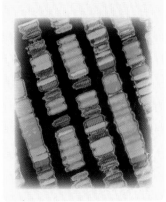

## 2.2.5 Bacteria and Disease

- Name several disease-causing bacteria and the diseases they cause.
- Determine where bacteria need to thrive and analyze how best to avoid being exposed to harmful bacteria.
- Describe methods of protecting food from being spoiled by bacteria.

VOCABULARY

- **pasteurization** the process of heating liquids to kill harmful bacteria
- **pathogen** a disease-causing organism
- **preservative** a chemical added to food to slow the growth of bacteria and mold

The world cannot function without bacteria, but there are some bacteria people would like to be without. These bacteria are **pathogens**, disease-causing organisms. Pathogens can cause diseases in humans, animals, and plants. Some of the diseases caused by bacteria are strep throat, cholera, Hansen's disease (leprosy), certain kinds of pneumonia, tuberculosis, typhoid fever, and whooping cough. These bacteria enter the body through natural openings (such as nose, eyes, or mouth) or through breaks in the skin. The bacteria destroy healthy cells, preventing the body from functioning properly. Other bacteria produce poisons called *toxins*, which cause diphtheria, scarlet fever, tetanus, and other diseases. Sometimes living bacteria produce toxins; other times toxins are released after the bacteria die. Some pathogens are anaerobic bacteria that produce powerful toxins causing serious diseases such as gangrene and botulism. Other bacteria cause food poisoning, infections, and blood poisoning.

Some bacteria that usually live in the body without causing harm may cause diseases when a person's resistance to infection is low. When resistance is low, the body cannot get rid of bacteria as fast as they reproduce. Bacteria in the throat, for example, might build up, causing a sore throat. That is why you are more likely to get sick when you are tired or under stress.

The primary method for destroying or stopping the growth of harmful bacteria is antibiotics. Different antibiotics are used to fight different bacterial infections. For example, penicillin is often used to cure strep throat, and sulfa antibiotics are often used for eye infections.

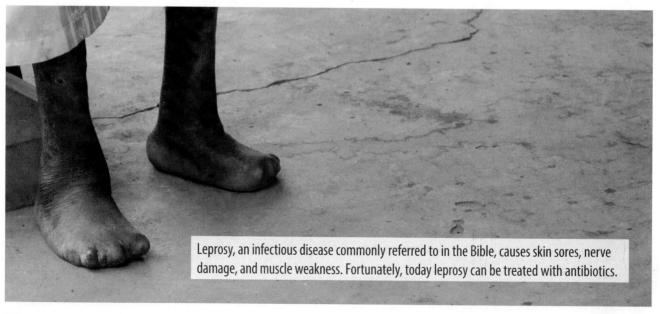

Leprosy, an infectious disease commonly referred to in the Bible, causes skin sores, nerve damage, and muscle weakness. Fortunately, today leprosy can be treated with antibiotics.

# Diseases Caused by Bacteria

| Disease | Bacteria | How Transmitted | Symptoms |
|---|---|---|---|
| **Cholera** | *Vibrio cholerae* | Untreated water, eating food prepared by someone with cholera | Severe diarrhea, vomiting, dehydration, possibly death |
| **Lyme disease** | *Borrelia burgdorferi* | Deer ticks | Skin rashes, headache, fever, and fatigue, possibly arthritis or death |
| **Gangrene** | *Clostridium perfringens* | Bacteria infection after blood supply to tissues is blocked following an injury or frostbite | Discoloration of skin tissue accompanied by a bad odor, death of tissues because of a lack of oxygen |
| **Botulism** | *Clostridium botulinum* | Food poisoning from improperly canned foods; botulism in babies comes from honey | Terrible stomach pains, partial paralysis, respiratory failure |
| **Tuberculosis** | *Mycobacterium tuberculosis* | Close contact with an infected person, inhaling bacteria | Bad cough that does not go away, fever, weight loss, night sweats, loss of energy and appetite |
| **Legionnaires' disease** | *Legionella pneumophila* | Inhaling contaminated water droplets in warm water environments | Loss of energy, nausea, fever, chest pains; later mental confusion can occur |
| **Bacterial meningitis** | *Neisseria meningitides* | Contact with saliva of infected person, coughing, or sneezing | Stiff or painful neck, fever, severe headache, nausea, vomiting |
| **Tetanus (Lockjaw)** | *Clostridium tetani* | Dirty cuts or wounds, especially deep cuts and puncture wounds | Headache and jaw stiffness followed by stiffness of neck, muscle spasms, sweating, and fever |

## TRY THIS

**Wanted**
Make a "wanted" poster for a specific harmful type of bacteria. Your poster should include a drawing, a description, information on how the bacterium attacks and spreads, its favorite victims, degree of danger, number of victims, hideout, most effective weapons against it, when it was first identified, and other interesting features.

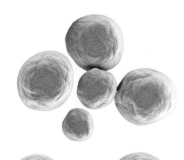

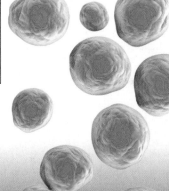

MRSA infection is caused by a strain of staph bacteria that have become resistant to the type of antibiotics commonly used to treat ordinary staph infections.

Canning helps to prevent bacterial contamination.

 **BIOGRAPHY**

### Alexander Fleming

Alexander Fleming (1881–1955), who was born in Scotland, is famous for discovering penicillin. Fleming and his brothers and sisters spent a lot of time exploring the streams, valleys, and moors of the Scottish countryside. Without even trying, he learned a great deal from nature. After Fleming graduated from Regent Street Polytechnic in London, he got a job at a shipping firm, but he did not like it. His brother Tom, who was a physician, encouraged Fleming to pursue medicine, so he went to medical school and studied bacteriology.

When World War I broke out, many of the staff of the bacteriology lab where Fleming was working went to France to set up a battlefield hospital lab. They found that scores of soldiers were dying from infections. Fleming believed that there must be a cure for these infections.

In the 1920s, Fleming searched for an effective antiseptic. He worked with so many things at once that his lab was often very cluttered and disorganized. Had he been neater, he might never have made his important discovery. In 1928, he was straightening up some old petri dishes in which he had been growing bacteria. He had left them piled in the sink when one of the dishes caught his attention. Mold was growing on it, and the bacteria around the mold had been killed, which was very unusual. Fleming took a sample of the mold and found it to be of the Penicillium family. He published a report on penicillin and its potential uses in the British Journal of Experimental Pathology. Chemists Howard Florey and Ernst Chain took over, refining and growing the mold so that it could be used as an antibiotic. When World War II broke out, interest in penicillin grew, and Fleming's work was further refined. Alexander Fleming was knighted in 1944, and he was awarded the Nobel Prize in 1945, along with Florey and Chain.

Have you ever swallowed sour milk or smelled rotten meat? If you have, you probably started paying more attention to expiration dates on foods. When foods spoil, bacteria are usually responsible. Bacteria break down food, making it taste bad or even poisoning it.

The food industry has devised many food packaging methods to prevent even one bacterium from reproducing. Many foods are processed by canning them. Canning is a process in which food is heated to kill the bacteria and sealed in airtight containers. **Pasteurization** is the process of heating milk or other liquids to kill harmful bacteria. Since most food-spoiling bacteria cannot grow without moisture and oxygen, many foods are preserved by freeze-drying, removing all of the moisture from a food before sealing it in an airtight package. Bacteria also need a warm

## TRY THIS

### Preservatives

Dissolve a bouillon cube in one cup of hot water. Divide the solution equally into three small glasses. Label the glasses *salt*, *vinegar*, and *control*. Add 1 tsp salt to the salt glass, 1 tsp vinegar to the vinegar glass, and add nothing to the control glass. Place the glasses in a warm spot for 2 days. Then observe the solution in each glass. Use a microscope to look for bacteria in each solution. Which substance is the best preservative against bacteria?

Milk is pasteurized to kill harmful bacteria.

environment to grow in, so keeping food cool in a refrigerator or a freezer slows down the growth of bacteria. Many foods are now treated with **preservatives**, chemicals added to food to slow the growth of bacteria and mold. Think of the foods you have eaten today. How were they preserved?

Disease-causing bacteria affect not only people but also animals and plants. Animals—especially cattle, horses, and sheep—can be infected with the bacterial disease anthrax. This disease can be passed to humans. Plants may contract fire blight, a disease occurring in apple and pear trees, or soft rot, a disease that decays some fruits and vegetables.

## LESSON REVIEW
**1.** What are some pathogens that cause diseases? What illnesses do they cause?
**2.** What do most bacteria need to thrive?
**3.** Name four methods of preserving foods.

Sodium benzoate is a preservative commonly used in carbonated drinks.

Thermophiles live in extremely hot environments such as Grand Prismatic Spring of Yellowstone National Park in Wyoming.

The term *bacteria* is commonly used to refer to all prokaryotic organisms found in both the bacteria and archaea domains. However, even though some similarities exist between the two types of microorganisms, there are enough significant differences that scientists categorize them into two separate domains. In 1977, microbiologist Dr. Carl Woese and his colleagues at the University of Illinois developed the three domain system (Bacteria, Archaea, Eukarya) to account for these differences as well as the eukaryotic organisms classified in the eukarya domain. Domains were officially added to the classification system in 1990.

Both bacteria and archaea are unicellular prokaryotes that have cell walls but lack interior membranes and organelles. Because of this they were originally classified together as monerans. Although archaea look similar to bacteria, they have a different cell structure and chemical makeup. Their genetic information (ribosomal RNA structure) and cell walls are not alike.

Archaea reproduce asexually using binary fission or budding. They are typically less than 1/1,000 mm long and have one or more flagella used for mobility. Like bacteria, archaea can be round (cocci), rod-shaped (bacilli), or spiral shaped (spirilla), although some archaea are unique in that they have unusual geometric shapes. Archaea live in a wide range of habitats and environmental conditions. Some archaea (as well as some bacteria) are **extremophiles**, which are microorganisms that live in extreme conditions.

Not all extremophiles are the same. For instance, halophiles are extremophiles that live in areas with high concentrations of salt,

 **FYI**

**Methanogens**

Methanogens are archaea that make methane. Methanogens are anaerobic archaea found in sewage and in marine and freshwater sediments; they also live in the intestines of animals, such as cows, sheep, and horses, to help them digest their food. Methanogens are also responsible for marsh gas produced in wetlands and sewage-treatment plants. Methane helps maintain the balance of gases in the atmosphere. Without methane, the atmosphere might have too much oxygen. But too much methane is harmful. Methane is a greenhouse gas that some researchers believe contributes to climate change. Because a large part of the methane in the atmosphere is emitted by livestock, scientists are developing animal feed that will reduce their methane output.

**"Ghost" Lights**

Have you ever seen strange, unexplained lights hovering over a marsh or a swamp? These lights have frightened many people. They have even been the source of rumors about ghosts or UFOs. The lights are known as *will-o'-the-wisp, ghost lights, or ignis fatuus* (Latin for "fool's fire"). The "ghosts" responsible for these lights are really archaea. The methanogens that live in the swamp produce methane, or marsh gas. The conditions that produce methane are also right for producing the gas diphosphane, which burns when exposed to air. The combination of methane and diphosphane create these "ghost lights."

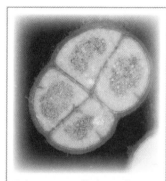

*Deinococcus radiodurans* has been called *a polyextremophile* because it can survive acid, hot or cold temperatures, and 3,000 times more radiation than a human being.

such as the Great Salt Lake and the Dead Sea. Thermophiles thrive in temperatures between 45°C–122°C. These organisms live in hot springs, volcanoes, and near hydrothermal vents. Acidophiles are extremophiles with optimal growth at pH levels of 3 or less. One specific species of acidophiles, which was discovered in Japan, is so acid-tolerant that it can grow in soils with a pH of 0! There are many other types of extremophiles, including some that are resistant to high levels of radiation, some that can tolerate high levels of heavy metals, and some that thrive in the deep ocean where there is great pressure.

## LESSON REVIEW

**1.** Describe the similarities between bacteria and archaea.

**2.** Describe the differences between bacteria and archaea.

**3.** What is an extremophile? List three types of extremophiles and describe the type of environment they live in.

Thermoacidophiles live in extremely acidic conditions and at very high temperatures.

## OBJECTIVES

- List several characteristics of fungi.
- Explain how multicellular fungi reproduce and how God designed fungi to help maintain environmental balance.

## VOCABULARY

- **hyphae** the threadlike filaments in fungi that produce enzymes
- **mycelium** the large mass of hyphae that forms the growing structure of fungi
- **spore** a small reproductive cell that can develop into an adult without fusing with another cell

When you hear the word *fungus*, what do you think of? Green fuzz growing on the leftovers in the refrigerator that somebody forgot to throw out? The itchy misery of athlete's foot? Black crud creeping across the walls of a shower? A fungus is something to get rid of. Or is it?

Before you ban fungi altogether, consider these questions. Do you enjoy looking at a forest-covered mountain? What kind of bread do you use to make your favorite sandwich? Do you like mushrooms on your pizza? Even if you do not, would pizza be the same without cheese? Has penicillin ever cured you of strep throat or another illness? Without fungi, these things would not be possible!

Fungi (singular: fungus) are a diverse group of organisms. They can look like bright miniature umbrellas, waxy blobs, green fuzz, white flowers, purple sponges, or red cups. Some are only one cell; others are almost 1.5 m long. Mycologists, scientists who study fungi, have named thousands of species of fungi. Fungi can be found almost anywhere on Earth. Although most fungi live on land, some live in the ocean. A few even grow in acid! Despite their differences, fungi have many common features.

Fungi are nonvascular organisms that do not have chlorophyll. Unlike most plants, they do not have vessels to transport water and nutrients. They cannot photosynthesize to make their own food. Fungi are heterotrophs that feed off other organisms. Many are also saprophytes, which are organisms that eat dead or decaying things. Animals first eat food and then digest it, but

Mold growing on a strawberry

Fungi growing on tree stump

fungi first release chemicals to digest the food and then absorb it. Because fungi do not have vessels like plants and animals do, they absorb their nutrients like a sponge. In fact, the word *fungus* probably originates from the Greek word *sphongos*, meaning "sponge."

Some fungi feed on small animals. For example, oyster mushrooms release a chemical that stuns roundworms so they can be ingested. Some fungi snare their tiny prey with threadlike nooses; then, they release chemicals that break down their prey's tissues. Other fungi feed off dead organisms. These fungi are called *decomposers* because they decompose (break down) dead plant and animal matter. These fungi along with decomposing bacteria help clean up the dead things in the world. They are an important part of God's recycling plan for the earth.

Some fungi, such as yeast, have only one cell, but most are multicellular organisms. The green mold on a piece of cheese, the mushroom on a pizza, and the black mildew in a shower all have more than one cell. Multicellular fungi are made of **hyphae**, the threadlike filaments that produce enzymes to break down living or dead organisms. Hyphae weave together to form different kinds of fungi. The large mass of hyphae that forms the growing structure of fungi are called **mycelium**. Hyphae grow very rapidly. That is why the leftover lunch stashed in your locker grows mold so quickly.

Both plant and animal cells are separate units, with each unit containing one nucleus separated from each other by a cell wall or cell membrane. But hyphae are continuous threads that contain many nuclei. They are not divided by cell walls or cell membranes, although some are divided by incomplete cross walls with many openings. Because hyphae do not have cell membranes, substances move freely throughout the hyphae. Some fungi reproduce sexually through conjugation. Hyphae from two mating strains connect. All fungi, however, can reproduce asexually by producing spores. A **spore** is a small reproductive cell that can develop into an adult without fusing with another cell. Because these spores are very light, they can easily be carried away by the wind or by animals. Spores are

Porcelain fungus

Oyster mushrooms come in a variety of colors—from white and tan to yellow and pink.

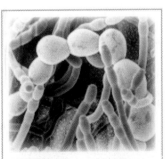
Fungal hyphae and fruiting bodies, magnified 1,500×

**TRY THIS**

**Fungus Structure**
Prepare three microscope slides: one with mold, one with a very thin slice of mushroom, and one with yeast. Observe the fungi under the microscope at both low and high power. Sketch your observations.

able to resist heat, dryness, and cold. Their toughness protects them until conditions become favorable for their growth. Ideal growing conditions vary from fungus to fungus, but generally fungi thrive in warm, damp places. If the spores land in a place where the growing conditions are right, they grow hyphae.

Spores are produced in a structure called *the fruiting body*. A fruiting body may be a stalk tipped with a cluster of spores or many hyphae packed close together. One fruiting body can produce trillions of spores!

If fungi can reproduce so quickly and if one fruiting body can produce so many spores, you might wonder why the world is not overrun with fungi. Remember that spores grow hyphae only when conditions—temperature, moisture, food, and space—are right. So of the few fungi that begin to grow, even fewer live long enough to produce spores. By creating fungi to produce an abundance of spores but limiting how many are able to grow in certain environmental conditions, God ensured balance would be maintained.

## LESSON REVIEW

**1.** What are the basic characteristics of organisms classified as fungi?
**2.** How do fungi obtain their nutrients?
**3.** Describe hyphae and their relationship to mycelium.
**4.** What is the purpose of a spore?
**5.** Why do fungi not overrun the world?

Fruiting body of a basket fungus

Think about the types of fungi with which you are familiar. They may include yeast, mushrooms, and mold. How would you classify them? People find it useful to group fungi according to their basic shapes. Yeasts consist of single cells, mushrooms are shaped like umbrellas, and molds are shapeless and fuzzy.

**Yeasts** are unicellular fungi. Does the smell of baking bread make your mouth water? If you love freshly baked bread, be thankful for yeast. And be thankful that years ago people learned how to combine yeast with other ingredients to make bread rise. Remember that in order to reproduce, fungi need certain environmental conditions. Yeasts need warm temperatures, moisture, and food such as sugar. To make bread, bakers mix yeast with sugar and warm water along with other ingredients. Then they allow the dough to sit while the yeast reproduces. As it grows, the yeast produces carbon dioxide gas, which forms bubbles and makes the bread rise.

Yeasts reproduce by budding. A small part of the yeast cell pushes beyond the cell's boundaries and forms a tiny bud that grows into a new yeast cell.

Unlike yeast, mushrooms are multicellular. You may have eaten mushrooms on pizza or in pasta dishes, and you've probably seen mushrooms growing in the woods or in your yard. **Mushrooms**

**FYI**

**Truffles**
Have you ever eaten a truffle? Truffles, the fruiting bodies of fungi that grow underneath the ground, are considered delicacies. They also have nearly twice the essential nutrients as most mushrooms. Because truffles grow underground, people cannot find them very easily. Truffle-hunters train pigs and dogs to sniff out these treats.

*Candida* is a common cause of fungal infections.

are fungi with a cap on top of a stalk. The cap acts as the mushroom's fruiting body; the mushroom's spores are produced on the underside of the cap. In many mushrooms the spores are produced in gills, which are thin sheets of tissue that stretch from the stalk to the cap's edge. Some mushrooms produce spores in tubes or flaps.

Although mushrooms can be great food, some species are poisonous—even deadly. Some poisonous species look a lot like edible species. Never eat a mushroom without being sure that it is safe to eat.

Most people enjoy bread made by using yeast, and many people like mushrooms. But most people are not too happy to see molds. **Molds** are shapeless, fuzzy fungi that grow on rotting organic matter or in damp places. Molds have a cottony mycelium, which give them a fuzzy appearance. Molds come in a wide variety of colors. Their spores give them their different colors. Black bread mold is a common form of mold. The green fuzz that pops up on old oranges is another form of mold.

### Parts of a Mushroom

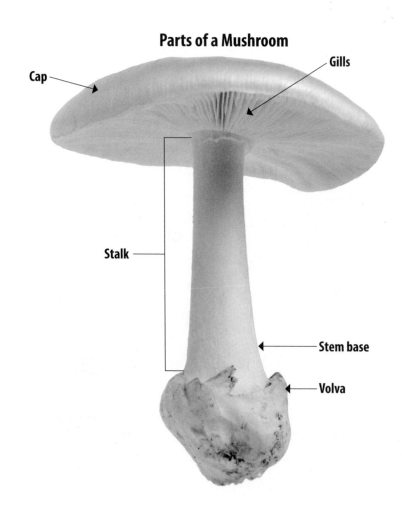

Cap

Gills

Stalk

Stem base

Volva

Mold on a rotting lemon

**TRY THIS**

**Mushrooms**
• Examine a mushroom. Write a detailed description. Name one difference between a mushroom and a plant. What does this tell you about the mushroom?
• Cut the mushroom in half from top to bottom. Draw the cross section of the cap, gills, and stalk. Notice how well protected the gills are beneath the cap. How do the gills relate to reproduction?
• Diagram the mushroom and its various sections.
• Look in a mushroom field guide to determine what kind of mushroom you examined.
• Wash your hands after this activity!

## LESSON REVIEW

**1.** Describe the three basic types of fungi.
**2.** Describe two ways that fungi can reproduce.
**3.** Draw and label the parts of a mushroom.
**4.** Explain the reproductive structures of mushrooms.

*Chlorophyllum molybdites* is one of the most frequent causes of mushroom poisoning in the United States. It has a large white cap with brown or pink scales on top, and it often grows in rings or rows in lawns. It looks quite similar to some edible mushrooms, but if eaten it causes severe abdominal pain, vomiting, and diarrhea.

### TRY THIS

**Fungi, the Great Recyclers**
Design and conduct an experiment comparing seed growth in untreated poor soil and seed growth in poor soil mixed with compost (organic matter broken down by fungi and other decomposers). What important role do fungi play in creation?

Many fungi live off dead or decaying materials. In addition, they also break down many substances in dead plants and recycle these substances back through ecosystems. Along with bacteria, fungi also recycle vital elements, such as nitrogen and phosphorus. So when you see fungi growing on a log in the forest, do not think that the log is dirty. Instead, marvel at the methods that God designed to recycle matter!

Many plants—including most trees—could not survive without fungi. Helpful fungi grow on the roots of 90% of plants. Some fungi send their hyphae into the roots of plants. This helpful interaction between fungi and roots is called *a mycorrhizal relationship. Mycorrhizal* means "fungus-root." The fungi transport nutrients, such as nitrogen and phosphorus, from the soil to the roots, and the plant or tree provides the fungi with moisture and sugar. Fungi also help protect trees and other plants against root diseases, cold, acid rain, drought, and certain insects. Scientists are beginning to realize how important these interactive relationships are. They believe that some reforestation attempts have failed because of a lack of understanding of the importance of the interaction between tree roots and fungi. By better understanding the relationship between fungi and trees, people can be better caretakers of the world.

Another example of fungi working together with another organism is found in lichens. Lichens are actually made up of two different organisms: fungi and algae or cyanobacteria. The way these organisms interact is called **mutualism**—the relationship

Dryad's saddle fungi on a log

between two organisms that live and work together for the benefit of both. The fungi support the lichen and provide water and nutrients. The algae or cyanobacteria in the lichen make food through photosynthesis and share it with the fungi. This relationship enables lichens to live on bare surfaces and in some very hostile environments.

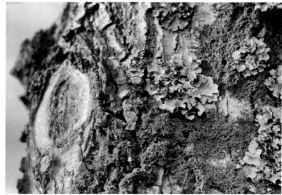

Lichens are most common in the Antarctic, Arctic tundra, high mountains, and the tropics, although they are frequently found in forests as well. In northern forests, lichens commonly grow on tree bark and rocks. Lichens are among the slowest growing organisms. The average lichen grows only a centimeter per year, and some lichens grow as little as a few millimeters per century. Lichens also live a long time. Scientists believe that some lichens are over 1,000 years old.

Lichen is the mint green leafy growth on this tree stump.

Lichens are an important part of many ecosystems. Lichens slowly wear down rocks. They prepare places for plants to grow by speeding up the weathering process and helping make new soil. Lichens are also the foundation of Arctic food chains because caribou and deer eat them.

## FYI

**Natural Insecticides**
Some fungi can be used as insecticides against gypsy moths and aphids. These insecticides are much safer for the environment than chemical insecticides are.

## How a Fungus Eats

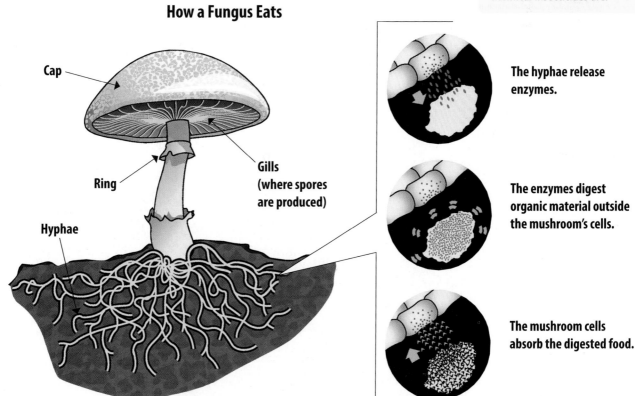

**Cap**

**Ring**

**Gills**
(where spores are produced)

**Hyphae**

The hyphae release enzymes.

The enzymes digest organic material outside the mushroom's cells.

The mushroom cells absorb the digested food.

## HISTORY

**Canary in a Coal Mine?**
Miners used to take canaries into mines with them—not as pets or to hear them sing but to serve as an early warning for harmful gases that could build up in the mine shaft. The canaries would be affected by the gases and collapse even before the miners could smell the toxic fumes. This would warn the miners to escape. How do lichens function for an ecosystem like a canary in a coal mine?

Although in many ways lichens are very hardy organisms, they die easily in polluted air. Lichens are especially affected by the sulfur dioxide produced when coal is burned. Because of their sensitivity to pollution, scientists use lichens to study the health of an ecosystem and to test for air pollution. In fact, scientists can even determine the amount of sulfur dioxide in the air of a certain area by determining how much lichen the area has. A decrease in lichen is a signal of increasing air pollution. Lichens also absorb more radioactive fallout than plants do. Unfortunately, lichens can pass along this fallout through the food chain. For example, years after the Chernobyl nuclear accident in 1986, caribou in that part of Ukraine tested positive for radioactivity.

Fungi can be seen as both helpful and harmful. For example, fungi are very important in the production of some foods. For thousands of years people have used yeast to make bread. Using molds to ripen certain cheeses such as Camembert, Brie, and blue cheese is also a common practice. And many people like to eat mushrooms and truffles. However, fungi such as mold can also ruin foods and make them inedible.

Some fungi are used in antibiotics. Fungi and bacteria are natural competitors, and scientists have discovered that fungi make powerful bacteria-killing drugs. Of these drugs, penicillin is the best known. Centuries ago people sometimes placed moldy bread or cheese on an infection in an attempt to cure it. It may surprise you, but sometimes this strange treatment worked!

Ergot fungus on barley

Blue cheese

In 1928, Alexander Fleming, a Scottish scientist, discovered why: the mold *Penicillium* was able to kill certain bacteria that cause disease. Since Fleming's discovery, the antibiotic penicillin has cured millions of infections and has saved countless lives.

But fungi also cause a wide variety of diseases and other ailments in plants, animals, and people. Certain fungi cause diseases in important food plants—in citrus trees, raspberries, avocados, and rye plants. Fungi are also responsible for destroying nearly all the chestnut and elm trees in North America.

Fungi can live as parasites in or on animals. Some lung diseases in birds and sores on mammals are caused by fungi. Microscopic fungi are even parasites on microorganisms such as amoebas. A few fungal parasites are so particular that they will feed off only one sex of an insect species or one particular body part. An irritating fungal parasite is *Trichophyton rubrum*, the most common cause of athlete's foot.

Some fungi cause serious diseases in people. Ergot is a fungus that grows on rye and other cereal plants. Ergot poisoning constricts the blood vessels, causing numbness, gangrene, and other symptoms. It can also affect the nervous system. Ergot poisoning was common during the Middle Ages. Some historians wonder if the strange symptoms that led to the 1692 witchcraft trials in Salem, Massachusetts, were the result of ergot poisoning. The strange symptoms of the 19 people who were hanged for witchcraft (and of the many others who were accused but not hanged) could have been from a disease caused by ergot fungi.

Ergot has value as a medicine, however. It is the source for a medication for migraine headaches. The medicine narrows the blood vessels in the brain, which helps to relieve the pain.

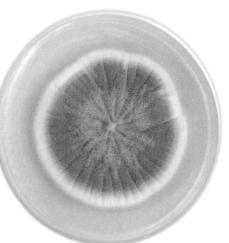

*Penicillium* mold growing in a petri dish

## LESSON REVIEW

1. Explain how fungi help maintain ecosystems.
2. How do fungi interact with plants?
3. In what ways are fungi helpful and harmful?
4. What might the presence of healthy lichen near a factory tell you about that factory?
5. How do fungi affect people?

### TRY THIS

**Preservatives in Food**
Choose several different kinds of packaged foods such as bread products, snack foods, and fruit. Place them in separate beakers or jars and label the contents. Set the jars on the counter. Observe the foods daily to determine which have more preservatives in them, measured by the amount of mold growing on the food. Some foods have a lot of preservatives. Would the mold on these foods grow any faster if you sprinkle water on them?

### TRY THIS

**Protist Temperatures**
Prepare a microscope slide with a drop of pond water. Observe it under a microscope. Record your observations. Then pour equal amounts of pond water into three beakers. Store one at room temperature, one in a refrigerator, and one under a heat lamp. After 24 hours, prepare slides with a drop of water from each beaker and observe them under a microscope. (Use a separate eyedropper for each beaker.) Record your observations. What effect does temperature have on the protists?

What are protists? Scientists generally define a protist as an organism with a nucleus and that is not an animal, plant, or fungus. All protists have cells with a nucleus, and most are **unicellular**, or one-celled, organisms. Nor is it a bacterium, because bacteria do not have nuclei. Most protists have a flagella and most are aquatic. Scientists have placed these organisms into a separate kingdom called *kingdom Protista*, which is one of the four kingdoms in domain Eukarya. Some protists are microorganisms, or organisms that can be seen only under a microscope. Other protists are much larger.

Scientists have not studied the creatures of kingdom Protista as extensively as those of the other kingdoms. Most protists are not major food sources or causes of human diseases, so they do not receive much attention. But they are still important. Many protists help produce the world's oxygen through photosynthesis. They are also important parts of food chains. By learning more about protists, which are some of God's smallest creatures, people can appreciate the important role God gave them to keep the planet healthy!

Protists are very diverse. It has been a challenge for scientists to classify them. When it comes to protists, it is important to remember that the classification system was made by humans, but the organisms themselves were created by God.

Most protists are unicellular eukaryotic organisms. Imagine what this means. Your body has many tissues and organs made from many, many cells that help you move and perform digestion, respiration, excretion, and other functions. But protists carry out all these life functions with only one cell!

Most protists are aquatic. They live in saltwater, in freshwater, and in the watery tissues of other living things. Some protists live in soil. With advancements in technology, scientists continue to find new protists in a variety of environments. It is possible that every species of animal, plant, and fungus has some association with a protist. Some of these protists are parasites, but some help the animal in which they live. For example, cows and other plant-eating animals have a symbiotic relationship with protists in their intestines. The protists, along with bacteria, help the animals digest plant matter.

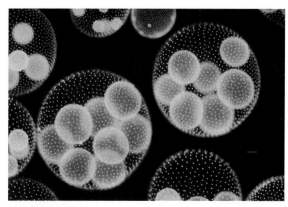
*Volvox aureus*

Some protists are autotrophs, or organisms that make their own food from simple substances. Others are heterotrophs, or organisms that cannot make their own food. Heterotrophic protists eat or absorb other organic substances, such as each other or bacteria. Other protists are able to both make their own food and eat existing food! These organisms use photosynthesis when light is plentiful but hunt for or absorb food when it is dark. Some protists live individually as unicellular organisms during their entire life cycle, and others form large colonies. Some protists have rectangular cells like those of plants, some form beautiful geometrical shells, and some look like shapeless blobs. Many protists float in the ocean their entire lives; others require one or more hosts to complete their life cycles. The wide variation found in protists is a wonderful show of God's limitless creativity!

*Trichia persimilis*, a slime mold, is classified as a protist.

The main characteristic that determines whether a protist is considered plantlike, animallike, or funguslike is how the organism feeds. Plantlike protists use photosynthesis, animallike protists eat other organisms, and funguslike protists eat decaying matter.

## LESSON REVIEW
1. Why can kingdom Protista be considered a catchall kingdom?
2. What do most protists have in common?
3. Describe the habitats of protists.
4. What does the wide variety among protists tell you about God's creativity?

Algae under a microscope

Among the variety of protists God created are some that share characteristics with plants. Plantlike protists include algae and **plankton**, which are tiny organisms that live near the surface of the water. Plankton are a vital part of the food chain. Small fish and other animals eat these protists, and then they are eaten in turn by larger fish or animals. These tiny creatures are the foundation of marine food webs.

The main characteristic that plantlike protists share with plants is the use of photosynthesis. Most plantlike protists are autotrophs that use photosynthesis to make their own food. Phytoplankton are tiny marine plants that use photosynthesis, and they actually produce much of the world's oxygen. These tiny marine protists are oxygen-producing powerhouses, which is incredible when you consider how very small they are!

Plantlike protists move with the help of one or more flagella. Flagella are thin, whiplike structures that help organisms move through liquids. (Flagella is the plural of flagellum.) Some of these protists have symbiotic relationships with other creatures. For example, protists live inside clams, sea anemones, and coral and help provide their hosts with food.

**FYI**

**Changing Color**
One species of euglena turns white if it spends enough time out of the light. It then begins hunting microorganisms for food. When placed back into the light, it will turn green within a few hours and be able to use photosynthesis to produce food again.

Photo taken from space of phytoplankton bloom in the ocean

Three of the autotrophic protists God designed are dinoflagellates, euglena, and diatoms. Another kind of autotrophic protist, algae, is presented in the next lesson.

Dinoflagellates are unicellular ocean organisms that are a type of plankton. Most dinoflagellates are phytoplankton, but some are zooplankton, which are small marine animals that feed on smaller phytoplankton. A few species are parasites of marine animals.

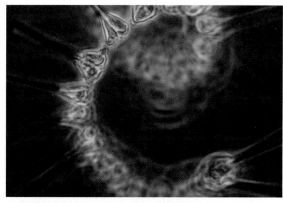

Phytoplankton are autotrophic plantlike organisms that form the basis of the marine food web.

All dinoflagellates have platelike cell walls that give them their regular shape. Dinoflagellates also have two flagella. One flagellum trails behind the organism like a tail, and the other wraps around the dinoflagellate like a belt. The dinoflagellate moves this second flagellum to spin like a top. Dinoflagellates also have an amazing and beautiful feature: they glow. Some nights the oceans glitter with tiny sparkles or glow with eerie, bluish lights produced by dinoflagellates.

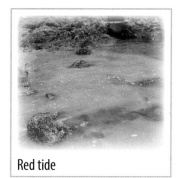

Red tide

Under certain conditions, ocean dinoflagellates can multiply so rapidly that they discolor the water. A rapid growth of dinoflagellates that colors a section of ocean water is called *a red tide*. Red tides are not always red. Depending on the color of the dinoflagellates, red tides can range from yellow-green to almost brown. Red tides can be dangerous because some dinoflagellates produce large amounts of toxins, or poisons. These toxins have been known to kill thousands of fish and to make people sick.

Euglena are also a type of plankton. These unique plantlike protists can be both autotrophs and heterotrophs. When they have sufficient light, they use photosynthesis to produce their own food. But when light is too scarce for photosynthesis, they hunt for bacteria and other small microorganisms to eat. All euglena have three common features. First, euglena have a pouch that holds two flagella: one short and one long. The long flagellum is used for movement. Second, euglena have chloroplasts, which are green structures that are used in photosynthesis. Third, euglena have a photoreceptor and an eyespot that are sensitive to light. Euglena use these organelles to determine if there is enough light for photosynthesis.

Diatoms are phytoplankton.

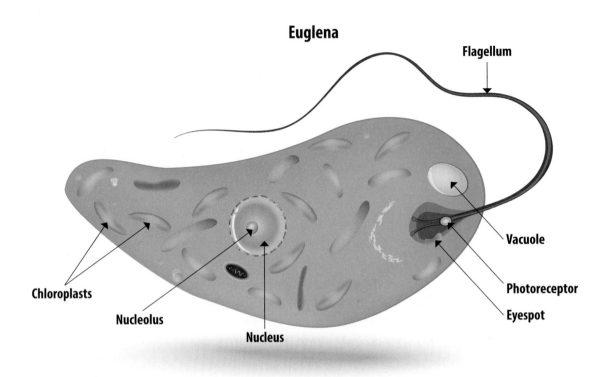

**Euglena**

**Flagellum**

**Vacuole**

**Photoreceptor**

**Eyespot**

**Nucleus**

**Nucleolus**

**Chloroplasts**

You probably brush your teeth with diatoms. They are an ingredient in many toothpastes.

Diatoms are a third type of plantlike protists that live as phytoplankton. These protists are found in freshwater, saltwater, or soil. They are a major part of ocean food chains—they feed both zooplankton and baleen whales! Diatoms produce 20% of the oxygen in the atmosphere.

Each tiny diatom is encased in a two-part shell made from silica, which looks like a glass box. The shell is wonderfully patterned with ridges, spines, or holes. Under a microscope the shells look like miniature stained-glass windows. When diatoms die, their shells are left behind. These tiny shells are used as polishing agents in some products you might use at home, such as toothpaste and car polish. Because they reflect light, the shells are sometimes added to the paint used to mark traffic lanes on roads.

## LESSON REVIEW

**1.** Describe the characteristics of dinoflagellates, euglena, and diatoms.
**2.** Do you think that the word *plantlike* is a good word to use to describe dinoflagellates, euglena, and diatoms? Why?
**3.** How do euglena adapt to thrive in a variety of conditions?
**4.** What role do plankton play in marine food webs?

If someone asked you what algae are, you might describe the green scum growing on ponds or creeping up the sides of your aquarium. If someone asked you what seaweed and kelp are, you might say that they are plants growing in water. But did you know that algae are not always green? Or that seaweed and kelp are kinds of algae? Or that algae are not actually plants?

In the past, algae were classified as simple plants because they use photosynthesis to produce their food. Some biologists still consider them to be plants. But algae do not develop from a seed or spores like plants do, and many types of algae possess one or more flagella at some stage in their life cycle and plants do not. Most algae reproduce through binary fission—cell division in which one cell splits into two identical daughter cells—but some algae reproduce sexually. All algae contain chlorophyll, the green pigment that captures light energy for photosynthesis, but most algae also have other pigments that give them a specific color. Most algae live in water. Some algae are unicellular; some are multicellular. Most multicellular protists are algae.

The most common algae on Earth, green algae, are the primary producers of oxygen in freshwater ecosystems. Green algae

## OBJECTIVES

- Defend why algae are classified as protists instead of plants.
- Identify the four main groups of algae.
- List the major features of each algae group.

## VOCABULARY

- **air bladder** a grapelike bulb that keeps an organism afloat
- **blade** a leaflike structure of algae
- **holdfast** a rootlike structure an organism uses to anchor itself

Green algae on a pond

Air bladders of brown algae

also help reduce carbon dioxide levels in the air; in freshwater and saltwater ecosystems, they store more than 6 billion tons of carbon (taken from the carbon dioxide in the air during photosynthesis).

Brown algae have brown pigments that mask the green color of chlorophyll. This algae group contains the largest known protist—the giant kelp, which can grow over 90 m long. Brown algae live in both warm and cold waters; most live in the ocean. They are common along rocky seashores where they form massive beds of seaweed that provide habitats for countless invertebrates and fish. Some brown algae float free in the ocean. Sargassum, for example, forms massive floating islands of seaweed in the Sargasso Sea, a region of the North Atlantic Ocean. Brown algae form different structures, including **blades**, which are leaflike; stipes, which are stemlike; and **holdfasts**, which are rootlike and anchor the algae to rocks and structures. Brown algae (and some other algae) also have grapelike bulbs called **air bladders** that keep the algae afloat in the sunlight. In many parts of the world people eat brown algae, and at low tide in some places, livestock graze on brown algae. Chemical extracts from brown algae are added to salad dressings and other foods to keep them smooth and thick.

Red algae growing on the Pearl and Hermes Atoll near Hawaii

Most of the world's seaweed are red algae. Most red algae are multicellular. Red algae can grow to be several meters long. These algae live along the beaches and rocky shores of tropical waters. Some form clumps of branching red threads, others grow in flat sheets, and still others form stiff branches that are rich in calcium carbonate—the same substance that forms the shells of coral and crabs. The green chlorophyll in red algae is masked by red pigments. You might think that this would hinder red algae from photosynthesizing, but it does not. In fact, red algae can live in deeper water than other algae and still use photosynthesis. The red pigment absorbs traces of light that reach these depths.

Whether you know it or not, you have probably eaten red algae—chemicals from red algae are used in foods like ice cream and frosting. On a food label, the chemicals from red algae are listed as *carrageenan* in the ingredients. Red algae is also used to make agar, the jellylike medium used in petri dishes in science labs.

Golden algae

Golden algae contain structures filled with golden-yellow pigment. Golden algae live in oceans, lakes, and streams. Some species form swarmer cells, which are cells with whiplike tails. Swarmer cells swim away to form new algae colonies. God gave golden algae an amazing capability—if the algae are not getting enough light, or if they are in the presence of many dissolved nutrients, they will stop photosynthesizing and act as heterotrophs, eating bacteria or unicellular protists.

By God's design, thousands of algae species, both in the oceans and in freshwater, help maintain Earth's health. Algae produce oxygen in aquatic environments. They also serve as the starting point for many aquatic food chains. Many algae are sensitive to changes in light, salinity (saltiness), pollution, acidity, and nutrients. Changes in any of these areas result in changes in the algae. Algae, for example, can multiply quickly in polluted water and throw the ecosystem off balance.

## LESSON REVIEW

**1.** How are algae like plants?
**2.** How are algae different from plants?
**3.** How are algae helpful?
**4.** What are the main groups of algae, and what are the features of each?

Carrageenan is used for stabilizing chocolate, milk, eggnog, ice cream, sherbet, instant pudding, frosting, creamed soup, and many other products.

Some protists have features similar to those of animals; these are called **protozoa**. They are heterotrophs, which means they eat other organisms instead of producing their own food. Most protozoa can move, and their cells do not have a cell wall.

Amoebas are one group of animallike protists. They are widely found in freshwater, saltwater, and soil. Although all amoebas lack flagella, they have a unique way of moving with a **pseudopod**. Amoebas use this flowing extension of cytoplasm to move and to obtain food. The word *pseudopod* means "false foot." To get food, an amoeba approaches a food particle and extends its pseudopod around the particle until the particle is completely surrounded. This process produces a food vacuole, a bubblelike structure that encases and digests the food. The digested food is used by the amoeba for growth and energy. The waste products are pushed out of the cell. Other substances do not have to be removed from the cell in this way. For example, water, carbon dioxide, and oxygen simply diffuse through the membrane.

Amoebas reproduce through binary fission—they make a copy of the hereditary material in the nucleus and split in two. Amoebas are sensitive to their environment. They move from light into dimmer areas, and they move toward some chemicals and draw away from others.

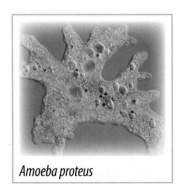

*Amoeba proteus*

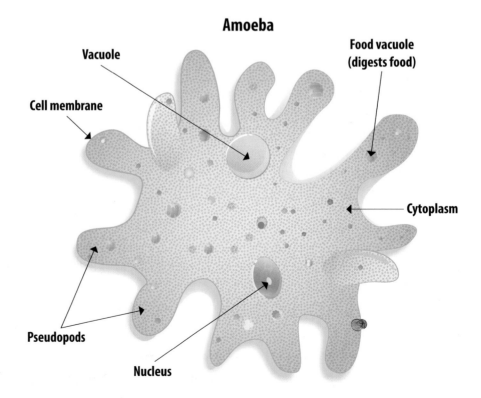

**Amoeba**

Vacuole · Food vacuole (digests food) · Cell membrane · Cytoplasm · Pseudopods · Nucleus

## Slipper-Shaped Paramecium

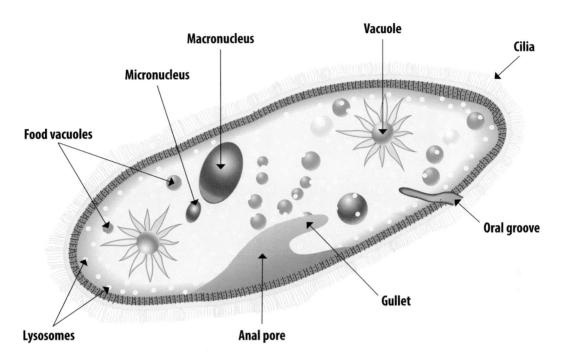

Micronucleus

Macronucleus

Vacuole

Cilia

Food vacuoles

Oral groove

Lysosomes

Gullet

Anal pore

Some amoebas are parasites of animals. In certain regions of the world where sanitation is poor and sewage is not treated, amoebas can get into the food and water supplies and then infect humans. One type of amoeba moves into human intestines and causes amoebic dysentery, a disease of the large intestines that results in severe diarrhea and abdominal pain.

Ciliates are another group of protozoa. Ciliates are named for their distinctive feature, **cilia**—tiny, hairlike projections that move rhythmically like oars to help ciliates move. Ciliates also use their cilia to sense their surroundings and to sweep food such as bacteria or dissolved nutrients toward themselves. Ciliates have two nuclei, one small and one large. The large nucleus controls the ciliate's life functions; the small nucleus controls reproduction. Ciliates reproduce by binary fission, but most ciliates also use **conjugation** as a step in their reproductive process. Conjugation is a process during which two ciliates attach themselves to each other and exchange genetic information.

The slipper-shaped paramecium is one commonly studied ciliate. This paramecium is covered

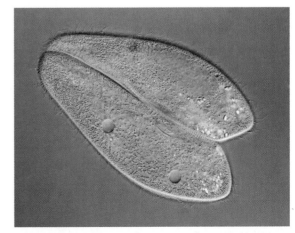

*Paramecium caudatum* in the process of conjugation

## FYI

**Amoebic Dysentery**

*Entamoeba histolytica*, a parasitic amoeba, causes amoebic dysentery and ulcers. It invades the intestine and could spread to the liver, lungs and other tissues, causing pus-filled sores to develop. Cysts are passed through the body in the feces. Infection is caused by the ingestion of food and water that has been contaminated. This species of amoeba is found throughout the world and is most common in areas with poor sanitation.

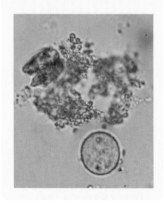

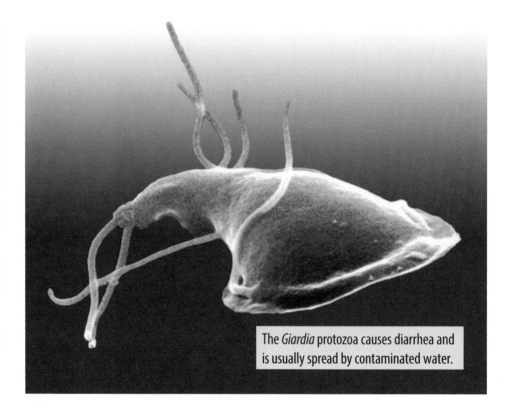

The *Giardia* protozoa causes diarrhea and is usually spread by contaminated water.

by a tough layer that consists of the cell membrane and some underlying layers. Its cilia sweep food particles into an indentation called *the oral groove*, which leads to a structure called *the gullet*. When food reaches the base of the gullet, food vacuoles form around it. After the food is digested, waste products are eliminated through the anal pore.

Zooflagellates are protists that have one to eight flagella. They use the flagella to move and to collect food. Zooflagellates reproduce through binary fission. They live in water or as parasites in fish, reptiles, and amphibians. Some have a symbiotic relationship with other organisms. For example, zooflagellates live in the intestines of wood-eating cockroaches and termites, and the insects rely on the zooflagellates to help digest the wood that they eat. Some zooflagellates cause diseases in people and animals; they are responsible for various intestinal diseases as well as African sleeping sickness.

## LESSON REVIEW

1. In what ways are amoebas, ciliates, and zooflagellates like animals?
2. Describe the major features of amoebas, ciliates, and zooflagellates.
3. What are cilia?

**Sporozoans** are heterotrophic protists that cannot move on their own. They have no cilia or flagella. You may wonder how they can survive if they can neither produce their own food nor move around to find it. They survive by being parasites. Sporozoans live in the bloodstream of animals. Most have complicated life cycles that involve more than one kind of host. Sporozoans form cells known as *spores*, which pass from one organism to another through food, insect bites, and other means.

Sporozoans can cause serious or even fatal diseases. The sporozoans of one genus (the *Plasmodium* genus) cause **malaria**, a serious disease that affects red blood cells and results in muscle aches, chills, and high fevers. A person with malaria first feels very cold and then very hot and sweaty. Even after the patient appears to be well, the illness can suddenly reappear. Malaria is especially a problem in tropical and subtropical areas of Africa, Asia, Central America, and South America. Although malaria can be treated with drugs, many people around the world die from malaria each year.

## OBJECTIVES

- Describe the main characteristics of sporozoans.
- Explain why the life cycle of the *Plasmodium* sporozoan is complicated.
- Diagram how malaria spreads.

## VOCABULARY

- **malaria** a disease caused by a sporozoan that is characterized by periodic attacks of chills and high fevers
- **sporozoan** a heterotrophic protist that cannot move on its own

## Life Cycle of Malaria Sporozoans

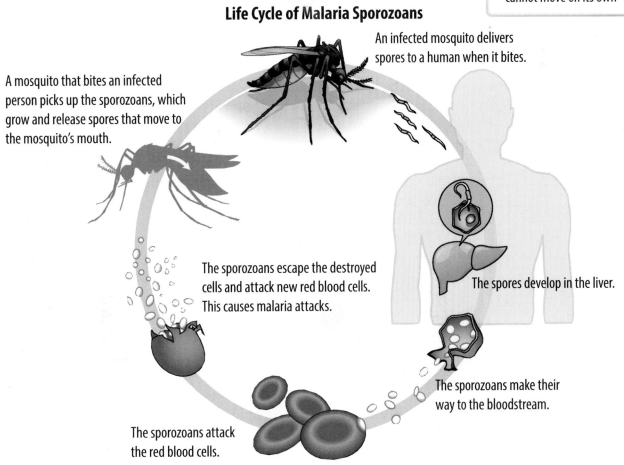

An infected mosquito delivers spores to a human when it bites.

A mosquito that bites an infected person picks up the sporozoans, which grow and release spores that move to the mosquito's mouth.

The spores develop in the liver.

The sporozoans escape the destroyed cells and attack new red blood cells. This causes malaria attacks.

The sporozoans make their way to the bloodstream.

The sporozoans attack the red blood cells.

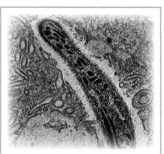

The sporozoan *Plasmodium falciparum* is shown gliding through a cell in the gut of its host mosquito. There are five different species of *Plasmodium* that can cause malaria. However, the *Plasmodium falciparum* causes the most severe symptoms.

 **HISTORY**

**Malaria Myths**
For thousands of years people noticed a connection between malaria and swamps or marshes. They once thought that the disease was caused by the air around the swamps. In fact, the word malaria means "bad air." But malaria is not caused by bad air. It is caused by sporozoans and spread by mosquitoes that live around swamps.

Since World War II, the World Health Organization has supported a worldwide effort to eliminate the disease by using pesticides to reduce mosquito populations. These methods have eliminated malaria from North America and Europe, but the disease remains a problem in Africa, Central and South America, and Asia.

The sporozoans that cause malaria have a complicated life cycle. They use two hosts: mosquitoes and human beings. These *Plasmodium* sporozoans are fertilized in the gut of a mosquito, forming many sporozoans. The sporozoans then migrate into the mosquito's saliva. When the mosquito bites someone, it injects sporozoans into the person's bloodstream. In the bloodstream these cells grow on a diet of iron taken from human blood cells. Then, the sporozoans travel to the liver, where they develop and grow before making their way back into the bloodstream. Inside the blood cells sporozoans divide rapidly, developing into many infection-carrying cells that swarm back into the bloodstream. These infection-carrying cells attack more blood cells, grow and develop, and repeat the cycle. The flood of these infected cells entering new blood cells causes malarial attacks in the human host. If a mosquito bites someone with malaria, it can spread the disease by biting someone else.

For centuries Native Americans used the bark of the cinchona tree to treat malaria. Quinine, a drug used to treat malaria, was developed from this bark. In fact, until 1930, quinine was the only drug available to treat malaria. Now several synthetic drugs are available.

People who live in countries where malaria is a problem often put netting around their beds to help prevent mosquito bites.

## LESSON REVIEW
**1.** Describe the characteristics of sporozoans.
**2.** Why is the life cycle of the *Plasmodium* sporozoan complicated?
**3.** Diagram how malaria spreads. Explain each step.

A funguslike protist changed the course of history when it caused the Irish potato famine. This protist invaded and destroyed Ireland's potato crop, a crop the people depended on to survive. The famine raged from 1845–1849. Ireland's population dropped from 8.6 million to 6.6 million because of death and another 2 million because of mass emigration.

Funguslike protists act or look like fungi. Protists like the one that caused the potato famine act like fungi. They extend filaments into a host and release strong enzymes that destroy the host. Similar types of protists can cause animal diseases. If you have an aquarium, maybe you have seen ich—white fuzzy growths—on your fish. Ich is caused by this type of protist.

Some funguslike protists and fungi share a basic structure: individual cells that form filaments, which are long, threadlike body structures. During different stages of their lives, some funguslike protists resemble amoebas. Others have flagella for part of their lives. Like fungi, most funguslike protists are **saprophytes**. These organisms eat dead or decaying materials. Some funguslike protists are parasites, and a few species even eat bacteria. Funguslike protists are heterotrophs, and they reproduce using spores.

Slime molds are funguslike protists. Slime molds reproduce through the use of spores, which are cells encased in a hard, protective covering. The spores are produced in a receptacle called *a sporangium*. Once the sporangium releases the spores, each spore develops into a new organism. It will either transform into an amoeba-like cell or a cell with flagella. In some species the cell may grow into a huge cell that produces spores. In one species, the amoeba-like cells live independently for a time and quickly reproduce. When the food supply is used up, the cells come together through fusion and act as one organism called *a zygote*. The zygote may travel for several centimeters to find food. After it locates food, it grows and becomes a plasmodium cell. The plasmodium cell then matures and develops a new sporangium, repeating the process of reproduction.

*Phytophthora infestans* is the funguslike protist that causes Late Blight, the potato disease that lead to the Great Irish Famine in the 1800s.

### OBJECTIVES

- Recount several basic characteristics of funguslike protists.
- Describe the structure and behavior of slime mold.

### VOCABULARY

- **saprophyte** an organism that eats dead or decaying materials

## Plasmodial Slime Mold Life Cycle

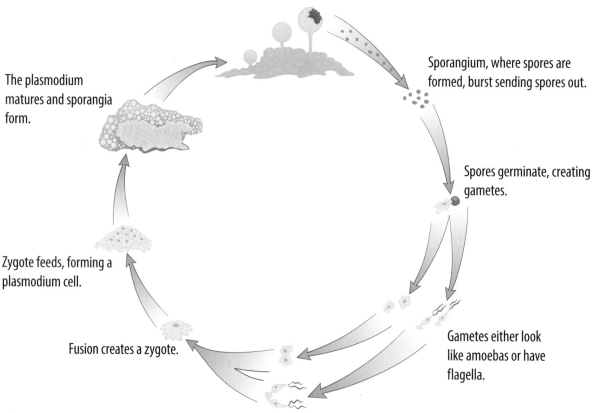

Sporangium, where spores are formed, burst sending spores out.

The plasmodium matures and sporangia form.

Spores germinate, creating gametes.

Zygote feeds, forming a plasmodium cell.

Gametes either look like amoebas or have flagella.

Fusion creates a zygote.

Fusion occurs and the gametes combine.

At one point in their life cycle, slime molds look like slimy blobs that ooze across fallen logs and leaves. They move and spend their entire life cycle within a path of slime that they form. Slime molds cannot move unless they are in their slime paths. They use a wavelike motion to move, but have no flagella or pseudopods. Slime molds feed using three different methods: they absorb decaying vegetation, they eat bacteria, or they absorb dissolved food particles.

A colorful slime mold

## LESSON REVIEW

1. What are the basic characteristics of funguslike protists?
2. How are some protists like fungi?
3. How are slime molds different from other protists?
4. Describe how slime molds reproduce.

108

**Chapter 1:** *Introduction to Plants*
**Chapter 2:** *Plant Activity*

*Vocabulary*

| | | |
|---|---|---|
| angiosperm | humus | sporophyte |
| annual | monocot | stamen |
| biennial | nonvascular | stigma |
| cambium | ovary | stomata |
| chlorophyll | ovule | style |
| dicot | perennial | taproot |
| embryo | phloem | thigmotropism |
| epiphyte | photosynthesis | transpiration |
| fragmentation | phototropism | tropism |
| fronds | pistil | vascular |
| gametophyte | pith | xylem |
| gravitropism | rhizome | |
| gymnosperm | sepal | |

**Plants**

*Key Ideas*

- Systems, order, and organization
- Evidence, models, and explanation
- Change, constancy, and measurement
- Form and function
- Abilities necessary to do scientific inquiry
- Understandings about science and technology
- Structure and function in living systems
- Regulation and behavior
- Diversity and adaptations of organisms
- Science as a human endeavor
- Science and technology in society
- Science as a human endeavor
- Nature of science
- History of science

**SCRIPTURE**

Then God said, "Let the land produce vegetation: seed-bearing plants and trees on the land that bear fruit with seed in it, according to their various kinds." And it was so.

Genesis 1:11

Most buildings are ordinary looking, but some buildings are beautiful. The architects and builders designed these buildings to be not only useful but also to be interesting or beautiful, even though they would be just as functional with a boring design.

God is the architect of the world. He created plants to be useful and beautiful. He planted the first garden, making "all kinds of trees grow out of the ground—trees that were pleasing to the eye and good for food" (Genesis 2:9). Plants beautify the world and provide food and oxygen for other living things. God gave Adam and Eve the responsibility of caring for the Garden of Eden, "to work it and take care of it" (Genesis 2:15). Today God continues to give people the responsibility to work the ground and to care for the earth. Learning about plants gives you a greater understanding of the complexity of creation, helps you better appreciate the important role of plants in the world, and enables you to take care of plants.

Imagine what life would be like without plants. If you used shampoo or toothpaste this morning, you have taken advantage of plants. If you are using a pencil and paper, you are using plant products. Many medicines are derived from plant products as are many fabrics. More importantly, plants produce oxygen through the process of photosynthesis. Plants also have the ability to use the sun's energy to turn minerals and carbon dioxide into compounds that form the foundation of the

Corn plants in a farmer's field

food supply. So, no plants—no people. As you study plants, remember the important role that they play. Notice the plants that surround you in the world. Remember that God created these plants for a special purpose. Be thankful for plants!

God created and provided for thousands of plant species. In many ways plants are the same, and in many ways they are different. Of the approximately 400,000 species of plants, all of them—cacti, ferns, oak trees, roses, dandelions, and all other plants—have several things in common:

Green, red, and purple grapes growing on a vine

**Plants make their own food.** Plant cells have organelles called *chloroplasts*. The chloroplasts contain chlorophyll, a green pigment that captures light energy from the sun. Plants use this energy to make food in a process called *photosynthesis*.

**Plant cells have cell walls outside of their cell membranes.** The cell walls protect and support the plant,

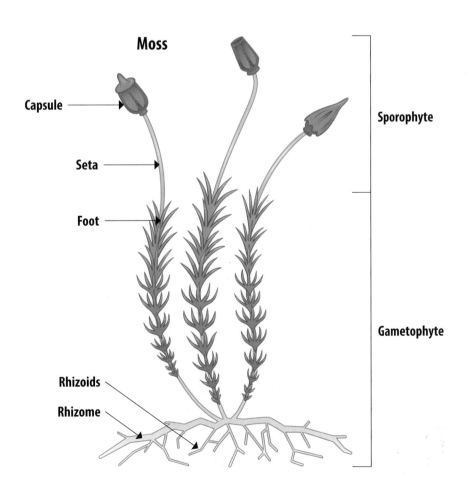

giving it the necessary stiffness to remain upright. The cell walls are made of complex carbohydrates and proteins that form a hard material.

**Plants reproduce with a two-part life cycle.** One phase produces spores, and the other phase produces egg and sperm cells, which are sex cells. The spore-producing phase of a plant is called the **sporophyte**. The phase of a plant that produces sex cells, or gametes, is called the **gametophyte**. The two phases are much more obvious in mosses and ferns than they are in flowering plants.

Although all plants share these basic characteristics, plants have many differences. Plants can be classified into two main groups—nonvascular and vascular. **Nonvascular** plants do not have vessels to transport materials throughout the plant. Water and nutrients move around the plant through diffusion and osmosis. Such transportation is possible because nonvascular plants are small. Mosses, hornworts, and liverworts are nonvascular plants.

**Vascular** plants have vessels to transport materials throughout the plant. They have vascular tissues that deliver materials such as water and nutrients throughout the plant—just as pipes carry water to the faucets in your house. Although some vascular plants, such as violets, are very small, other vascular plants, such as oak trees, grow very large.

Vascular plants can be divided into two groups—those that do not produce seeds and those that do. Ferns and club mosses are

| Nonvascular | Vascular | | |
|---|---|---|---|
| | No seeds | Seeds | |
| | | Nonflowering | Flowering |
| | | | |
| Mosses and liverworts | Ferns, horsetails, and club mosses | Gymnosperms | Angiosperms (annuals, biennials, perennials) |

### George Washington Carver

If you like peanut butter with your jelly, thank George Washington Carver. Carver (1860–1943), who was born a slave in Missouri, was an agricultural scientist who helped southern farmers restore vitality to their soil. Years of growing only cotton had drained the nutrients from the soil, so Carver advised crop rotation: planting peanuts and sweet potatoes to restore the necessary nitrogen to the soil and enrich it again. Since there was no market for peanuts and sweet potatoes, Carver invented hundreds of uses for these crops so that farmers could use these plants for crop rotation and still make a profit.

Carver attended Simpson College in Iowa and Iowa State College of Agriculture and Mechanic Arts (now Iowa State University), earning his master's degree in agriculture in 1896. That same year, Booker T. Washington, founder of Tuskegee Institute in Alabama, noticed Carver's achievements and asked him to head the institute's agriculture department. Carver stayed with the institute for 47 years, conducting important work in scientific agriculture and chemurgy (the industrial use of raw products from plants).

Carver invented over 100 products from soybeans—including flour, breakfast food, oil, and soy milk. Carver launched a multi-billion dollar industry using peanuts in over 300 products such as peanut butter, beverages, pickles, Worcestershire sauce, bleach, wood filler, washing powder, metal polish, paper, ink, plastics, shaving cream, linoleum, shampoo, and axle grease. He also developed many materials from sweet potatoes, including synthetic rubber and a material used to pave roads.

Carver also noticed that "nature has a way of evening things out" because it produces no waste. He was ahead of his time as a recycler and found ways to transform discarded materials into new, valuable resources. He turned industrial waste products into useful materials such as imitation marble, paint, wallboard, and paving materials.

With all of his inventions, Carver could have been very rich, but he did not want recognition. He only wanted to help people. Despite having hundreds of inventions, he sought patents on only three, saying, "God gave them to me—how can I sell them to someone else?" In his will, Carver left his life savings to Tuskegee University to establish the George Washington Carver Foundation for research in agricultural chemistry.

seedless plants. Seedless plants reproduce by using spores rather than seeds. A spore is a small reproductive cell that can develop into an adult without fusing with another cell. Plants that produce seeds are divided into gymnosperms and angiosperms.

**Gymnosperms** are plants that produce seeds but not flowers. In Greek, *gymnosperm* means "naked seed." Gymnosperm seeds are exposed rather than enclosed in a seed case. The four groups of gymnosperms are cycads, ginkgoes, gnetophytes, and conifers.

Cycads are tropical gymnosperms that are often mistaken for palm trees. The large leaves of a cycad resemble feathers. They grow in clusters on top of the trunk. Cycad plants are either male or female. The large seeds develop in cones.

The ginkgo tree has fan-shaped leaves. Unlike most gymnosperms, the ginkgo is deciduous: it drops its leaves each year. The ginkgo tree is the only living species of ginkgoes. It is very resistant to pollution, so it is often planted in urban areas. Individual ginkgoes are either male or female. Usually only male trees are planted in public areas because female trees produce seeds that smell terrible.

Gnetophytes are widely varied gymnosperms that have large, leathery leaves. Their seeds develop in cones. The *Welwitschia mirabilis* species is a very unusual plant—most of it is buried in sandy soil, and the exposed part produces only two leaves. This plant can live for centuries—one specimen is 2,000 years old!

Conifers are the most familiar gymnosperm. The tallest plant, the redwood tree (*Sequoia sempervirens*), is a conifer. Conifers produce cones. They normally have needles, and they are usually evergreen. The bristlecone pine, the longest-lived tree, can retain its needles for up to 45 years.

The other group of vascular plants that produce seeds are flowering. Plants that reproduce seeds in flowers are called **angiosperms**. In Greek, *angiosperm* means "enclosed seed."

Bristlecone pines can be found at the timberline, often between the elevations of 2,280 m and 3,500 m.

These plants produce seeds in protective seed cases called *fruit*. The fruit may not be the kind you think of when you hear the word *fruit*. Even grasses and weeds produce fruit. All agricultural crops, weeds, broad-leaved shrubs, garden flowers, and trees are angiosperms. Flowering plants are divided into groups according to their life spans: annuals, biennials, and perennials.

**Annuals** are plants that complete their life cycle in one growing season. These plants produce seeds during their single growing season and then die. The word *annual* comes from the Latin word *annus*, meaning "year." Petunias and many other garden flowers are annuals, as are crops such as wheat, corn, rye, and rice.

**Biennials** are plants that complete their life cycle in two growing seasons. The Latin word *bi* means "two." In the first year, biennials sprout and grow roots, stems, and leaves. The stems and leaves may die during the first winter, but the roots survive. In the second growing season, the plants produce stems, leaves, flowers, and seeds. Biennials include carrots, celery, lettuce, and dianthus.

**Perennials** are plants that live through many growing seasons and produce reproductive structures each year. The Latin word *per* means "through." Most perennials are woody plants such as trees. Because these plants live for many years, they are able to build up the woody layers in their stems. The leaves and stems of nonwoody perennials, such as chrysanthemums and peonies, die each winter and grow again in the spring.

## LESSON REVIEW

**1.** Explain the difference between nonvascular and vascular plants.
**2.** What is the difference between seedless plants, gymnosperms, and angiosperms? Give an example of each type of plant.
**3.** What is the difference between annuals, biennials, and perennials?
**4.** How might life be different if all plants were annuals?

Banana plant

### OBJECTIVES
- Identify the basic needs of plants.
- Explain the importance of each basic need of plants.

Humus—sometimes referred to as *compost*—is often added to soil to make it more fertile.

Plants need certain conditions in order to live. Although plants have varying specific needs, all plants have the same basic needs. These basic needs include soil, water, sun, air, space, and the correct temperature. Have you ever planted a flower garden? Most seed packets tell you if the plants need full sun, part sun, or shade. Houseplants often come with a tag indicating how much sunlight, water, and what type of soil they need.

Plants need soil because it provides minerals and nutrients; soil also anchors the plants and holds water around them. Microorganisms such as bacteria and fungi live in soil and make nutrients available to the plants. Soil is made of little pieces of rock mixed with **humus**, which is partially or totally decayed organic matter. Humus contains large numbers of microorganisms. The amount of humus in soil determines the nutrients in the soil (the more humus, the more nutrients) and how well the soil holds water. The size of the rock particles also determines how much water the soil can hold. Sandy soil is composed of particles that are 0.05 mm to 2 mm. Because the spaces between sand particles are quite large, water drains from sandy soil fairly quickly. Silt is soil composed of particles that are 0.002 mm to 0.05 mm. Smaller particles make up clay, which becomes waterlogged easily. Most soils are made up of a mixture of humus, sand, silt, and clay. Different plants thrive best in different types of soil, but the best kind of soil for plant growth is generally a blend of these four ingredients.

Seedlings

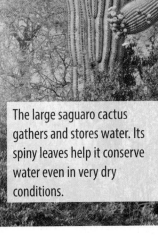

## TRY THIS

**Plant Ingredients**

Remove a plant from the soil, brushing off any remaining soil particles. Determine the plant's mass. Then hang the plant upside down until it has completely dried out. Determine the mass again. What percentage of the plant was water?

Take the plant outside, and (with adult supervision) burn the dried plant material in a beaker. Then determine the mass of the ashes. About how much of the dry mass was carbon compounds? How much was minerals? How do these things relate to the plant's needs?

Water is necessary for plants to make their own food through photosynthesis. Plants use water to carry moisture and nutrients from the roots to the leaves and then to carry food from the leaves back down to the roots. Amazingly, 80% to 90% of the total weight of a growing plant is water. God created plants with varying needs for water, so plants can live in different habitats all around the world. For example, willow trees usually grow where a lot of water is available, often along rivers and on the shores of lakes. On the other hand, saguaro cactuses thrive in hot, dry deserts.

Water conditions within a plant depend on water intake by the roots and water loss from the leaves. Plants are equipped with structures that help them conserve water. For example, many plants have a waxy coating that slows down water loss from leaf surfaces, and many desert plants have spiny leaves that help them conserve water.

Plants use the sun's energy to manufacture food through photosynthesis and to grow. Some plants require lots of sunlight, but others need very little. For example, wheat needs a lot of sunlight, but ferns do not.

Because they need sunlight, plants require clean air. Polluted, smoggy air blocks out sunlight. Plants also need air to provide the carbon dioxide used in photosynthesis to make carbon compounds. Carbon compounds make up 85% to 95% of the dry matter of the plant. The rest of the dry weight of a plant is minerals. If a plant is burned, the ashes that remain are the plant's minerals.

The large saguaro cactus gathers and stores water. Its spiny leaves help it conserve water even in very dry conditions.

Plants also need space to grow. Plants that are crowded together will be small and stunted;

trees and other large plants need more space for their roots and branches.

Temperature also affects plant growth. If the temperature is too cold, the plants may not grow at all. If the temperature is too hot, the heat may damage the plants. Each plant species has an ideal growth temperature, which can vary with a plant's age.

## LESSON REVIEW

**1.** Name the basic needs of plants, and explain why plants need each of these things.

**2.** What do you think would happen if all plants needed the same amount of sun and water and the same type of soil?

Most nonvascular plants are small and grow in moist habitats like wetlands, forests, and meadows. They need the water provided by such places to live and reproduce. Nonvascular plants are grouped into three categories: mosses, liverworts, and hornworts.

Have you ever been told that if you are lost in the woods, you can orient yourself because moss always grows on the north side of trees? You may have an extremely long walk home if you follow this advice! Moss can grow on any side of a tree. However, moss often does grow on the north side of a tree because it is usually damp and shady there—the perfect living conditions for moss.

Mosses, liverworts and hornworts prefer a damp and shady environment because of their structures. Because these plants lack sturdy stems with a vascular system and depend on the stiff cell walls of their cells for support, they are small and grow close to the ground. These plants survive without a system of vessels to transport water and nutrients. They use their entire body surface to absorb water and nutrients by diffusion and osmosis. These methods of water collection will not carry materials very far, but they do not have to because these plants are small.

Mosses and liverworts usually live in large groups. They cover rocks, tree bark, or soil with a mat of tiny plants.

## OBJECTIVES

- Identify the basic characteristics of nonvascular plants.
- Name the three types of nonvascular plants.
- Describe sexual and asexual reproduction processes in nonvascular plants.

## VOCABULARY

- **fragmentation** the separation of a parent plant into parts that develop into whole new plants

Moss growing on a tree

Moss growing on rocks near a river in Japan

Their ability to hold water helps rocks break down and helps logs decay. They do not have true roots, stems, or leaves that contain vascular tissue, but they have structures that carry out the functions of roots, stems, and leaves. Instead of stems, they have stalks called *setae* (singular: seta). Instead of roots, they have threadlike cells called *rhizoids*. And instead of leaves, they have scales.

Mosses, liverworts, and hornworts all live in moist places, so they do not need a waxy coating like that of vascular plants to prevent water loss. Although their reproductive cells have a tough, waterproof coating to help them survive dry periods, moisture is a necessity for nonvascular plants to reproduce.

Mosses are the most familiar nonvascular plants. The most conspicuous and dominant phase of the moss life cycle is the gametophyte, or the sex-cell producing phase of the plant. Sometimes the same plant produces both male and female organs; sometimes separate plants are male and female. First, the moss plant produces sex cells—either eggs or sperm. Then, the sperm swim through a film of moisture to fertilize the eggs, producing a cell called *a zygote*. Next, each zygote divides, creating a sporophyte (the spore-producing phase of the plant), which grows on top of the gametophyte. This sporophyte usually

Moss with identifiable gametophytes and sporophytes

## Moss Life Cycle

The sporophyte releases spores into the air.

Spores that land in a damp area grow into gametophytes.

Gametophytes

The gametophytes produce gametes (sex cells).

Capsule

Spores

Sperm cells

Sporophyte

Egg cells

The fertilized egg grows into a sporophyte.

Sperm cells swim through moisture to fertilize the egg cell.

Fertilized egg

consists of a capsule that is raised up on a stalk. The spores produced within the capsule are released and carried away by the wind. When the spores land in a damp area, they grow into a new gametophyte. Because moss sporophytes take 6–18 months to mature, it is very important not to pick up or step on moss.

The liverwort is liver-shaped and was named in the Middle Ages when it was believed that the shape of the plant related to its medicinal value. Most liverwort gametophytes are leafy and mosslike, but others are broad and flattened. Some liverworts produce leafy, stemlike structures, which grow flat along the ground. The gametophytes of most liverworts are either male or female. The plant parts that produce sperm are raised on stalks that are disk-headed. The parts that contain eggs grow on umbrella-headed stalks. The life cycle of liverworts is similar to that of mosses.

Some mosses and liverworts can reproduce not only sexually but also asexually by **fragmentation**—the separation of a parent plant into parts that develop into whole new plants. Fragmentation can occur at any point in the gametophyte phase.

## FYI

**Peat Moss**
Peat moss gametophytes have clusters of branches that form moplike heads, creating bright green or reddish clumps in boggy areas. Peat moss has special hollow leaf cells that can hold up to 20 times their dry weight in water. Gardeners often use this moss to make the soil more capable of holding water. Throughout history people have used peat moss to dress wounds and to diaper babies because it can hold so much liquid.

Thalloid liverwort

Mosses and liverworts play a very important role in the world. They are often the first plants to grow in a new environment, such as newly exposed rock. When these plants die, they form a thin layer of soil, which can serve as a habitat for new plants. Mosses and liverworts also help prevent erosion by covering the soil and holding it in place. Some birds use mosses for nesting material.

Hornworts are the least common nonvascular plants. The most familiar hornwort grows around the world in moist, shaded habitats. They usually live in moist soil and are seldom found on tree trunks or rocks. Each cell of a hornwort has a single, large chloroplast, so hornworts look a lot like algae.

## LESSON REVIEW

**1.** What are the characteristics of nonvascular plants?
**2.** How are nonvascular plants different from vascular plants?
**3.** What are three types of nonvascular plants?
**4.** Explain how nonvascular plants reproduce.

Hornwort

If you have ever walked in damp woods, you have probably seen ferns. Ferns are leafy plants that grow in areas that are moist at least part of the year. Ferns usually do not grow in hot, dry areas because they need moisture to reproduce. Ferns are vascular plants, but they do not reproduce using seeds. They reproduce using spores.

Like all plants, ferns have a two-part life cycle. The sporophyte stage of the fern is the most familiar stage. Most ferns have a horizontal underground stem called a **rhizome**. The rhizome produces large, divided leaves called **fronds**. Roots grow out of the rhizomes near the base of the leaves. Most garden and woodland ferns have rhizomes that produce new sets of fronds each year. Young fronds are called *fiddleheads* because they are tightly coiled and look like the end of a violin. On the underside of the fronds are little clusters of spore cases called *sori*. The sori contain the spores. Each spore is designed to grow into a new fern under the right conditions.

The fern releases its spores and they land on the soil. Under suitable conditions (including the correct amount of moisture and the right temperature), a spore will grow into a heart-shaped plantlet—the gametophyte. The gametophyte has two sets of reproductive organs on its underside—the male set that contains

## OBJECTIVES

- Examine and identify the structures of ferns.
- Distinguish between ferns, horsetails, and club mosses.

## VOCABULARY

- **epiphyte** a plant that grows on another plant, which it uses for support but not for nutrients
- **frond** a large, divided leaf usually found on a fern
- **rhizome** a horizontal underground stem

Club moss

Fiddleheads on an ostrich fern

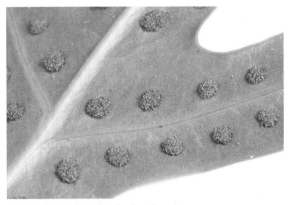

Sori on the underside of a fern frond

sperm cells and the female set that contains egg cells. The male organs and female organs are separate from each other, but if water is present on the gametophyte, the sperm cells swim to the egg cells. Fertilization occurs when the sperm and egg fuse. Their genetic material makes a cell with a full set of genes. This cell is the beginning of an adult fern, located in and protected by the gametophyte structure. As it grows to become the sporophyte, it takes over from the gametophyte and becomes the adult fern.

Horsetails and club mosses have the same life cycle as ferns, but they lack the stem and leaf structure of ferns. Horsetails belong to the genus *Equisetum*, from the Latin words for "horse" and "tail." They are most often found in moist or damp places, by streams, or along the edge of woods. Horsetails have tiny scales and short branches that grow in swirls around the stem. Horsetail stems are hollow and gritty; people once used

## Fern Life Cycle

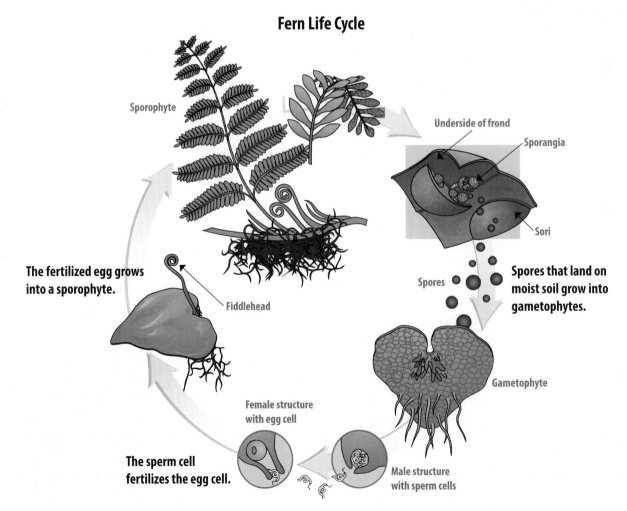

these abrasive stems to clean pots and pans. Because of this, horsetails are sometimes called *bottlebrushes*. If you find a horsetail, try polishing a penny with it by pulling apart a section of the stem and rubbing the end of it on the surface of the penny. People also call this plant *the puzzle plant* because its segments can be taken apart and put together again.

Club mosses are simple plants that are similar to true mosses except that they have vascular tissues in their stems. There are over 300 species of club mosses, and they live in moist areas from the Arctic to the tropics. The tropical species are mainly **epiphytes**, plants that grow on other plants, which they use for support but not for nutrients. This temperate species often form evergreen mats on forest floors. The spores of club mosses can live in the soil up to 7 years as they develop into new plants, and it might take 10 years before the new plants grow aboveground. It takes almost 20 years for a single club moss plant to complete its cycle from spore to mature plant. Club moss spores were once used to coat pills and to make fireworks that gave off a yellowish glow.

Horsetails

Ferns, horsetails, and club mosses help form soil and prevent soil erosion. Ferns are used in yards and as houseplants. Coal was formed from fossilized ferns, horsetails, and club mosses that lived many years ago.

## LESSON REVIEW
**1.** Name and describe the three types of vascular seedless plants.
**2.** Why do seedless plants need moisture to reproduce?
**3.** Describe the life cycle of a fern.

 **TRY THIS**

**Vascular Seedless Plants**
Gather a variety of seedless vascular plants such as ferns, horsetails, and club mosses. Carefully sketch each one. Find the parts that are like roots, stems, and leaves. Look for reproductive structures, especially on the undersides of the fern leaves.

Create a chart to record how the leaflike, stemlike, and rootlike structures of each plant are alike or different from plants with seeds.

**FYI**

One genus of club mosses is *Lycopodium*, which means "wolf foot" in Greek. This name comes from the plant's appearance. It has several "toes" branching from the central stem that look like the toes and claws on a wolf's foot.

## OBJECTIVES

- Describe the functions of the cambium, phloem, and xylem.
- Explain the primary function of the three main parts of all seed plants.
- Compare and contrast the structures of monocots and dicots.

## VOCABULARY

- **cambium** the layer of a woody stem that produces new vascular tissue
- **dicot** a plant that produces two seed leaves
- **monocot** a plant that produces one seed leaf
- **phloem** the vascular plant tissue that transports food from the leaves to the rest of the plant
- **pith** the very center of a woody stem
- **stomata** the openings in a leaf through which gases pass
- **taproot** a large root that grows downward and has smaller roots branching off from it
- **xylem** the vascular tissue that transports water and minerals from the roots throughout a plant

Have you ever enjoyed a great big slice of watermelon? Even as you savored the red fruit, you had to spit out the seeds. Vascular seed plants are among the most familiar plants. For example, beans, dandelions, cotton, tomatoes, strawberries, grasses, and trees all grow from seeds. Seed plants have three main parts: roots, stems, and leaves.

Each seed plant gets its nutrients from underground and from aboveground through an extraordinary system of roots and shoots. The root system is made up of roots; the shoot system is made up of stems and leaves. These two systems depend on each other and are connected with two kinds of vascular tissue: xylem and phloem. **Xylem** is specialized plant tissue that transports water and minerals from the roots throughout the plant. The **phloem** is the plant tissue that transports food from the leaves to the rest of the plant.

Although they are hidden from sight, roots are a very important part of the plant. Root systems can be very extensive—roots often extend underground as far as the aboveground part of the plant extends. Remember this the next time you see a large tree! Roots have three main functions. Roots absorb water and dissolved minerals from the soil; these materials are transported by the xylem throughout the plant. In addition, roots support and anchor the plant in the soil. Roots also store extra food that the plant makes during photosynthesis; this food, which is usually stored as sugar, is carried from the leaves to the roots in the phloem.

Roots are carefully structured to do their work. The outer layer of the root, or its skin, is called the *epidermis*. Cells called *root hairs* extend out from the root and increase its surface area so that the root can absorb more water. The root hairs draw water and minerals up into the next layer, which is the cortex. In many plants, the cortex stores food. Water and dissolved minerals are then diffused into the inner layer of the root. The inner root contains xylem and phloem that carry water and dissolved minerals throughout the plant. The tips of roots are covered by a root cap, which protects the tip of the root as it pushes through the soil as the root grows. Older roots, such as those of trees, become tough and woody and serve as anchors for the tree.

There are two types of roots: taproots and fibrous roots. A **taproot** is a large root that grows downward through the soil and has smaller roots branching off from it. The small roots branching off a taproot are called *lateral roots*. Some taproots store sugars and starches. A fibrous root is a mass of small roots that spread out from the base of the stem. Tomatoes and grasses have fibrous roots. Although taproots and fibrous roots

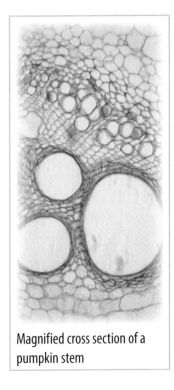

Magnified cross section of a pumpkin stem

**Detail of a Root**

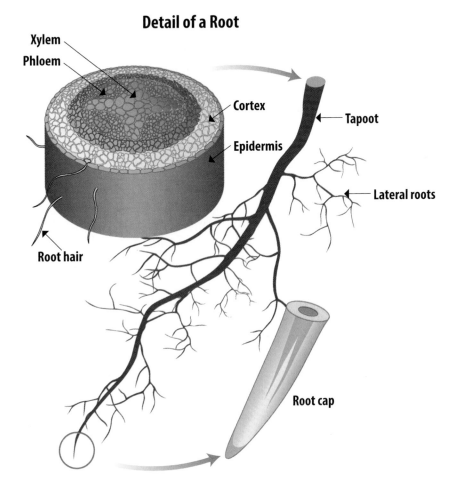

Xylem
Phloem
Cortex
Epidermis
Tapoot
Lateral roots
Root hair
Root cap

Dandelions grow from a long taproot that, when cut, can grow into multiple new plants. When removing dandelions, make sure to cut at least four to five inches of the taproot and the aboveground growth. Otherwise, new shoots will develop and the plant will continue to flourish.

Tapioca is made from the starch of the cassava root. What type of root system do you think it has?

Fibrous tree roots

have different designs, they both anchor plants into the ground, support them, and absorb water and minerals from the soil.

For thousands of years people have eaten root vegetables such as parsnips, carrots, and turnips. Root products like tapioca are ingredients in food, and roots such as licorice and horseradish are used as spices. Some roots are also used to make medicines and dyes.

The shoot system of stems and leaves is another important part of the plant. Although some plants have stems underground, stems are usually located aboveground. A stem has three main functions. Stems support the plant by holding up the leaves so they can absorb sunlight for photosynthesis and holding up flowers so that they can be pollinated. Stems also transport materials through the xylem and phloem between the roots and the rest of the plant. Some stems also store materials such as water or food. For example, the potato is an underground stem (not a root, as many people think) that stores starch, and cactuses store water in their thick stems.

Plants can be classified into two groups on the basis of the structure of their stems—woody and herbaceous. Woody plants have tough, strong stems made of wood. Woody plants include trees as well as smaller plants such as roses. Herbaceous plants have soft, green stems. Grasses, peas, petunias, and carrots all have herbaceous stems.

The tough outer layer of a woody stem is called *bark*. The outer bark waterproofs the plant and protects the plant's inner tissues from disease, insects, and harsh weather conditions. The inner bark is made of phloem. The next layer under the bark is the **cambium**, which is the layer of a woody stem that produces new phloem and xylem. This layer produces a new layer of wood each year, which causes the plant to grow in diameter. The cambium, located between the outer and inner bark, produces a new outer bark each year by separating the two bark layers. As new bark is produced and pushed outward, the older layers split, creating a pattern on the bark. Every species of tree has a different bark pattern that can be used for identification.

There are two layers of xylem within the cambium. *Sapwood* is the outer layer of xylem that transports water and minerals from the roots to the leaves of woody plants. Sapwood is lighter and softer than *heartwood*, which is the inner nonliving xylem

## Specialized Stems

A rhizome is a horizontal underground plant stem that often can produce the root and shoot system of a new plant. When a rhizome is cut, it does not die but grows additional plants. Certain weeds, such as crabgrass and Canada thistle, are very hard to get rid of because they have rhizomes and just keep spreading. Other plants with rhizomes are ginger, trillium, irises, and water lilies.

A tuber is an enlarged tip of a rhizome that stores food. Although tubers do not look like stems, they have all the usual stem parts. For example, they have tiny scale leaves, each with a bud that has the potential for developing into a new plant. A potato is a tuber, and the "eyes" of the potato are the buds.

A tendril is a slender, sensitive structure of many climbing plants. It responds to touch and supports the plant. Tendrils can be vines, stems, leaves, or roots. Most young tendrils turn slowly as they grow so that they coil around a structure. For example, pea plants and grapes have tendrils that coil around a slender support. Other plants, such as Boston ivy, have tendrils that attach to brick, stone, or wood, allowing the plant to crawl up walls and buildings. Vines are weak, flexible stems that require support. By using tendrils, some vines climb trellises, walls, or other plants. Others, such as cucumber and squash vines, creep along the ground.

A runner is a slender, creeping stem that takes root where its buds touch the ground, producing new shoots. The runner usually dies at the end of the season, leaving new separate plants. Strawberries, black raspberries, and some grasses reproduce by runners.

A bulb is an underground stem that stores food and water and plays a part in asexual reproduction. The bulb sustains the plant from one blooming season to another. Some bulbs have many fleshy layers, such as the onion and hyacinth. Others have thin dry scales, such as lilies. Both the layers and the scales are highly modified leaves.

## TRY THIS

### Colored Pigments

Tear three fresh, green sugar maple leaves (or any other leaves that change colors dramatically in the fall) into pieces as small as possible, and place them in a glass container about 15 cm tall. Add 70% rubbing alcohol to cover the leaf fragments. Tape a 2 cm × 15 cm strip of filter paper to a pencil, and lay the pencil across the top of the glass container. Adjust the length of the filter paper by rotating the pencil so that the end of the paper strip just touches the alcohol. When the alcohol has moved up the strip about 10 cm, remove the filter paper strip, and hang it to dry. Describe the different pigments in the leaves.

**Cross Section of a Tree**

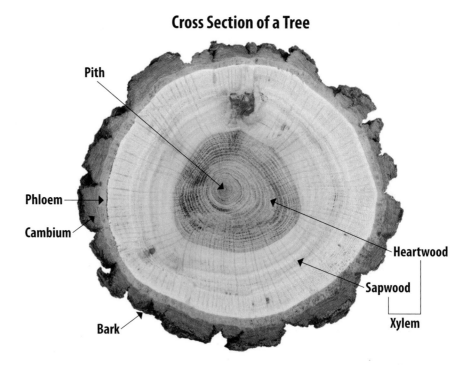

layer at the center of a tree. No water transport occurs in the heartwood. As the plant grows, the xylem tubes press together to form the heartwood. Heartwood is very strong and provides support for the tree. The very center of a woody stem is called the **pith**. The cells in the pith store water and food in young plants.

A tree adds new layers of wood during spring and summer each year. In a cross section of a tree trunk, these layers appear as rings. Each ring includes both a light and a dark section. In the spring, the growth is fast, which produces lighter colored wood because the cells are larger. In the summer, the growth is slower, so the wood is darker and has smaller cells. To determine the age of a tree, count only the light or the dark rings.

A tree's rings can tell the story of its life. For example, wide rings indicate years with a lot of rain or snow, and narrow rings indicate dry years. The tree may also develop a thin ring during a year in which insects infest the tree's leaves. The amount of sunlight a tree receives also affects the size of its rings. Wider rings form when the tree receives plenty of sunlight, and thinner rings form when it does not. Other factors can also affect growth rings, such as exposure to strong winds that slow the tree's growth on one side, causing misshapen rings.

Woody plants grow as their twigs lengthen. At the end of every twig is a terminal bud. The terminal bud is where primary growth takes place. Each terminal bud contains the beginning of a twig, leaf, or flower and is protected from the weather by thick, overlapping scales. To tell how much a woody plant has grown in a year, look at the distance between the current terminal bud and the scale scar left by the previous year's terminal bud. If the terminal bud dies or is removed, a lateral bud (a bud at the side of the twig instead of at the top) will develop. If a terminal bud or an end flower is clipped, the plant will grow bushier; otherwise it will become tall and spindly.

Like woody stems, herbaceous stems have xylem and phloem and buds where leaves grow. The cambium in herbaceous stems is not as active as the cambium in woody stems, so herbaceous stems do not grow as much as woody stems do. Most herbaceous stems die off every year.

Stems are used to make a variety of products. Wood, of course, is commonly used to make many objects ranging from buildings to matches to furniture. Paper is also made from wood. Other stems are used as food, such as potatoes, onions, and sugar cane. In addition, many stems are used to make medicines and dyes. Flax stems are even used to make linen cloth.

Each species of seed plant also has specially designed leaves. In fact, leaves can be useful in identifying plants if you know the structure of leaves. Most leaves have a stalk and a blade. The stalk is the part that connects the leaf to the stem of the plant; the blade is the flat, thin part. There are two kinds of leaves. A simple leaf has a one-piece blade; maple and apple trees have simple leaves. A compound leaf has a blade that is divided into separate parts. Clover, for example, has compound leaves.

Leaves show God's amazing creativity. You might not think of the needles of pines and firs as being leaves, but that is what they are. The leaves of maples and oaks, in contrast, are wide and flat. Some leaves are spines or prickles to protect the plants. Other leaves are tendrils that help to support the plant. Some plants, such as the poinsettia, have brightly colored leaves that help attract pollinators to the flowers. Despite all this variety, all leaves function to make food for the plant. Leaves capture the sunlight that starts the process of photosynthesis. You will learn more about photosynthesis in the next chapter.

Despite their different shapes, all leaves have features in common. The outer layer of a leaf is the epidermis. The

A terminal bud of *Acer pseudoplatanus*

## TRY THIS

**Twig Growth**

Examine the twigs on trees. Check out the terminal buds on different trees. Look at the length of the twigs from the terminal bud to the previous year's bud scale scars. How much did the tree grow last year and in previous years? Compare branches that are low on the tree with those that are higher. Which branches have the longest yearly growth zones? Why?

Simple leaf of an oak tree

Compound leaf from a chestnut tree

epidermis cells are coated with a waxy covering that helps prevent water loss. The inner cells of leaves are known as *the mesophyll*; there are many air spaces in the mesophyll to allow water vapor, oxygen, and carbon dioxide to flow. The openings in a leaf through which gases can pass are called **stomata** (singular: stoma). The stomata come up from the mesophyll to the epidermis. Carbon dioxide enters a leaf and oxygen and water vapor exit the leaf through the stomata. Guard cells open and close the stomata.

Angiosperms are plants that produce seeds in flowers. The flower is the reproductive structure of the angiosperm; it produces the egg cells and the sperm cells within the pollen. An egg cell that is fertilized by a sperm cell becomes a seed. The ovary of the plant enlarges to form a fruit. You will learn more about this reproductive process in Chapter 2 of this unit.

Flowering plants are grouped into two categories depending on the number of seed leaves produced once the seed germinates or sprouts. A seed leaf, or cotyledon, is a leaf that is attached to the plant embryo; it provides food for the young seedling. Plants

## Detail of a Leaf

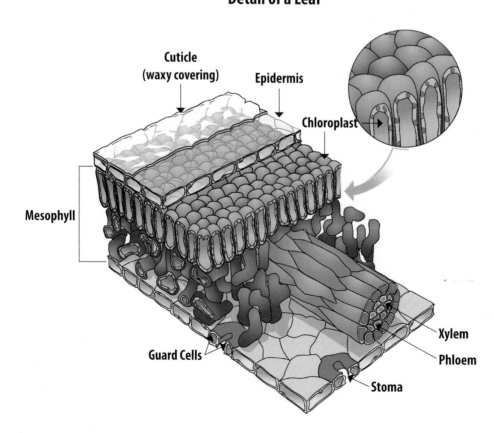

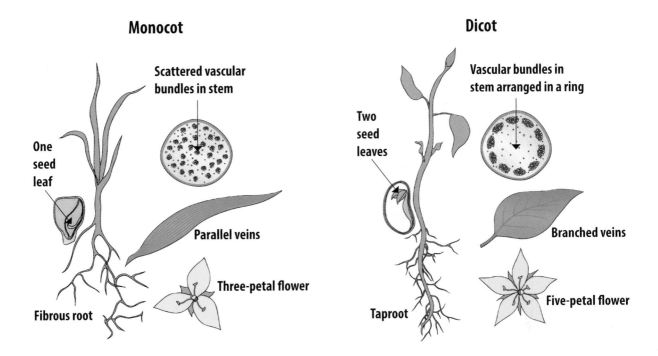

**Monocot**

Scattered vascular bundles in stem

One seed leaf

Parallel veins

Three-petal flower

Fibrous root

**Dicot**

Vascular bundles in stem arranged in a ring

Two seed leaves

Branched veins

Five-petal flower

Taproot

that produce one seed leaf are called **monocots**, which include grasses, lilies, irises, orchids, cattails, and palms. Monocots have leaves with parallel veins, and their flowers and petals grow in groups of three (or multiples of three). The stems of monocots have scattered vascular bundles for transporting water and nutrients. Monocots usually have fibrous roots and soft stems.

The dicots include almost all flowering trees and shrubs and many herbs. **Dicots** are plants that produce two seed leaves. Dicots have flowers with four or five (or multiples of four or five) petals and leaves with a network of branched veins. The stems of dicots are more organized than those of monocots. They have a ring of vascular bundles that transport water and nutrients. The bundles of vascular tissue surround the pith in the center of the stem, in which food is usually stored. Most dicots have taproots. Roses and pansies are examples of dicots.

## LESSON REVIEW
1. What are the three main parts of seed plants and their functions?
2. Describe the functions of the cambium, phloem, and xylem.
3. Describe how a tree grows.
4. Compare and contrast monocots and dicots.

The *Orchis simia* (monkey orchid) is a great example of God's creative hand.

## TRY THIS

**See What Transpires**
Carefully seal a plastic bag around a leaf on a live plant. Wait 24 hours. What happens? Why? Why do plants need so much water? Design an experiment to estimate how much water a tree transpires in one day.

God created your body in a wonderful way. It is sustained by many processes working together. The blood traveling through your blood vessels brings water, nutrients, and gases to every part of your body. You breathe in oxygen so that your cells can break down food into energy that you can use. Your cells get rid of waste products. Your body responds to cycles such as day and night, and hormones tell your body to grow and develop.

God designed similar systems to sustain plants. Factors inside a plant work with factors outside so that the plant can grow. In this chapter you will learn about some of the systems and processes that help plants survive and reproduce.

Factories are structured places that produce products such as appliances, paper, clothing, or furniture. What goes on in a factory? Factories take in raw materials and combine them with energy to produce a product. Think of a chocolate factory, for example. A chocolate factory takes in raw materials such as milk, sugar, and cocoa beans, combines them with heat, and produces chocolate.

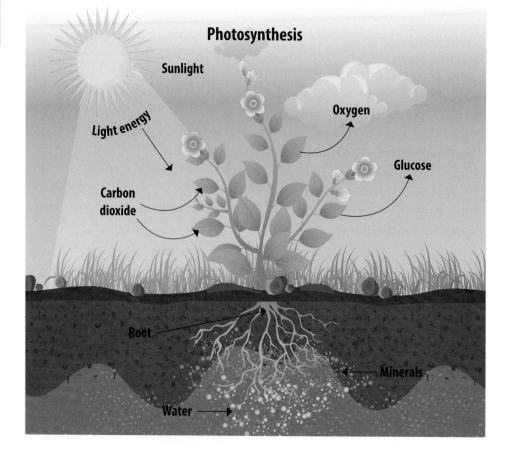

## Photosynthesis and Cellular Respiration

$$6CO_2 + 6H_2O + Energy = C_6H_{12}O_6 + 6O_2$$

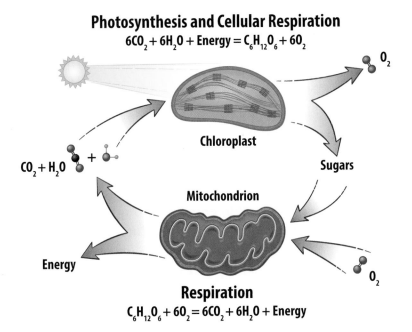

Chloroplast

$O_2$

$CO_2 + H_2O$ +

Sugars

Mitochondrion

Energy

$O_2$

### Respiration

$$C_6H_{12}O_6 + 6O_2 = 6CO_2 + 6H_2O + Energy$$

In many ways plants are like factories. Plants take in raw materials and use energy from sunlight to produce a finished product (food). **Photosynthesis** is the process by which plants use energy from the sun to make their own food. During photosynthesis, plants use energy from sunlight to make food in the form of glucose ($C_6H_{12}O_6$), a simple sugar made from the raw materials of carbon dioxide ($CO_2$) and water ($H_2O$).

**Chlorophyll** is the green pigment in each plant cell's chloroplast that captures light energy for photosynthesis. The formula for this process is written as follows:

$$\underset{\text{Carbon dioxide}}{6CO_2} + \underset{\text{Water}}{6H_2O} \underset{\text{(Light energy)}}{\longrightarrow} \underset{\text{Glucose}}{C_6H_{12}O_6} + \underset{\text{Oxygen}}{6O_2}$$

During photosynthesis:
- Light energy strikes the leaf, passes into the leaf, and reaches a chloroplast inside an individual cell.
- The chlorophyll inside the chloroplast captures the light energy.
- Some of the energy is used to split water ($H_2O$) into hydrogen (H) and oxygen (O).
- The oxygen is released into the air because it is not needed.
- Carbon dioxide ($CO_2$) enters the leaf through the stomata and passes into the chloroplast.
- Using more of the light energy, hydrogen combines with carbon dioxide in the chloroplast to make glucose ($C_6H_{12}O_6$), which is a sugar.

## TRY THIS

### Transport

Fill a jar about three-fourths full of water. Add a few drops of food dye to the water. Make fresh cuts on one end of two celery stalks. Record your observations. Are the parts of the celery uniform in color? Make a deep cut about halfway up one of the celery stalks. Stand the celery stalks in the water. Wait a few hours and observe the plants again. Record any differences in the color of the celery stalks. After a few more hours, cut the top of the celery stalks, and record your observations. Was the dye transported to the top of the stalks? Did the notch affect the transport?

- Glucose is carried in the phloem to other plant parts.
- The plant uses the glucose for energy and to carry out plant functions.

The phloem transports much of the glucose to other parts of the plant because every plant cell needs sugar or it will die. The plant can combine the simple sugars it produces into starch, which is often stored in the plant. Potatoes, for example, are high in starch. Plants need glucose to provide the energy needed to carry out processes such as growth and repair. Plants break down food molecules into usable energy through a process called *cellular respiration*. During this process, which takes place in each cell, a plant uses oxygen and releases carbon dioxide and water. Photosynthesis and cellular respiration are nearly the reverse of one another.

For photosynthesis to occur, a plant needs to get carbon dioxide from the air into the chloroplasts of its cells. Carbon dioxide gets

## Transport and Transpiration

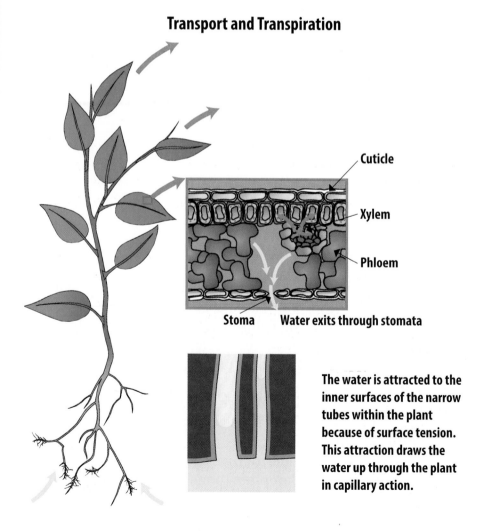

Cuticle

Xylem

Phloem

Stoma    Water exits through stomata

The water is attracted to the inner surfaces of the narrow tubes within the plant because of surface tension. This attraction draws the water up through the plant in capillary action.

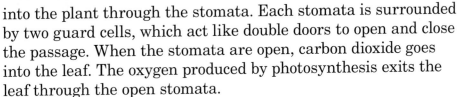
Grass creates oxygen for the earth through the process of cellular respiration.

## FYI

**Transpiration and Surface Area**

Transpiration presents a challenge for plants. For transpiration, a plant needs as little surface area as possible to reduce water loss. But for photosynthesis, a plant needs as much surface area as possible to collect sunlight and carbon dioxide. In order for carbon dioxide to be absorbed, the cell surfaces must be moist. This presents a dilemma. But God designed plants to carefully balance these needs. Many plants in dry, sunny environments have small, light-colored leaves that reduce water loss. Plants in shady, moist habitats have large, dark green leaves that photosynthesize more easily. Some plants have specialized leaves; tropical jade plants have waxy, thick leaves that slow transpiration but allow for maximum photosynthesis. Plants that have lots of water are designed so that excess water can run off; many rain forest plants have drip tips so water can fall off their leaves. Some rain forest plants get little sun, so they have large leaves that give them more sun-absorbing surface areas.

into the plant through the stomata. Each stomata is surrounded by two guard cells, which act like double doors to open and close the passage. When the stomata are open, carbon dioxide goes into the leaf. The oxygen produced by photosynthesis exits the leaf through the open stomata.

Plants move the water that they absorb and the food that they produce from place to place within the plant through the xylem and phloem. The movement of materials throughout a plant is called *transport*. Transport involves a combination of processes. Plants transport water from the roots to the stem through a process called *root pressure*. Water molecules enter the root cell faster than they leave, which causes pressure to build up in the xylem and forces water to move upward to the stem. Then, the process of capillary action transports the water in the stem to the top parts of the plants. Water molecules move up the stem in capillary action because the water molecules are attracted to each other and to the walls of the narrow xylem passages; this attraction draws the water up through the stem to the top parts of the plant.

© *Life Science*

Water moves continuously through a plant from the roots to the leaves. When light falls on a plant, the stomata usually open so that the plant can photosynthesize, which also allows water vapor to escape from the leaves. The loss of water vapor by plants through stomata is called **transpiration**. Plants wilt when they lose more water through transpiration than they have absorbed through their roots. Under these conditions, the stomata will close to prevent further water loss. When it is dark, the stomata usually close, which helps the plant conserve water. Transpiration is believed to cause negative pressure at the leaves. This negative pressure then pulls the plant's water up from the roots and through the stem in much the way you would drink water through a straw. However, some scientists believe that transpiration is just how a plant "sweats."

## LESSON REVIEW

1. Name the raw materials, energy, and products involved in photosynthesis.
2. What role do leaves and chlorophyll play in photosynthesis?
3. How do water and nutrients move up a plant?
4. Explain how damage to the roots, stem, and leaves of a plant can harm the entire plant.
5. Explain the process of transpiration.

One reason that rainforest climates are so humid is that rain forest plants transpire a lot of water.

Think about the many flowers that decorate the world: roses, lilacs, daisies, bird-of-paradise flowers, cactus flowers—even dandelions and other "weeds" that add interest to the landscape. Now think about all your favorite fruits—perhaps apples, oranges, bananas, kiwis, grapes, peaches, and raspberries. If you appreciate these flowers and fruits, then be thankful for angiosperms.

It is hard to choose a favorite among the wide array of lush flowers and fruits. But God did not create flowers and fruits just so that people can enjoy their sight, scent, and taste; God created flowers and fruits of all shapes, sizes, and colors as food for other living creatures. He also designed them to produce seeds that ensure the birth of a new generation of angiosperms. The seeds of angiosperms are produced because of the intricate design of their flowers.

**Sepals** are the leaflike structures that cover and protect an immature flower. As the flower blossoms, the sepals fold back. Sepals are often green like leaves.

Petals are the often brightly colored parts of a flower that surround the reproductive structures. Together they form the corolla. Like sepals, petals are broad, flat, and thin, but they

## OBJECTIVES

- Identify the reproductive structures of angiosperms.
- Describe the conditions necessary for seed germination.
- List the ways seeds are dispersed and the designs that make this possible.

## VOCABULARY

- **embryo** an organism in its earliest stages of development
- **ovary** the rounded base of a pistil
- **ovule** the structure that contains the egg cell of a seed plant
- **pistil** the female reproductive structure of a flower
- **sepal** the leaflike structure that covers and protects the immature flower
- **stamen** the male reproductive structure of the flower
- **stigma** the tip of the pistil
- **style** the long, slender part of the pistil

Flowers display an amazing variety of shapes and colors.

Catkins of a birch tree

The common blue violet has a blue and white pattern of nectar guides on its petals.

Corn is planted in large fields. The wind can carry pollen from the male tassels to the silk, which is the female stigma.

Columbine flowers have a blue and yellow color pattern to attract butterflies or birds. The have long nectar spurs for hummingbirds.

The wind also pollinates many plants, trees in particular. Willows, birches, and oaks have long clusters of flowers called *catkins*. The stamens on the catkins shed pollen, and the wind scatters it. The stigma of a flower pollinated by the wind has a sticky substance to capture pollen as it blows past. Because they do not need to attract insects, wind-pollinated flowers seldom have attractive scents or bright colors. Instead they produce great quantities of dry pollen grains, which can drift for hundreds of kilometers.

New plants grow best if they have been cross-pollinated, or pollinated by pollen from different plants rather than from their own pollen. If plants self-pollinate, their offspring are often genetically weaker. Flowers have many features that increase their chances of cross-pollination. Some plants produce both male and female flowers on the same plant, but the pistils and stamens are located in different flowers. Sometimes male and female flowers grow on different plants. In this case, pollination and fertilization can occur only if both male and female plants are present. In other plants, the pistils and stamens mature at different times, so the pollen is dispersed before the pistils of that plant are ready for fertilization.

Pollination leads to fertilization. Fertilization begins when a pollen grain sticks to a flower's stigma. The protective coat on the pollen grain ruptures, and a tube called *a pollen tube* grows from the grain through the stigma to an opening in an ovule in the ovary. The pollen grain contains two nuclei. One nucleus disintegrates; the other divides to form two sperm. The sperm move down the pollen tube and enter the ovule. One sperm fuses with the egg to form a zygote; the other helps develop the tissue that supplies food for the growing **embryo**, which is an organism in its earliest stages of development.

Following fertilization, the ovule wall develops into a protective seed coat, the zygote develops into an embryo, and the ovary develops into a fruit. The embryo grows and begins to form vascular tissue and cotyledons, or seed leaves. When the embryo is complete, the seed coat hardens, and the seed becomes dormant. The fruit, or ripe ovary of the plant, protects the enclosed seeds.

After seeds have finished forming, they are usually scattered. Some seeds depend on animals to spread them. For example, animals that eat fruits such as raspberries, dates, or cherries will later drop or expel the seeds in a different place. Such seeds are designed not to be broken down during the digestive process. Animals and humans also scatter sticky seeds or seeds with burrs that get caught in fur or clothes. Some seeds are carried by water. Willow trees have lightweight seeds that float. Coconuts produce woody, waterproof seeds that can float for long periods in salty water. Some seeds depend on the wind to spread them. For example, cottonwood and dandelion seeds have fluffy heads that catch the wind. Others, such as maple tree seeds, have wings that help them blow in the wind. The fruits of some plants, such as geraniums, poppies, milkweeds, and sensitive plants, burst open and throw their seeds out.

This showy lady's slipper orchid has a pouch that holds a fragrant chemical to attract insects. Insects climb inside the pouch, which is designed to direct them past the flower's anthers. When they climb out, pollen sacs are stuck to the insects' backs. As the insects visit another orchid, that flower's stigma removes the pollen sacs from the first orchid and pollination is achieved.

Milkweed pods burst open, scattering their seeds.

A stand of conifer trees

Gymnosperm means "naked seed." The seeds of gymnosperms are not enclosed in a flower or fruit. Gymnosperms reproduce differently than angiosperms. To understand how gymnosperms reproduce, examine the reproduction of conifers, the most familiar gymnosperms.

Conifers grow both male and female cones, which usually grow on the same tree. The male cones are quite small, from 1 to 2 cm long. These cones produce male spores that develop into pollen, the dustlike particles that carry the sperm cells. Pollen grains are the male gametophytes. The tree sheds the yellowish pollen grains in huge quantities. The wind carries the pollen to female cones on the same tree or to female cones on a nearby tree.

Female cones are much larger and more complex than male cones. Two ovules (undeveloped seeds) are found on the upper surface of each scale of a female cone. In the spring, the scales

### Gymnosperm Pollination and Fertilization

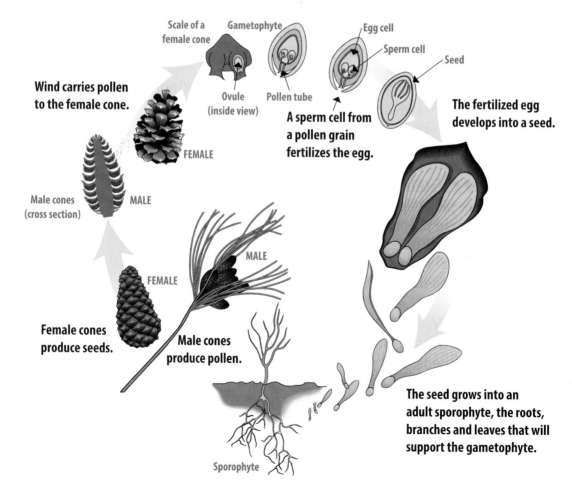

Scale of a female cone

Gametophyte

Egg cell

Sperm cell

Seed

Ovule (inside view)

Pollen tube

**Wind carries pollen to the female cone.**

**A sperm cell from a pollen grain fertilizes the egg.**

**The fertilized egg develops into a seed.**

FEMALE

Male cones (cross section)

MALE

MALE

FEMALE

**Female cones produce seeds.**

**Male cones produce pollen.**

**The seed grows into an adult sporophyte, the roots, branches and leaves that will support the gametophyte.**

Sporophyte

Female cones on a Norway spruce

of the female cone spread open. Pollen from male cones drifts through the air and sticks to a fluid near the opening of an ovule. As the sticky fluid evaporates, the pollen grain is drawn into the ovule, where the embryo will develop. One month after pollination, the scales of the female cone thicken to close up the open spaces between the scales. The female gametophyte, which will produce the eggs, begins to grow.

About 15 months later, the cone has enlarged, the female gametophytes have produced eggs, and the pollen tubes have discharged sperm cells into the egg. Almost two years later, the cone has grown to full size, but it is still tightly closed. The seedling embryos have developed inside the ovules. Two years after pollination, the cone opens to allow seeds to fall out. Each scale contains two ripe seeds that have wings formed from a thin layer of the scale. The winged seeds flutter through the air or are carried by the wind. Seeds germinate in soil, and seedlings begin to grow. When each tree is mature, the cycle starts again.

## FYI

### Scattering Pine Seeds

The seeds of some pine species are scattered by birds. For example, the seeds of limber pines, whitebark pines, and pinyon pines of western North America are eaten by large crowlike birds called *nutcrackers*. The birds digest the seeds and expel them with other waste. Other pines, especially the jack pine, require extreme heat for the scales to separate and the seeds to release. When a forest fire sweeps through jack pine forests, the parent trees burn, but the cones are heat resistant and open to release their seeds. A forest fire can be helpful because it clears out dead needles and underbrush while preparing cones to begin a new generation of jack pine trees.

Seeds (or nuts) from a cedar pine cone

Male cones on a pine tree

Male cones usually grow on lower branches, and female cones grow on upper branches. Some pine trees bear male and female cones on the same branch, but the female cones are closer to the end of the branch. This design limits self-pollination, which happens when the pollen from a cone fertilizes an egg cell from another cone on the same tree, because pollen is usually not blown straight upward. As a result, the tree's eggs are usually pollinated by the pollen from another tree (cross-pollination). Cross-pollination is more desirable than self-pollination because self-pollination can produce weakened plants.

## LESSON REVIEW

**1.** Compare and contrast male cones and female cones.
**2.** Briefly describe the steps in a conifer life cycle.

Male and female cones on the same branch

Imagine your surprise if you stepped on a plant and it shrieked in protest. What if you reached down to pick a wildflower and it pulled itself out of the ground and fled? Plants do not defend themselves in these ways, but God did give plants ways of responding to their environment.

Plants are able to move toward or away from external stimuli by growing in a particular direction. A change in the growth of a plant in response to a stimulus is called a **tropism**. The direction of the movement is usually determined by the direction from which the most intense stimulus comes. Growth toward a stimulus is a positive tropism; growth away from the stimulus is a negative tropism.

**Phototropism** is the growth of a plant in response to light. If you place a houseplant in such a way that it gets light from only one direction, the shoot tips will bend toward the light. This response happens because the cells of the shaded side of the plant grow longer. Growth toward a light source is positive phototropism. In contrast, ivy growing on walls moves away from the light toward the darker, solid walls. This response is an example of negative phototropism.

Some plants orient themselves to the sun's direct rays, following the sun's movement throughout the day. This movement is called *solar tracking*. Some common plants that do this are cotton, soybeans, and sunflowers.

A conifer seedling pushing up through the forest floor

Notice how all these sunflowers are pointing in the same direction—toward the sun.

**Gravitropism** is the growth of a plant in response to gravity. Gravitropism demonstrates both a positive and a negative tropism. The roots grow downward, a positive response to gravity; the shoots grow upward, a negative response to gravity.

Contact with solid objects causes some plants to respond. A plant's growth response to touch is called **thigmotropism**. Tendrils demonstrate thigmotropism by wrapping around objects to provide support for a plant as it climbs.

Other types of plant movements can be observed in daily cycles such as the flowering time of flowers or the folding and unfolding of leaves. These regular rhythms of growth and activity that occur in approximately 24-hour cycles are called *circadian rhythms* after two Latin words that mean "approximately a day." Although most scientists now believe that circadian rhythms are controlled internally by a plant's biological clock, external factors may also influence these plant movements. Circadian rhythms are important for plants because they help plants adjust to changing seasons and changing environments. Circadian rhythms also affect plant growth and reproduction.

Photoperiodism is the biological response of a plant to changes in the amounts of lightness and darkness within a 24-hour cycle. Understanding this phenomenon can answer some confusing questions. For example, photoperiodism explains why ragweed does not grow in far northern Maine: ragweed plants begin

The roots of this orange tree grow downward, a positive response to gravity. The stem and branches grow upward, a negative response to gravity.

## TRY THIS

### Rotating the Light

Set up four geraniums along a sunny window ledge for four weeks. Leave one in its original position, rotate the second plant one-quarter of a turn each week, rotate the third half a turn, and rotate the last plant three-quarters of a turn.

Set up a certain time every week to rotate the plants. Draw or describe your observations.

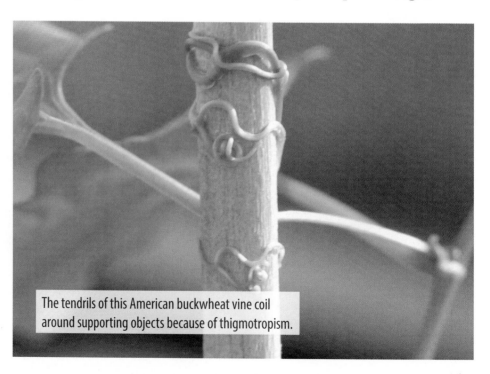

The tendrils of this American buckwheat vine coil around supporting objects because of thigmotropism.

making flowers when the day is about 14.5 hours long; the long summer days in that area do not shorten to 14.5 hours until August, so there is not enough time for the plant to seed before the first frost. For similar reasons, spinach does not grow naturally near the equator: spinach requires 14 hours of light a day for at least two weeks to flower, and because of the tilt of the earth, that latitude receives only 12 hours of sunlight per day.

Experiments with tobacco plants and soybean plants have helped scientists discover the influence of day length on plant growth. In their experiments, American scientists W. W. Garner and H. A. Allard discovered that neither tobacco nor soybeans will flower unless the day length is shorter than a certain number of hours. Garner and Allard defined three categories for flowering plants on the basis of their response to light: short-day plants, long-day plants, and day-neutral plants. Short-day plants are those that flower during the short days of spring or fall. In order to flower, these plants require a period of daylight that is shorter than a particular length. Chrysanthemums, poinsettias, and strawberries are short-day plants. Long-day plants need longer light periods, so they usually flower in the summer. Spinach and some potatoes fall into this group. The photoperiodic response varies for different species; some species need only one exposure to the critical day-night cycle but others require weeks of exposure. The flowering of day-neutral plants does not relate to a certain day length. Petunias are day-neutral plants.

## TRY THIS

**Follow the Light**
Use cardboard or index cards, scissors, and tape to set up a simple maze through which a plant can grow inside a shoe box.

Cut a hole in the box at the end of the maze, and place the plant inside the box at the beginning of the maze. Sketch the plant and the maze. Place the lid on the box, and seal it with a few strips of tape. Each day dim the lights and open the box to check the plant's progress and to make sure that the plant has enough water. Limit the amount of time the plant is exposed to light when the box is open. Sketch the plant as it grows through the maze, recording the date of each observation.

You could also grow plants in four shoe boxes: one with a slit cut in the lid, one with a slit cut on one end, one with a slit cut on one side, and one with no slit. Observe how the plants grow in each box.

Morning glory flowers (left) open in the morning and close their petals at night. The Moon flower (right) blooms during the evening and closes its petals during the day.

The poinsettia plant requires a short amount of daylight to flower, and therefore blooms during seasons when the days are shorter.

The leaflets of the *Mimosa pudica* plant close up by folding inward when they are touched.

Nastic movements are more rapid plant movements that are not related to the direction of a stimulus. The most common nastic movement is the night-closing or "sleep" movements of leaves. Other movements result from touch. For example, the leaflets of the sensitive plant, *Mimosa pudica*, droop suddenly when touched. A Venus flytrap's nastic movement results from a fly landing on one of its traps, causing the trap to close and capture the fly.

## LESSON REVIEW

**1.** Give an example of a positive and a negative tropism.
**2.** Describe the three tropisms and give an example of each.
**3.** What are circadian rhythms?
**4.** Define a nastic movement and give an example.

UNIT 4

# Animals

## Chapter 1: *Invertebrates*
## Chapter 2: *Arthropods*
## Chapter 3: *Vertebrates*

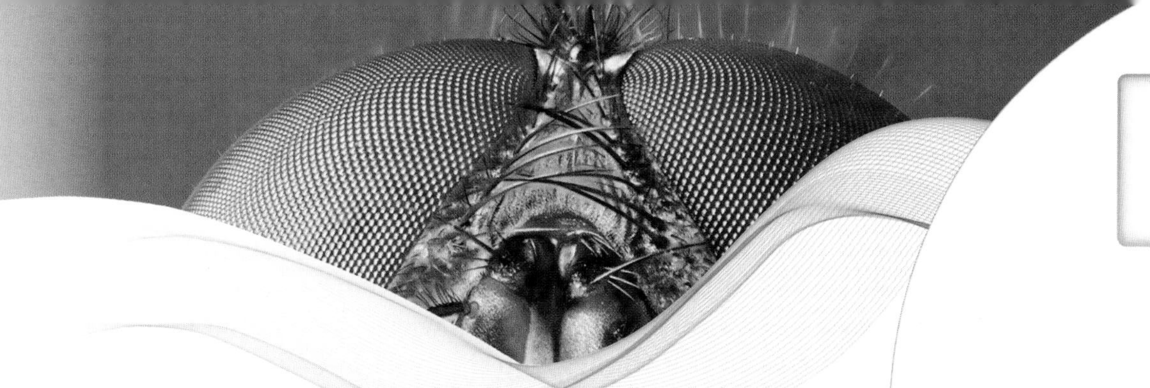

*Vocabulary*

abdomen
arachnid
arthropod
bivalve
centipede
cephalopod
cephalothorax
collar cells
constrictor
crustacean
echolocation
ectotherm
endoskeleton
endotherm
entomologist
exoskeleton
gastropod
herpetology
keel
mammary gland
mandible
mantle
marsupial
maxilliped
medusa
metamorphosis
migrate
millipede
nematocyst
nymph
operculum
placenta
polyp
radula
regeneration
spicule
swim bladder
thorax
tube feet

*Key Ideas*

- Systems, order, and organization
- Evidence, models, and explanation
- Change, constancy, and measurement
- Form and function
- Abilities necessary to do scientific inquiry
- Understandings about science and technology
- Structure and function in living systems
- Regulation and behavior
- Populations and ecosystems
- Diversity and adaptations of organisms
- Risks and benefits
- Science and technology in society
- Science as a human endeavor
- Nature of science
- History of science

SCRIPTURE

Your righteousness is like the highest mountains, Your justice like the great deep. You, Lord, preserve both people and animals.

Psalm 36:6

## OBJECTIVES

- State the three basic groups of invertebrates by symmetry.
- Distinguish between asexual and sexual reproduction.

We see animals every day—family pets, squirrels and birds in the trees, and insects or spiders crawling in the grass. Animals are a familiar part of the world. When you think of the importance of animals, you might think of pets that offer friendship; cows, chickens, and fish that provide food; or insects that pollinate crops and other vegetation. But all animals, including sponges, clams, worms, and sea stars, are part of the goodness of creation. God designed all animals to contribute to life on Earth, even the ones that we do not see often or have never seen at all!

Think of an animal. What kind of animal did you think of? Was it a mammal such as a tiger or a dolphin? Maybe you thought of a reptile, such as a rattlesnake, or a bird, such as a beautiful parrot, or perhaps a fish, such as the great white shark. When most people think of animals, they think of animals that have a backbone. Zoologists, who are scientists who study animals, have named more than 1.5 million species of animals, but only about 3% of these species have backbones. Approximately 97% of all animal species are known as *invertebrates*, which are animals without backbones. Animals with backbones are called *vertebrates*. Even more amazing is that zoologists believe that millions of invertebrates have not even been discovered yet!

God designed invertebrates with many functions. Invertebrates are an essential source of food for many animals. Some invertebrates, such as shrimp and clams, are a food source for people. Most flowering plants depend on invertebrates like insects for pollination. Without insects to pollinate them, many trees, crops, flowers, and weeds could not reproduce.

Most animals produce offspring through sexual reproduction, which involves both a male and a female parent and results in offspring that have characteristics of both parents. In sexual reproduction a sperm cell fertilizes an egg cell to

produce a new individual. But many invertebrates reproduce through asexual reproduction, which involves only one parent and results in offspring that are identical to the parent. These animals may reproduce by breaking off small parts that grow into a new individual.

Scientists use symmetry to group invertebrates. Invertebrates can be asymmetrical or they can have radial or bilateral symmetry. Sponges are asymmetrical. They cannot be divided into equal parts along a line of symmetry. Jellyfish and coral have radial symmetry. These animals' bodies are arranged around a center point. They do not have a right and left side or a head and rear end. Most invertebrates have bodies with bilateral symmetry. Their bodies can be divided into two equal parts that are mirror images of each other. Worms, crustaceans, and insects have bilateral symmetry.

This beetle has bilateral symmetry. The two halves of its body mirror each other.

## LESSON REVIEW

**1.** What is an invertebrate?

**2.** What are the three types of body symmetry that invertebrates can be divided into? Explain each type.

**3.** Explain the difference between sexual and asexual reproduction.

Sea sponges are asymmetrical. It is not possible to draw a line down the middle of a sponge to divide it into equal parts.

This sea anemone has radial symmetry. Its body is organized around a center, like spokes on a wheel.

When you hear the word *sponge*, you probably think of the colored rectangle that people use to wash dishes. But in the sea, and even in some freshwater lakes, sponges are living creatures. Because sponges move so slowly and show no visible reaction when they are touched, sponges were once thought to be plants.

All sponges live in water. Most sponges live in the ocean. The smallest sponges are a few millimeters long. The largest sponges can be 1–2 m tall as well as 1–2 m or more in diameter. Sponges grow in a wide variety of shapes, sizes, and colors, all of which vary depending on their species, age, and the location and the depth of the water that they live in.

Sponges belong to the phylum Porifera. The word *porifera* comes from the Latin words *porus* meaning "pore" and *ferre* meaning "to bear." Sponges have many pores, or tiny openings. Unlike other animals, sponges do not have tissues or organs. Their cells have specific roles, and they cannot work without the other types of cells. Their individual cells work together so that the entire sponge may feed, grow, and reproduce. If a sponge is squeezed through a fine screen, the cells will reassemble back into a single sponge within a few hours! No other animal can do this. Sponges are also capable of **regeneration**, which is the regrowth of lost body parts.

Sponges have no mouths, intestines, or major systems for moving food throughout their bodies. A sponge is basically a water filtering system made of three basic types of cells: collar cells, epithelial cells, and amoebalike cells. **Collar cells** are cells that line the inner chamber of a sponge. These cells have hairlike structures called *flagella* that move

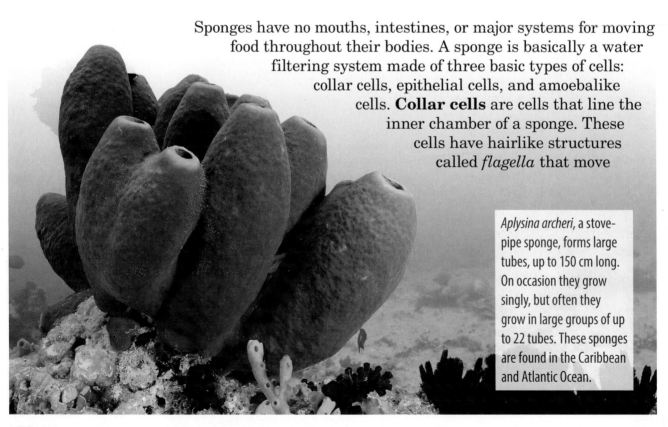

*Aplysina archeri*, a stovepipe sponge, forms large tubes, up to 150 cm long. On occasion they grow singly, but often they grow in large groups of up to 22 tubes. These sponges are found in the Caribbean and Atlantic Ocean.

## Sponge

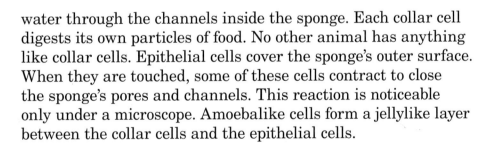

Water out
Amoebalike cell
Osculum
Flagella
Collar cell
Ostium (Pore)
Epithelial cell
Spicule
Water in
Water in

Barrel sponges are some of the largest sponges—up to 2.5 m in diameter!

## FYI

**Natural Sponges**
People often use natural sponges to scrub themselves. Using natural sponges is nothing new. Ancient Greeks and Romans padded their helmets and armor with sponges. In the Middle Ages, people burned sponges and used the ash in their medicines. People use them in medicines today, too. Some sponges are a source of antibiotics that destroy disease-causing bacteria.

water through the channels inside the sponge. Each collar cell digests its own particles of food. No other animal has anything like collar cells. Epithelial cells cover the sponge's outer surface. When they are touched, some of these cells contract to close the sponge's pores and channels. This reaction is noticeable only under a microscope. Amoebalike cells form a jellylike layer between the collar cells and the epithelial cells.

What do sponges eat? Sponges eat plankton, organic particles, and disintegrated debris; even the largest sponge cannot eat anything larger than microscopic organisms. How does a sponge eat? The sponge sweeps water into its body through tiny pores called *ostia* on the outer surface. The collar cells move the water through the central cavity. Each collar cell filters food particles from the water and digests them, a feeding method known as *filter feeding*. At the same time, oxygen in the water moves into the cells. Undigested food and carbon dioxide are released back into the water through a larger opening called *an osculum*.

Sponges have needlelike skeletal materials inside them called **spicules**. Viewed through a microscope, spicules look like slivers of glass. Spicules link together to form a skeleton. They also help

## TRY THIS

**Look Closely at a Sponge**
Examine the surface of a natural sponge with a hand lens. Sketch what you see. Then use a razor blade to cut off a thin slice and examine it under a microscope. Sketch what you see.

## FYI

**Loofahs**
Loofahs are often thought of as natural sponges. In actuality, even though they are natural, they are not the animal sponges classified as Poriferans. Loofahs come from fibrous plant seed pods, such as gourds, pumpkins, and cucumbers. The outer skin is removed to reveal the loofah, or xylem fibers, inside.

protect the sponge against predators. Spicules are very sharp and taste terrible! Other sponges have skeletons made of a softer, fiberlike material. The cleaned and dried skeletons of these type of sponges are the natural sponges you sometimes see in stores.

Sponges play an important role in coral reef ecosystems. They filter and use particles that are too small for other living things to use. Because they do not need light to live, sponges can live deeper than coral, providing homes for many animals. Sponges provide safe habitats for small ocean invertebrates and fish. Inside the sponge, these animals find a steady stream of food, and the sponge's spicules keep predators from eating them.

Sponges can reproduce both sexually and asexually. When they reproduce sexually, one sponge produces eggs and another produces sperm. The cells join in the water, and a new sponge begins to grow. Sponges reproduce asexually by budding. A part of the parent sponge falls off and that part grows into a whole new sponge.

## LESSON REVIEW

1. How are sponges different from most other animals?
2. How do sponges eat?
3. Imagine that you are snorkeling and encounter a 1.8 m sponge. You know that sponges are animals. Should you be afraid of it? Why?
4. How are sponges important to reef ecosystems?

Red finger sponge and a bowl sponge growing in a coral reef

Have you ever had smarting skin and a rash on your ankle after running through a field or vacant lot? If you have, you might have brushed up against a nettle, a plant with stinging bristles. Corals, sea anemones, and jellyfish are cnidarians, marine animals with stinging bristles. The name *cnidaria* comes from the Greek word for "nettle," which is *knide*. This is a fitting name because cnidarians may sting. Cnidarians come in a wide range of sizes. Some are so small that you can see them only with a microscope, but the giant lion's mane jellyfish can be up to 2.4 m wide and has tentacles that can stretch to over 30 m. Unlike sponges, cnidarians' bodies are organized into tissues and organs.

All cnidarians have radial symmetry. They have a central cavity with only one opening, which is the mouth. Surrounding the mouth are tentacles, which are long flexible appendages. Cnidarians' tentacles have special stinging cells called **nematocysts**, which cnidarians use to stun or kill prey. Cnidarians also use their tentacles to pull prey into their mouths. They digest prey in their hollow cavity and spit the waste products back out through their only opening—their mouth.

Cnidarians come in two main body shapes: the polyp and the medusa. A **polyp** is a cnidarian with a vase-shaped body. A **medusa** is a cnidarian with a bowl-shaped body.

## OBJECTIVES

- Identify the major types of cnidarians.
- Describe the general characteristics of cnidarians.

## VOCABULARY

- **medusa** the bowl-shaped body plan of a cnidarian
- **nematocyst** the special stinging cell of cnidarians
- **polyp** the vase-shaped body plan of a cnidarian

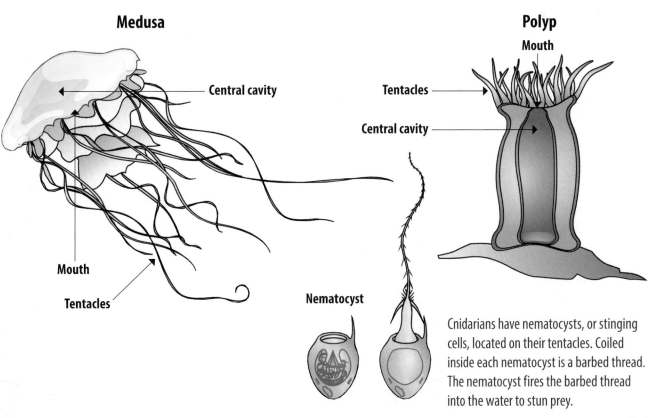

**Medusa**
- Central cavity
- Mouth
- Tentacles

**Polyp**
- Mouth
- Tentacles
- Central cavity

**Nematocyst**

Cnidarians have nematocysts, or stinging cells, located on their tentacles. Coiled inside each nematocyst is a barbed thread. The nematocyst fires the barbed thread into the water to stun prey.

**FYI**

## Coral Reefs: The Rain Forests of the Sea

Coral reefs are one of the world's most diverse ecosystems. Marine biologists (scientists who study ocean life) often refer to coral reefs as underwater rain forests because they are home to thousands of species. Coral reefs make up 0.1% of the earth, but scientists believe that more than 400,000 species may live in and on the world's coral reefs.

Coral reefs are easily thrown out of balance. For example, in the 1980s, fishers took too many parrot fish and sea urchins from the coral reefs surrounding the Cook Islands in the South Pacific. Without the parrot fish and sea urchins to keep the algae under control, algae smothered the reef, killing the corals. In Thailand's waters, nutrients from factory wastes caused a huge increase in the crown-of-thorns starfish population. This starfish species eats corals and is slowly destroying the coral reefs in these waters.

Thousands of ocean species depend on coral reefs for their homes and food. Many of the fish that grow up on the reefs are later caught and eaten by humans. Each year millions of tons of fish are caught near coral reefs. These fish are the main source of animal protein for people in Asia. Scientists believe that the majority of all coral reef animals have not even been discovered yet!

People have destroyed 10% of the world's reefs, and 60% of all coral reefs could be lost in 20–40 years. Some reefs are overfished with nets and dynamite, torn up to be used as

foundations for new buildings, and polluted with cyanide, which is used to stun and collect fish for pet shops and restaurants.

Sediment from logging, farming, mining, and building construction also kills coral reefs. The sediment blocks out sunlight so that the zooxanthellae (essential algae in the coral) cannot photosynthesize; it also smothers the coral's polyps. Careless tourists and divers also harm reefs by kicking fragile soft corals with their swim fins. Oil spills can destroy shallow reefs by smothering the homes of the coral polyps. Pollution and sewage dumped on coral reefs causes too much algae to grow, which blocks sunlight and deprives the corals of oxygen.

God assigned humans to be caretakers of His creation. He formed the coral reefs in the ocean for a purpose, and people should protect the reefs from harm so they can function as God intended. Australia's Great Barrier Reef Park is a coral success story. Park managers have divided the reef into different sectors for scientific research, tourism, and commercial fishing. Park rangers monitor and protect the reef.

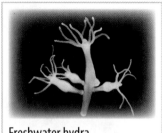

Freshwater hydra

There are four major groups of cnidarians: hydrozoans, jellyfish, sea anemones and corals, and box jellyfish. Hydrozoans include the fire corals, the Portuguese man-of-war, and the small freshwater hydras. Fire corals are not true corals; they are polyps that live on coral reefs. They can inflict a painful sting. The Portuguese man-of-war is a medusa found in warm oceans. Its tentacles average 10 m long. Hydras, which are common in freshwater habitats around the world, have long bodies with tentacles. Hydras are the only freshwater cnidarians. Unlike most other polyps, hydras can move themselves around.

Jellyfish, the second group of cnidarians, are medusas. They spend much of their life as plankton, the microscopic organisms that float near the water's surface. Under certain conditions, however, they can grow quickly. Jellyfish come in all different sizes and colors. Some are transparent blobs floating in the water; others look like giant vegetables and spend their lives swimming upside down. Jellyfish are not just harmless clumps of jelly—all jellyfish can sting. The sting of some jellyfish, such as the moon jelly, is harmless to people, but the sting of other jellyfish can kill a person. If you are swimming or diving and see a jellyfish, stay away from it.

Sea anemones and corals make up the third group of cnidarians. Sea anemones are polyps that look like flowers because their tentacles resemble petals. When fish swim through the tentacles of sea anemones, the nematocysts sting and stun the fish.

The lion's mane jellyfish can have over 800 tentacles. The largest recorded specimen has tentacles that reached over 36 m long!

The slow-growing brain coral is solid and strong enough to withstand storms that break up more delicate coral.

The coral animal is a polyp with a soft body, and is only a few millimeters long. Trillions of them work together to build coral reefs like Australia's Great Barrier Reef, which is over 2,000 km long. 6yCoral polyps make reefs by building small homes for themselves out of limestone (calcium carbonate), which they take from seawater. Coral reefs grow in a wide variety of colors and shapes. Some look like fans, some look like antlers, and some look like human brains.

Coral animals extend their tentacles to capture food.

Each generation of corals builds on the hard layer left by the previous generation. Most coral species build these layers very slowly. Brain coral, for example, adds only 1 cm to a reef every year. Because corals grow slowly, reefs damaged by storms may take as long as 50 years to recover. One careless diver can cause decades' worth of damage in just a few minutes.

Painted or Christmas anemone

Corals eat plankton and exist in a cooperative relationship with zooxanthellae, a type of algae. The zooxanthellae live inside the bodies of coral polyps, which provide protection for the algae. The zooxanthellae give corals their beautiful color and help make food for the corals. Because algae need sunlight to photosynthesize, corals must live in shallow water where sunlight can reach them. Many coral polyps around the world are spitting out the algae, turning white, and dying. This phenomenon has been named *coral bleaching*. Some marine biologists blame warmer ocean temperatures, but scientists are not sure why bleaching is happening.

Box jellyfish

Box jellyfish are similar in form to the true jellyfish, but they have some different characteristics. Box jellyfish have a square shape, four evenly spaced out tentacles or bunches of tentacles, and well-developed eyes that contain lenses, corneas, and retinas. The Australian box jellyfish *Chironex fleckeri*, also known as *the sea wasp,* is among the deadliest creatures in the world.

## LESSON REVIEW
**1.** Name the four groups of cnidarians.
**2.** What do cnidarians have in common?
**3.** What are some differences between polyps and medusas?
**4.** Why must corals live in shallow, clear water?

## 4.1.4  *Worms*

### OBJECTIVES

- Distinguish among the three major groups of worms.
- Observe and describe the behavior of earthworms.

Most people think of worms as the squirmy brownish things used to bait fishhooks. But God created a wide variety of worms—in fact, over one-fourth of all the animal phyla are worms. Most of these worms live in the ocean, although worms thrive in a wide variety of habitats. Worms live in some of the most hostile environments on Earth—inside pockets of methane gas deep on the ocean floor, in oil spills, inside deep ocean vents of near-boiling water, and inside other organisms.

God created worms in all colors, shapes, and sizes. One species of ocean worm is bright pink with brushlike appendages along its body. Feather duster worms look like beautiful white or purple flowers and live on coral reefs; they pop back into long tubes when you touch them. In contrast, fire worms have long stinging bristles and swim through the water. Christmas tree worms live on coral reefs and look like miniature Christmas trees. Worms come in a variety of sizes. Many worms are so small that you can see them only with the help of a microscope. The largest worm is the deep ocean bootlace worm, which can grow to 60 m long—longer than a blue whale. Try baiting a hook with that!

The three groups of worms most commonly studied are flatworms, roundworms, and segmented worms. (Worms that are not studied much include ribbon worms, velvet worms, spoon worms, spiny-headed worms, phoronid worms, peanut worms, tongue worms, beard worms, arrow worms, and feather duster worms.)

*Lineus longissimus*, the bootlace worm, is a type of ribbon worm. The bootlace worm is commonly found along the coasts of Britain and is one of the longest known animals.

Flatworms have soft, ribbon-shaped bodies, a mouth, and a gut. Flatworms reproduce sexually or asexually. These worms live in

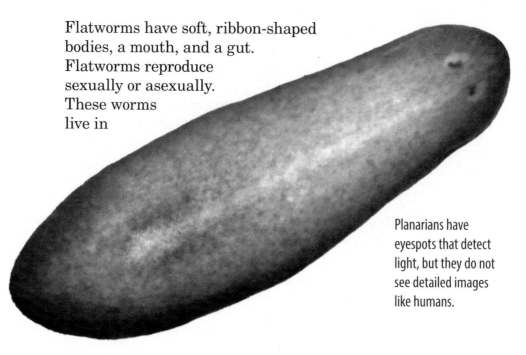

Planarians have eyespots that detect light, but they do not see detailed images like humans.

some odd places—such as in bat feces and human intestines. Flatworms are capable of regeneration. If you cut one flatworm into pieces, each piece will regenerate into a complete new worm. A common flatworm is the planarian, which lives in freshwater streams. A planarian can digest its own body parts when food is scarce and regrow them when food is available again. The largest flatworm is the tapeworm, which can grow to nearly 15 m long. This parasite can live in the intestinal tracts of people and animals. A tapeworm's head has hooks that attach to the host's tissues. The tapeworm takes the host's food and water, which makes the host sick. People can become infected with tapeworms by drinking contaminated water or from eating undercooked meat.

Scientists have identified tens of thousands of roundworms, but there are probably many more undiscovered species. Scientists believe that millions of roundworms live in the oceans. They also live in soil, beach sand, salt flats, lakes, and hot springs. Roundworms look like spaghetti with pointed ends. They do not have body segments, and they move by bending or flipping their bodies. Roundworms have a head and a tail, and they have a tubelike digestive system with a mouth and an anus.

The head of a beef tapeworm has a ring of hooks and four suckers to attach to a host's intestine. It is the most common tapeworm found in humans. People who eat infected or insufficiently cooked beef are at risk. The parasite most often occurs in areas where beef is a major food source and sanitation is poor.

The colorful tentacles on the Christmas tree worm are used for feeding on suspended particles and for respiration. Much of the worm is anchored in its coral burrow.

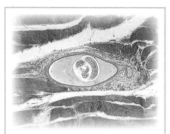

*Trichinella spiralis* is a roundworm that can infect pork. The larvae form small cysts in the muscles. When a human eats the infected meat, the larvae are released and migrate to the intestine, where they burrow into the intestinal lining, mature, and reproduce.

Some roundworms are so small that you can see them only with a microscope; others can grow to over a meter long. Most roundworm species reproduce sexually. Some females hold millions of eggs in their bodies and lay nearly 200,000 eggs a day!

Have you ever been warned that eating undercooked pork can make you sick? A roundworm is responsible for the disease trichinosis. This worm lives in the muscles of pigs, cats, dogs, rats, and bears, but you can only get the disease by eating the undercooked meat of these animals. Trichinosis is a very rare disease in North America, but you can avoid getting it by making sure that pork has been thoroughly cooked before you eat it.

The third type of worm is the segmented worm. These worms have ringlike segments on the outsides of their bodies. Segmented worms display a wide variety of colors, including pink and purple, and patterns such as stripes and spots. They live on land, in freshwater, in saltwater, and even in the seas of the Antarctic. The various species of segmented worms range in size from half a millimeter to the 3 m *Megascolides australis*, the Australian Giant Gippsland earthworm. The most familiar segmented worms are earthworms and leeches.

Earthworms have a closed circulatory system; their body fluids are contained within small tubes. They have no special respiratory organs because oxygen enters and carbon dioxide exits through the earthworm's skin. For this to happen, the

Some flatworms have beautiful designs.

**Earthworm**

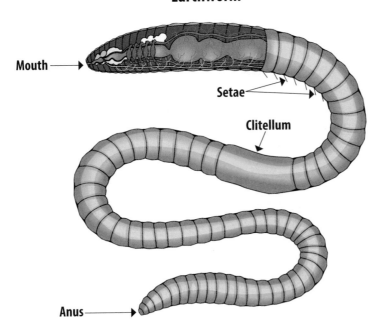

Mouth

Setae

Clitellum

Anus

earthworm's skin must remain moist and must be in contact with air. Earthworms suffocate if their skin dries out, but they will drown if too much water passes through their skin. The nervous system of an earthworm is simple. It consists of a brain, two nerves around the intestine, and a nerve cord on the underside. Earthworms reproduce sexually.

Leeches are segmented worms that have a sucker on each end. They breathe through their skin, and they have a digestive system that contains a pouch in which food can be stored for several months. Leeches reproduce sexually. Leeches are known for sucking blood from animals or people, but some leeches eat organic debris.

Worms play an important role on Earth. Besides being an important food source for many animals, they help make and add air to the soil. Earthworms, roundworms, and other worms help produce a rich upper layer of soil. Worms swallow large amounts of soil, which is enriched with nutrients as it passes through their digestive system. After digestion, the worms expel the enriched soil in the upper layer of soil. Over 1 million worms can be found in an acre of land.

## LESSON REVIEW

**1.** Describe the three main types of worms. Give at least three characteristics of each type.

**2.** How do worms help gardeners?

## FYI

**Get It Off Me!**
Has a leech ever attached itself to you when you waded in muddy water? Why did you see it before you felt it? Leeches are very careful when they attach themselves to their host. They secrete a local anesthetic before making their incision. They do not want you to notice them before they are finished eating!

## TRY THIS

**An Earthworm's Habits**
Pack hard soil as tightly as you can into two coffee cans, leaving about 8 cm of space on top of the soil. Pour about 2.5 cm of sand over the soil. Moisten the sand and soil in each can slightly with water. Place 8–10 earthworms on top of the sand in one can and cover the can with black paper. Secure with a rubber band. (The other can is the control.)

Check the cans the next day. Where are the earthworms? Let the cans sit for another day. Pour out the sand and the soil. What do you observe about the soil from each can?

Have you ever eaten clams, snails, squid, or oysters? If so, you have eaten mollusks. Phylum Mollusca contain over 100,000 species. Mollusks live in water or moist habitats. Some mollusks produce ink when they are startled to help them escape predators. Other species produce eerie lights deep beneath the ocean. A few mollusks are no larger than a grain of sand, but the largest mollusk, the giant squid, grows up to 14 m long. Giant squids are the largest invertebrates. Although they are not as long as the thin bootlace worm, overall, giant squids are much larger.

The name *mollusk* comes from the Latin word *molluscus*, which means "soft." Mollusks have several common characteristics. They have soft bodies with bilateral symmetry. Most have inner or outer shells. Mollusks have a hard, strong band called a **radula** that is used to scrape food or to bore into objects. Most mollusks also have a **mantle**, an organ in the wall lining the inner shell. The mantle secretes the material to make the shell. Most mollusks have a thick, muscular foot, which they use to move, to open and close their shells, or to bury themselves in sand or mud.

Mollusks can be divided into three major groups: bivalves, gastropods, and cephalopods. **Bivalves** have two shells that are held together by powerful muscles. Clams, mussels, scallops, and oysters are all bivalves. Bivalves live in the water and most are filter feeders. They pull seawater into their bodies and filter out plankton. Bivalves have a muscular foot that they use to attach themselves to surfaces or to burrow into the sand. If a grain of

**Bivalve**

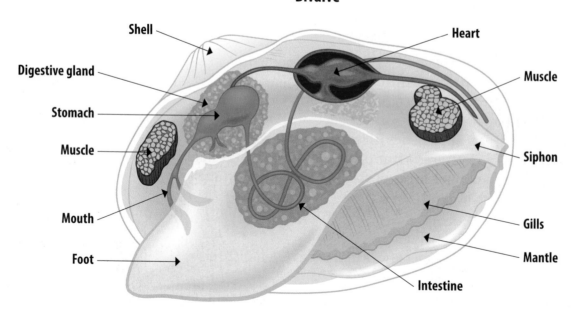

sand becomes caught within the mantle of some bivalves, such as oysters, the bivalve covers the grain with a secretion. After a few years, the secretions produce a pearl.

The second group of mollusks is gastropods. **Gastropods** are mollusks that have either a single coiled or cone-shaped shell, or no shell, and a foot for movement. *Gastropod* means "stomach foot." A gastropod's foot is on the same side of its body as its stomach. Gastropods live in freshwater, saltwater, or moist land environments. The snail is a familiar gastropod. Snails have a single, coiled shell and leave a trail of mucus, which helps them glide across different surfaces. Slugs are gastropods without a shell. Because they do not have a shell to protect them, they usually make their homes under logs or rocks.

Scallops do not burrow in the sand. Instead they move by using their abductor muscle to rapidly open and close their valves, ejecting water around the hinge and propelling themselves in the opposite direction.

**Cephalopods** are ocean mollusks with heads, tentacles, and beaklike jaws. Cephalopods include octopuses, squids, and nautiluses. Octopuses, which are eight-armed cephalopods, can change their skin color to camouflage themselves on the ocean floor. The eye of an octopus, much like a human eye, can see clear images. This good eyesight helps the octopus match its skin color to its environment. Octopuses do not have shells for protection, but God gave them other ways to protect themselves. Because they do not have shells, octopuses can squeeze into crevices in rocks. Surprised octopuses often squirt an inky substance or turn black to frighten off predators. If that does not work, they can become nearly transparent, almost vanishing in the water. Octopuses are considered the most intelligent invertebrates.

Nautilus shell

Gastropods have antennae that are used for finding food, detecting danger, or for feeling their way around.

## TRY THIS

### It's a Snail's Life

Examine a snail. Gently touch its eyes with a toothpick. Record your observations. Bore four holes in a strip of poster board—two holes that are large enough for the snail's body but not the shell and two that are large enough for the entire snail and shell to fit through. Use the poster board to make a bridge between two books, and position the snail so that it will encounter the holes. How does the snail select the right-sized holes to go through?

Lay a board flat on a table, and time how fast a snail moves 10 cm. Then time how fast the snail moves across 10 cm of sandpaper and then a piece of glass. Prop the different surfaces at different angles, and time how fast the snail moves up or down each angle. Translate the speed into kilometers per hour.

 **FYI**

### Boy or Girl?

Although most mollusks are either male or female, some have both male and female sexual body parts. Some fertilize each other; others fertilize themselves. Some sea slugs and oysters can even change their sex. One species of limpet gives birth to all males, which change into females after they grow to a certain size.

Experiments with octopuses have shown that they can even learn to remove the cork from a jar to reach food that is inside.

Throughout history people have used mollusks and their shells. People have carved tools and musical instruments from mollusk shells, and for thousands of years people have harvested oysters for their pearls. The Phoenicians used the yellow secretion of the murex, which is a gastropod, to make their purple royal dye. West Africans used cowry shells as currency. And mollusks have always been an important food source for people in regions around the world.

Many invertebrates' bodies contain chemicals that can be used to make medicines. One mollusk example is cone snail venom, which is being developed for use as a powerful painkiller. God's world holds millions of yet-unknown chemicals, many of which could one day prove to be valuable medicines.

## LESSON REVIEW

**1.** What are the characteristics of mollusks?
**2.** Name a mollusk from each of the three groups.
**3.** How can octopuses protect themselves from predators?

The cuttlefish, a cephalopod, injects saliva into its prey to paralyze it.

The Greek word *echinos* means "spiny," and *derma* means "skin." All echinoderms live in the ocean and have spiny skin. Echinoderms include sea stars (starfish), brittle stars, sand dollars, sea urchins, sea cucumbers, sea lilies, and feather stars. Many extinct echinoderm species are known only from their fossil remains, but several thousand species exist today.

Echinoderms are different from the other animals you have studied so far. Unlike sponges, cnidarians, worms, and mollusks, echinoderms do not have soft bodies. Instead, they have an **endoskeleton**, or internal skeleton. The endoskeleton of an echinoderm is made of plates. Tough, spiny skin covers these plates.

Adult echinoderms are wheel-shaped; they have five-part radial symmetry. All echinoderms have **tube feet**, which are bulblike suction structures that line their appendages. Each tube foot acts like a small, sticky suction cup. Echinoderms use water pressure inside their appendages to move and to pull things apart. Sea stars, for example, use their tube feet to pull apart bivalves, such as clams and oysters, so that they can eat them. Echinoderms reproduce sexually or asexually.

## OBJECTIVES

• Give examples of echinoderms.
• Describe the common characteristics of echinoderms.

## VOCABULARY

• **endoskeleton** an internal skeleton
• **tube feet** the bulblike suction structures lining the arms of an echinoderm

## Echinoderm

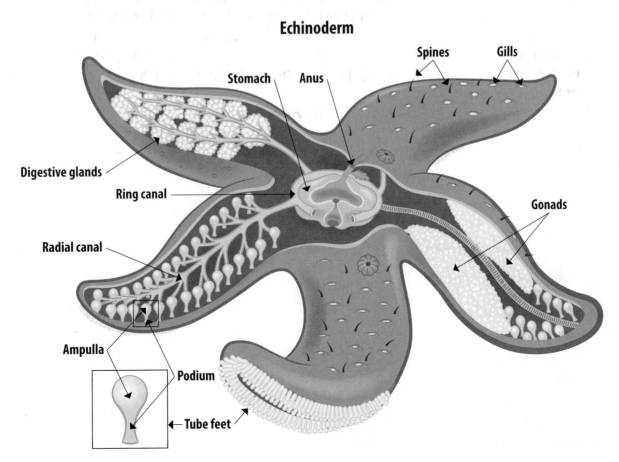

Spines, Gills, Stomach, Anus, Digestive glands, Ring canal, Radial canal, Gonads, Ampulla, Podium, Tube feet

The sea star's endoskeleton is made of calcium plates held together by tissues and muscles. A sea star does not have a head or a brain but does have a mouth and a stomach, which are located at the center of its body. When eating clams, which are one of its favorite foods, a sea star uses the tube feet on its arms to pry open the clam's shell. It then inverts its stomach through its mouth and into the clam. A sea star can push its stomach through cracks in the clam's shell as narrow as 0.1 mm. The sea star uses its stomach to digest the clam's tissues from inside the clam's own shell.

Like most echinoderms, sea stars regenerate, or grow back, lost body parts. Since both people and sea stars like to eat clams, oysters, and other mollusks, some people used to cut up any sea stars they caught to reduce the competition for mollusks. Then they threw the sea star pieces back into the ocean, thinking they were dead. But since sea stars can regenerate, people were actually making the competition worse instead of better.

Brittle stars and basket stars look like sea stars with long, very thin arms. Their arms bend as they crawl across the ocean floor. Brittle stars can easily snap off one of these arms to escape predators. In contrast, sea urchins and sand dollars are round and armless. Sand dollars are flat, and they burrow into the sand or mud. Sea urchins look like pincushions. Some sea urchins use their long spines to walk across the ocean floor. Sea urchins feed on decayed matter, algae, or underwater grasses.

Sea cucumber

The long-spined black urchin is found in tropical waters throughout the world. Its spines, which are covered with poisonous mucus, grow to 30–40 cm long. When disturbed, this urchin can point its spines toward the source of the disturbance and wave them about.

Brittle star

Sea cucumbers, which look like large underwater logs, can grow over a meter long. They exhibit bilateral as well as radial symmetry. They wander over sand and coral, vacuuming up decayed matter, plankton, and algae. God gave the sea cucumber an interesting weapon to use against predators: a sea cucumber can spew out all of its intestines when something threatens it. This behavior either repels a predator or lures it into eating the intestines, which gives the sea cucumber a chance to escape. The sea cucumber then grows back its intestines.

Sea lilies and feather stars spend most of their lives clinging to boulders or coral. They extend feathery arms out into the water, and their tube feet secrete mucus to trap plankton and other food particles.

## LESSON REVIEW
**1.** List the common characteristics of echinoderms.
**2.** Name two ways that echinoderms defend themselves.
**3.** Name six examples of echinoderms.

Red sea urchin

## OBJECTIVES

- Identify three arthropod characteristics that distinguish them from other invertebrates.
- Give examples of each type of arthropod.

## VOCABULARY

- **abdomen** the hind body segment of an insect or other arthropod
- **arthropod** an invertebrate that has a segmented body, jointed appendages, and an exoskeleton
- **cephalothorax** the body segment of crustaceans and arachnids that consists of the head and thorax fused together
- **exoskeleton** an external skeleton
- **thorax** the center body segment of an insect or other arthropod

Have you ever dreamed of having superpowers? Imagine being able to jump over the Statue of Liberty—that is 93 m—or to leap farther than the length of a football field. You would be the star of the track team. Imagine being so strong that you could carry 50 times your own weight—you would have no problem finding a summer job. Imagine having your entire body armored with thick plates—you would be able to fall great distances without being hurt. Imagine being able to fly 56 kph or to spin your own balloon to float 5 km up into the air or travel hundreds of kilometers across land or water. What would it be like to be such a super creature? It would be just like being an arthropod.

The Greek word *arthron* means "joint," and the word *pous* means "foot." **Arthropods** are invertebrates that have a segmented body, jointed appendages, and an **exoskeleton**, which is an external skeleton. The appendages of arthropods include legs, jaws, egg depositors, sucking tubes, claws, antennae, paddles, and pincers. Arthropods include a wide variety of animals, such as crabs, shrimp, centipedes, insects, and spiders.

Arthropods make up the largest animal phylum. Scientists do not know the number of arthropod species that are alive today, but estimates range from 2 to 30 million. Many arthropods are now extinct. Some extinct arthropods are known from their fossils, but many did not leave fossils.

So far biologists have named more than 1 million species of arthropods. Most are insect species. Scientists estimate that 10 quintillion (10,000,000,000,000,000,000) individual insects are alive on the earth at any given time. They have not even guessed at the total number of individual arthropods. Arthropods inhabit nearly every ecosystem on Earth. Even the air is filled with them.

Arthropods thrive in huge numbers and different ecosystems because God gave them variety. Arthropods have specially developed features. For example, their different types of legs and claws help them do a variety of tasks. Arthropods that fly can colonize faraway places. Arthropods also have very specific diets, which allows them to live in small habitats, such as a single plant or a few square centimeters of soil. These specialized diets cut down on competition between arthropod species. Two species can live near each other without needing the same food.

The most unusual feature of arthropods is the exoskeleton, which acts like armor. The exoskeleton protects the arthropod from predators and injury and is also waterproof. It also keeps the arthropod from losing too much water. The arthropod's muscles are attached to the inside of its exoskeleton, which gives the arthropod tremendous strength in relation to its small size.

## TRY THIS

**Compare the Arthropods**
Choose three arthropods (such as a crab, a grasshopper, and a spider), and use a hand lens to observe them closely. How are they alike? How are they different?

Ants often work together to accomplish tasks. What kinds of things could you accomplish if you worked together as a team?

**Spider**

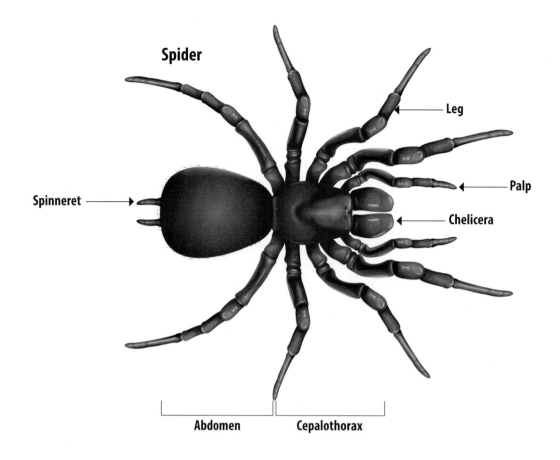

Spinneret

Leg

Palp

Chelicera

Abdomen

Cepalothorax

A molting cicada

Ants, for example, can carry objects that are 50 times their own weight. If you could lift 50 times your own weight, what kinds of objects could you lift?

The exoskeleton also keeps arthropods from growing very large. The exoskeleton does not grow along with the arthropod like your skeleton does, so the arthropod has to molt, or shed, its exoskeleton as it grows. The average arthropod molts its exoskeleton four to seven times throughout its life. Molting is a dangerous time for arthropods because they are vulnerable to predators and water loss at that time. Many arthropods hide until their molting is finished.

Arthropods have either two or three major body segments. Insects have three segments: head; **thorax**, which is the center body segment; and **abdomen**, which is the hind body segment. Crustaceans, such as lobsters and shrimp, and arachnids, such as spiders and scorpions, have one body segment called the **cephalothorax** that consists of the head and thorax fused together.

Instead of being controlled by their brains, arthropods are controlled by their body segments. Many arthropods such as grasshoppers, for example, move, eat, and jump even after their brain has been removed.

## Grasshopper

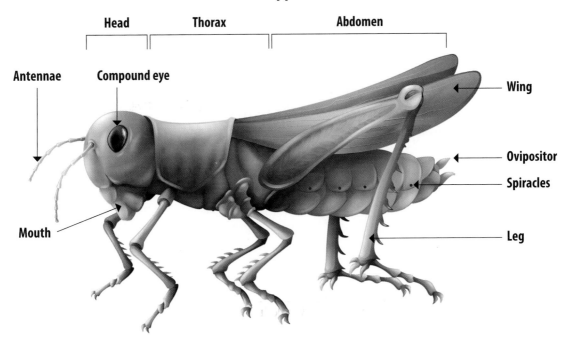

Head    Thorax    Abdomen

Antennae    Compound eye    Wing

Ovipositor

Spiracles

Mouth    Leg

**TRY THIS**

**Amazing Arthropods**

Match the arthropod to its amazing feat.

1. Grasshopper
2. Shrimp
3. Ant
4. Spider
5. Dragonfly
6. American cockroach

a. This arthropod can make its own balloon and float as high as 5 km.

b. If you were proportionately as fast as this arthropod, you could run 113 kph.

c. If you could proportionally jump as well as this arthropod, you could jump 93 m into the air.

d. This arthropod can live deep on the ocean floor under tremendous water pressure.

e. This arthropod can carry 50 times its own weight.

f. This arthropod can fly 58 kph.

## LESSON REVIEW

1. What are three characteristics of arthropods that distinguish them from other invertebrates?
2. How can arthropods live in such a wide variety of habitats?
3. Name each type of arthropod and give an example of each type.

### OBJECTIVES

- Describe the physical characteristics common to crustaceans.
- Give several examples of crustaceans.

### VOCABULARY

- **crustacean** an arthropod that has five or more pairs of legs and two pairs of antennae
- **mandible** a mouth part
- **maxilliped** the appendage that helps a crustacean eat

The Latin word *crusta* means "hard shell." **Crustaceans** are arthropods with hard shells. Crabs, lobsters, crayfish, shrimp, barnacles, prawns, water fleas, pill bugs, and sow bugs are among the over 45,000 crustacean species.

The head and thorax of a crustacean are not separate body parts; they are combined to form a cephalothorax. The cephalothorax often has a shieldlike cover called *a carapace*. Crustaceans have ringlike segments in their abdomen and jointed appendages, which help them move quickly. Crustaceans also have grinding mouth parts called **mandibles**.

Crustaceans have a variety of appendages for different uses. For example, crustaceans have two pairs of antennae. They use the longer pair to touch, taste, and smell and the shorter pair to touch, taste, and balance. Some crustaceans, such as crayfish and lobsters, have compound eyes at the end of a flexible stalk. Crustaceans also have five or more pairs of walking legs. The claws are considered the first pair of walking legs. **Maxillipeds** are the appendages that help a crustacean eat. Swimmerets are small leglike appendages attached to the abdomen. They are designed for swimming and for carrying eggs.

Crustaceans live in watery or damp environments and obtain oxygen from the water through respiratory organs called *gills*. As water passes over the gills, oxygen passes from the water into the gills. From the gills, the oxygen moves into the blood.

Pill bugs, or roly polies, are not classified as insects. They are crustaceans. Do you know why?

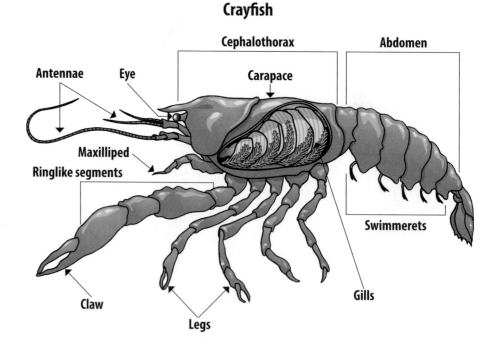

**Crayfish**

Meanwhile, carbon dioxide moves out of the blood through the gills and into the water.

Crustaceans reproduce sexually. Most crustacean species include males and females, but individual barnacles have both male and female body parts. A few shrimp begin their lives as males before turning into females.

Most crustaceans live in the ocean, such as lobsters, crabs, and shrimp. Tiny ocean crustaceans are a food source for fish and whales. Some crustaceans, such as crayfish, live in freshwater. Other crustaceans live on land. If you have ever walked in a forest, you have probably seen some of these: pill bugs, sow bugs, and rock slaters. All of these crustaceans live in moist leaf litter and in damp soil under rocks and logs. Several species of hermit crabs and other crabs that live in tropical areas like the Caribbean live mainly on land.

## LESSON REVIEW

**1.** What characteristics do all crustaceans have in common?
**2.** Why can crustaceans move rapidly?
**3.** What is a maxilliped?
**4.** Name five crustaceans.

## FYI

**Barnacles**

If you take a walk along the seashore, you might see barnacles clinging to solid surfaces that get covered by water. Barnacles are crustaceans that cling to docks, boats, rocks, and even whales.

After the barnacle larva selects a home, it attaches itself to the surface with a strong glue. This glue is so strong that scientists are researching how it could be useful to dentists. They then secrete limestone plates and grow into an adult. They live in hard, conelike houses, and they peek out when water covers them so that they can filter food from the water.

Ghost crab

# 4.2.3  Centipedes and Millipedes

## OBJECTIVES

- Compare and contrast centipedes and millipedes.
- Describe where centipedes and millipedes are found.

## VOCABULARY

- **centipede** a flat-bodied arthropod with one pair of legs on each body segment
- **millipede** a round-bodied arthropod with two pairs of legs on each body segment

Not all centipedes are black or brown. The *Scolopendra hardwickei*, or Indian Giant Tiger Centipede, is known for its vivid, unique coloring. This species of centipede can be as long as 25 cm!

The name *myriapod* means "many footed." The two major kinds of myriapods are centipedes and millipedes. **Centipedes** are flat-bodied arthropods with one pair of legs on each body segment, and **millipedes** are round-bodied arthropods with two pairs of legs on each body segment. Both of these myriapods have many feet. Centipede means "one hundred feet," and millipede means "one thousand feet," but they do not actually have that many feet. You do not have to count their legs to tell which is which—just look at how many pairs of legs they have per body segment.

Scientists have identified several thousand species of centipedes. The largest centipede grows 30 cm long. Depending on the species, centipedes actually have between 30 and 342 legs. The most common centipedes have 30 legs (15 pairs) that they use to move rapidly. Centipedes have a long pair of antennae on their heads, and they have mandibles for chewing. On the first segment behind their heads, centipedes have a pair of poisonous pinching claws. Centipedes rarely pinch people, however.

Most centipedes live in the tropics, but there are quite a few species that live in North America, Europe, and Asia as well. Centipedes generally live beneath rocks or logs. They are nocturnal, coming out at night to hunt earthworms and insects, which they kill with their poisonous claws. Some tropical species are even large enough to kill lizards and mice! Most centipedes live outdoors, but the house centipede, which can grow to 7.5 cm long, sometimes lives in basements and bathrooms. At night it hunts household insects such as cockroaches and other pests.

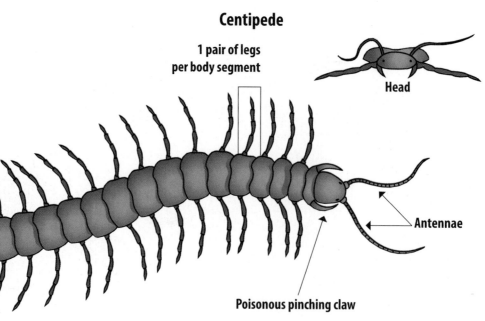

**Centipede**

1 pair of legs per body segment

Head

Antennae

Poisonous pinching claw

Scientists have also identified several thousand species of millipedes, most of which are found in the tropics; however, some species are found in North America and Europe. Most millipedes have between 100 to 300 legs. However, the *Illacme plenipes* has an amazing 750 legs! Even though they have so many legs, millipedes travel slowly, moving with a wavelike motion.

Millipedes spend most of their time beneath rocks and rotten logs, avoiding bright light. Unlike centipedes, which kill their prey, millipedes are scavengers—they eat dead animal and plant materials. When they are threatened by predators, some millipedes roll up into a tight ball to protect themselves. Others give off a bad smell with their stink glands to chase the predators away.

## LESSON REVIEW

**1.** How are centipedes and millipedes different?
**2.** How do centipedes and millipedes defend themselves?
**3.** Describe where centipedes and millipedes can be found.

## TRY THIS

**Centipede or Millipede**
Examine a centipede with a hand lens. Record the number of body segments and the number of legs per segment. Describe the centipede's mouth. Describe the centipede's body shape. Describe the way the centipede moves. Sketch the centipede. Repeat the process with a millipede.

If you look closely, you can see that the millipede has two pairs of legs per body segment.

## FYI

**Arachne of Lydia**

There is a myth in ancient Greek culture about Arachne of Lydia, a Greek woman who was a skilled weaver. She challenged the goddess Athena to a weaving contest. Arachne wove a tapestry more beautiful than Athena's, and the enraged goddess tore up Arachne's tapestry. In despair Arachne hung herself, but Athena took pity on her and turned the noose into a cobweb and Arachne into a spider.

What do you think of spiders? Some people are fascinated by them. Others are afraid of them. Even though the chance of being harmed by spiders is very slim, arachnophobia, the fear of spiders, ranks second among animal phobias—right behind the fear of snakes.

The word *arachnid* comes from the Greek word for "spider," which is *arachne*. But spiders are not the only arachnids. Scorpions, daddy longlegs, mites, and ticks are also arachnids. **Arachnids** are arthropods with six pairs of appendages. The first two pairs are mouth parts that are used for feeding, such as fangs or pincers. The other four pairs of appendages are legs. Arachnids have two major body segments: the cephalothorax, which is the head and thorax fused together, and the abdomen.

Spiders are probably the most familiar arachnids, and there are well over 30,000 species throughout the world. Spiders feed mainly on other arthropods, such as insects, but a few large spiders eat small mice, birds, frogs, and fish. All spiders use poison glands near the tips of their fangs to inject a digestive enzyme into their prey. This enzyme turns the victim's tissues into liquid. The spider then sucks up the liquid and discards whatever is left.

Most spiders produce long strands of silk that they can weave into webs. Orb webs, which have a circular design, are the most

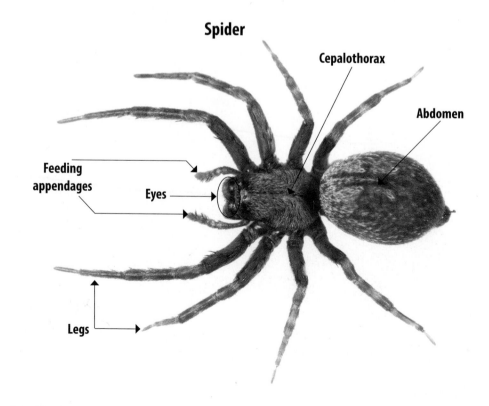

**Spider**

Cepalothorax

Abdomen

Feeding appendages

Eyes

Legs

common type of spider web. House spiders make irregular webs called *cobwebs*. Some spiders produce closely woven sheets of webbing, which they scatter across the grass. These webs often glisten with dew in early morning sunlight. Most spiders sit on their webs, waiting for passing insects to become trapped. When spiders sense the vibrations of their struggling prey, they descend on the thin strands to kill the animal with venom.

Spiders also use their silk to form long threads that they use to float long distances through the air, a practice called *ballooning*. Because female spiders lay an egg mass containing hundreds of young, ballooning prevents young spiders from overcrowding an area. The spiders climb to a high point, spin long threads of silk, and wait for the wind to lift them up into the atmosphere, where they can travel hundreds of kilometers.

Interestingly, most daddy longlegs, or harvestmen, are not considered spiders. Unlike spiders, daddy longlegs do not have a narrow waist that separates the cephalothorax from the abdomen. Daddy longlegs do not have fangs so they do not bite, but they do have stink glands that give off a bad odor. Some species eat insects; others eat dead matter.

Black widow

A lynx spider making an egg case

Common orb spider on a web

Most spiders are harmless to humans, but some spiders do have a painful bite. The two North American species most likely to bite people are the black widow spider, which has a red hourglass marking on its shiny black abdomen, and the brown recluse spider, a pale brown spider with a violin-shaped marking on its cephalothorax. The Brazilian wandering spider is perhaps the most dangerous spider in the world. Its venom is potent, but bites are rare. Australia is home to funnel-web spiders, whose bites can be fatal. However, since an antivenin was developed in the 1980s, no deaths have been recorded.

Scorpions are arachnids that have four pairs of legs, large pincers, and a segmented tail with a bulblike tip that contains a stinger. Scorpions live mainly in warm, dry climates. All scorpions are predators that feed mostly on other arthropods. Some larger species eat small lizards, snakes, and rodents. Many other arthropods, such as centipedes, spiders, and ants, eat scorpions. However, the greatest predators of scorpions are other scorpions. All scorpions are poisonous, but their bites are rarely fatal.

Mites are the tiniest arachnids, and they can live in almost any environment—on land, in freshwater, and in the oceans. They live in soil, in treetops, inside plants, in bird nests, on the bodies

Unlike true spiders, daddy longlegs do not have two distinct body segments. The cephalothorax and abdomen are fused together so the separation is not noticeable. These arachnids belong to the order Opiliones. True spiders belong to order Araneae.

of insects and vertebrates, in dried fruit and cheese, in beds or upholstery, in dust, and even on the bodies of humans. They are one of the few arthropods that live in the Arctic and Antarctic. Mites are the only arachnids that eat plants. Mites are helpful because they break down dead matter into nutrients, and they eat other pesky mites and insects. On the other hand, mites are parasites of humans and animals. They may also damage stored grain, attack crops, and transmit diseases. Chiggers are a group of microscopic mites that are parasites to all vertebrates except fish. They live on the outside of their host's body, burrow under the skin, and release enzymes that digest the host's skin cells.

The Arizona bark scorpion is only 5–8 cm long. It is the most dangerous scorpion in the United States.

Ticks are blood-sucking arachnid parasites that live on land, reptiles, birds, and mammals. Hard ticks are covered with a hard shield; soft ticks have a leathery, soft covering over their bodies. Ticks transmit more diseases to humans than any other arthropod except mosquitoes. For example, ticks spread Lyme disease, a bacterial infection that results in joint swelling, fever,

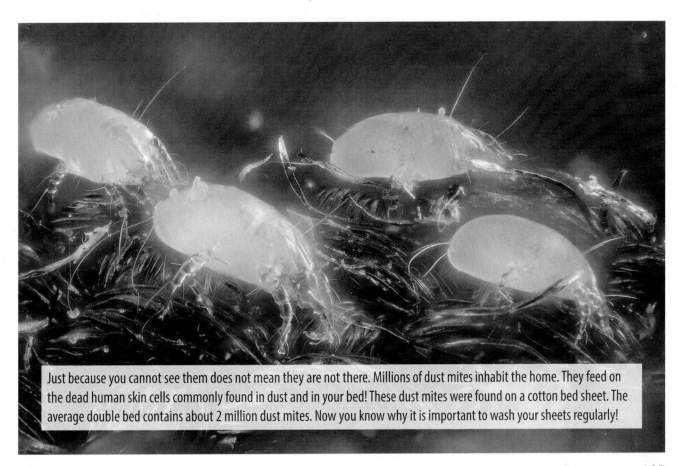

Just because you cannot see them does not mean they are not there. Millions of dust mites inhabit the home. They feed on the dead human skin cells commonly found in dust and in your bed! These dust mites were found on a cotton bed sheet. The average double bed contains about 2 million dust mites. Now you know why it is important to wash your sheets regularly!

**FYI**

**Spider Records**

- Wolf spiders are among the most common species of North American spiders. Instead of weaving a web, they lie in wait for their prey. Wolf spiders stalk their prey, much like a lion stalks an antelope.
- One of the oddest spiders is the European water spider, which spends its life beneath the surface of a pond. It lives in an air-filled diving bell that it makes from air bubbles, which it traps underneath the water inside a small sheet of silk. When it senses water vibrations, the water spider rushes to the surface to capture struggling insects; it then returns to its diving bell to eat them.
- The largest spider is the Goliath birdeater, a spider found in South American rain forests. One caught in Venezuela had a leg span of 27.5 cm.
- The heaviest individual spider, which was also the Goliath, was captured in Surinam and was 127.5 g. Most birds are much smaller than this spider. For example, wood warblers are 10 g and blue jays are about 100 g.

A Goliath birdeater

Tick

and a rash. Hard ticks that live on infected deer or small birds spread this disease to people.

Arachnids are part of God's design for the earth. They eat many insects that are pests to humans and other animals, such as mosquitoes and flies. They also eat other arachnids. Arachnids are also a source of food for other animals, such as birds, frogs, lizards, and other arachnids.

## LESSON REVIEW

1. What do arachnids have in common?
2. Explain why daddy longlegs belong to a different order and are not considered true spiders.
3. What important role do arachnids play in the world?
4. How are arachnids different from other arthropods?

Which weighs more—insects or people? If you compare the total weight of the world's insects with the total weight of the world's people, insects weigh a lot more. That is because the total mass of insects may be as much as 300 times the total mass of all the people on Earth!

God's world is home to an amazing variety of insects. The longest insect is the walkingstick of Borneo. The Natural History Museum in London has a walkingstick specimen that measures 56.7 cm long. The Goliath beetle of equatorial Africa is the heaviest insect at 100 g, and the smallest insect is the fairyfly wasp, which is one-fifth of a millimeter long. **Entomologists**, scientists who study insects, have identified over 900,000 species of insects—organisms that are classified in class Insecta. Scientists believe that there may be as many as 30 million species that have not even been discovered yet.

Insects have three body segments—a head, a thorax, and an abdomen. The insect's head is formed from many hard plates called *sclerites*. These plates join together to form a solid casing that houses the insect's eyes. Insects have two types of eyes. Their simple eyes help them focus on objects, and their compound eyes detect movement. The number of details an insect can see depends on the number of facets, or ommatidia, in its eyes. Ommatidia are the parts that make up an insect's

## OBJECTIVES

- Distinguish insects from other arthropods.
- Explain the life cycle of an insect and differentiate between complete and incomplete metamorphosis.

## VOCABULARY

- **entomologist** a scientist who studies insects
- **metamorphosis** a series of physical changes that certain animals undergo
- **nymph** a stage of incomplete metamorphosis that resembles the adult insect

Notice the three distinct body segments on this asparagus beetle.

compound eye. Each facet resembles a single simple eye, so the more facets an insect has, the better it can see. Houseflies have about 4,000 ommatidia, and dragonflies have 28,000. Some insects, such as ants that live in dark places, have no ommatidia. These insects rely on smell instead of sight. Insects also have one pair of antennae, which they use to feel and to smell.

Insects have three pairs of legs and most have one or two pairs of wings on the thorax. Flies, for example, have only one pair of wings; many ants and other simpler insects do not have any wings at all. The wings of most insects are laced with veins that allow blood to flow through the wings. Insects also have flying muscles inside their bodies to enable them to beat their wings. The flying muscles are attached to the walls of the thorax. These muscles give insects incredible flying power. For example, large-winged butterflies flap their wings 5–12 times per second. Mosquitoes beat their wings up to 600 times per second, creating the high-pitched whine that you hear when a mosquito flies close to your ear. Most insects do not fly faster than 32 kph, although dragonflies have been clocked at 58 kph.

Insects' mouth parts vary widely depending on their use. Insects that eat other insects have mandibles for grinding. A mosquito

Have you ever wondered why it is hard to catch a fly? The fly's compound eye is composed of a large number of individual facets, or ommatidia. Each facet sends a signal to the brain, which then constructs a mosaic view of the world. The image is poorly focused, but it can help the fly detect movement.

has a mouth part that pierces skin and injects a substance that prevents blood from clotting. A butterfly has a long tube to suck nectar from flowers. Insects also have varied diets. For example, flies eat blood, nectar, plant sap, and the liquid from rotting organic matter. Butterfly larvae eat leaves, and adult butterflies eat nectar. Mosquitoes eat blood.

An insect's abdomen is its longest part. It is flexible, and it contains the trachea, a tube that carries air so the insect can breathe. Each section of the insect's abdomen has a pair of openings called *spiracles* that allow the oxygen that the insect breathes in through its trachea to get to all the insect's organs and tissues. Carbon dioxide also exits through the spiracles.

The life cycles of most insect species are very similar. After mating, female insects lay hundreds or thousands of fertilized eggs. The eggs then undergo a **metamorphosis**—a series of changes that transform an egg into an adult.

Most insects undergo complete metamorphosis, which includes four stages: egg, larva, pupa, and adult. The egg hatches into a larva, which eats and grows until it becomes a pupa. The pupa stage can last from four days to several months. The insect does not eat during this stage. Some insects, such as moths and

## TRY THIS

**Water Strider**
Place a water strider in a filled aquarium. Place a drop of dishwashing detergent into the water and observe the water strider. (Do not let it drown.) Determine if other human-made substances have the same effect.

How can water pollution upset the balance that God intended for water insects?

**Complete Metamorphosis**

**Incomplete Metamorphosis**

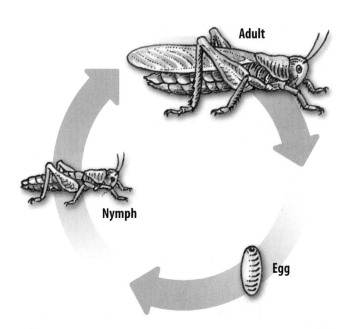

butterflies, form either a cocoon or a chrysalis during their pupa stage. The pupa finally emerges as an adult.

Incomplete metamorphosis includes three stages: egg, nymph, and adult. The eggs of crickets and grasshoppers, for example, hatch into nymphs. **Nymphs** are young insects that look like miniature adults without wings. The nymphs grow into adults.

Many insects are social insects; they stay close to their own species and work together in a group. Termites, which eat wood, have a social structure made up of three different kinds of individuals: workers, soldiers (who defend the colony), and a few who are solely responsible for reproducing. The known species of honeybees are also social insects. Honeybees are divided into queens, drones, and workers. Queens lay the eggs, drones mate with the queens, and worker bees do most of the colony's work. Ants are also social insects. A colony of ants usually consists of one queen ant and 150,000–700,000 worker ants. Soldier ants

A queen honeybee and attendant worker bees

are males with enlarged heads and powerful mandibles. They use their powerful jaws to defend the colony. Ants are the only known animals that domesticate other animals. For example, some ant species tend and protect herds of aphids. The ants will "milk" the aphids by stroking them with their antennae, which causes the aphids to produce a sweet liquid that the ants eat. Ants sometimes enslave other ants and force the slaves to collect food for them.

Some people think that all insects are pests, but insects do much more good than harm. Only about 1% of all known insect species are considered pests. Insects are a very important part of God's world; in fact, people could not live without them. Insects pollinate most flowering plants, including crops. Without insects, these plants would not reproduce. Insects break down dead organisms and waste products, till and aerate the soil, and are food for many animals. Insects provide products such as silk and honey. Even medicine has found a use for insects. Doctors use maggots (fly larvae) to clean out deep-rooted infections.

## LESSON REVIEW

1. Explain how an insect's structure is different from that of other arthropods.
2. What structures does an insect use to sense its environment?
3. Compare complete metamorphosis with incomplete metamorphosis.

 **FYI**

**Mosquito Meal**
Only female mosquitos require a blood meal. The reason for this is because they need protein for their eggs, and they get it from blood. Since males do not bear the burden of producing young, they head for the nectar of flowers instead. And when they are not trying to produce eggs, females are happy to stick to nectar too.

## OBJECTIVES

- Identify the basic characteristics of a vertebrate.
- Distinguish between ectotherms and endotherms.
- Model how a closed circulatory system works.

## VOCABULARY

- **ectotherm** an organism that uses the environment to regulate its body temperature
- **endotherm** an organism that internally regulates its body temperature

Judges 14:14 records a riddle that Samson told his companions: "Out of the eater, something to eat; out of the strong, something sweet." The solution was that bees had made honey in the carcass of a lion.

Here are some more animal riddles. What kind of animal can be any color or pattern—black, white, red, blue, yellow, purple, silver, gold, striped, or spotted? What kind of animal can swim, fly, climb, run, or jump? What kind of animal can grow to 3 cm long or to 30 m long? What kind of animal looks like a rock or a leafy plant or a small island in the ocean? What kind of animal can weigh as little as a couple of paper clips or as much as 150 small cars? What kind of animal roams the frozen ice of the Arctic, dives deep beneath the ocean, prowls desert sands, flocks in the middle of cities, and hides in the most remote rain forest? What kind of animal might eat grass or slugs or whales or even you? Sometimes people are horrified by this animal, sometimes they ride it, other times they cuddle next to it, and still other times they eat it. What is this animal?

The answer to all these riddles is vertebrates. The more than 50,000 species of vertebrates in the world display the creativity and the imagination of God, who created vertebrates of every color, shape, and size. He created bright yellow or purple frogfish that look like beautiful rocks on the ocean floor and sea dragons that look like leafy vegetables. He fashioned the 2 g tiny gobie fish and the 146,000 kg blue whale. He created polar bears to roam the Arctic, viperfish to generate their own ghostly light deep in the sea, lemurs to hide in the highest trees of the deepest rain forests, and pigeons and rats to thrive in cities.

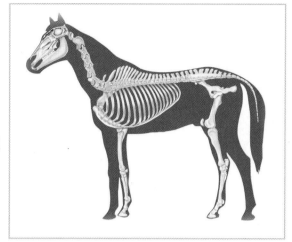

A vertebrate is an animal with a backbone. You are probably more familiar with vertebrates than you are with invertebrates. Think of the vertebrates that you see every day—your pet dog or the birds that fly around your yard. What do these animals have in common?

Vertebrates have many similarities. Each vertebrate has a backbone, which supports and protects its spinal cord. All vertebrates have endoskeletons that are formed of either bone or cartilage. They also have brains that are enclosed and protected by craniums, or skulls. Vertebrates have closed circulatory systems with hearts that are connected to blood vessels. The vertebrate circulation system contains veins and arteries, which are large blood vessels that transport blood. It also contains capillaries. These small blood

**Circulatory System**

**Blood vessels**

**Capillaries**

Annelids, cephalopods, and vertebrates have closed circulatory systems. This type of circulatory system is more efficient than open circulatory systems. Since blood circulates only inside blood vessels, it reaches farther distances between the organs and increases the oxygen supply to tissues. The greater efficiency of the closed circulatory system gives these animals faster movements.

 **CAREER**

## Veterinarian

Imagine how challenging it is to be a doctor who understands and treats the complex human body. Now think of the variety of animals God designed and imagine how challenging it is to be a doctor who knows enough to treat lizards, cats, cows, birds, zebras, pigs, fish, or even whales! That is the challenge that veterinarians face.

Most veterinarians specialize in the treatment of certain types of animals. Many vets in cities specialize in small animals and treat mostly cats and dogs. Out in the country, vets generally specialize in large farm animals such as cows, pigs, sheep, and horses. Some veterinarians specialize in specific animals. Poultry farmers rely on vets who understand chickens and turkeys. Horse trainers and ranchers need vets who understand equine (horse) medicine. Zoos, of course, need vets who can treat a wide variety of exotic animals.

To become a vet requires more than just a love of animals, although that is a good place to start. Vets must understand the biochemistry of animal systems, or how certain chemicals are made and broken down in their cells. They must study the genetics and the anatomy of many different animals. They must be able to recognize many different diseases and know the best way to treat them. Vets must also be surgeons who can anesthetize animals and cut and stitch tissues back together.

Animal medical emergencies happen at all hours of the day and night, so vets are often called out during the night to treat animals. Being a vet is a demanding profession, but it is satisfying to help hurting animals.

vessels keep fluids moving throughout the entire body and bring oxygen to every cell. The heart beats continuously to keep the blood flowing throughout a vertebrate's body.

Vertebrates can generally be divided into two groups: ectotherms and endotherms. **Ectotherms** are organisms that use the environment to regulate their body temperatures. For example, ectotherms control their body temperatures by basking in the sun or burrowing themselves in soil, leaves, or mud. In general, reptiles, fish, and amphibians are ectotherms. Because ectotherms do not generate their own body heat, they do not need as much food as endotherms do.

Ectotherms are active during the day when temperatures are warm. At night when temperatures are colder, they become less active.

**Endotherms** are organisms that internally regulate their body temperatures. Their temperatures remain in about the same range whether they are in hot environments or cold ones. In general, birds and mammals are endotherms.

The tissues of vertebrates are formed mostly of slightly salty water. Most living tissues freeze at a temperature of −0.5°C. Because living tissues die if they freeze, most vertebrates live in ecosystems that have a temperature range of 0°C–50°C. A few vertebrates, however, can survive in freezing temperatures. Some fish have chemicals in their blood that act like antifreeze and allow them to live in cold Arctic waters. Polar bears can survive in freezing temperatures because thick layers of fat insulate them from the cold.

## LESSON REVIEW
1. What do vertebrates have in common?
2. What are the two general groups of vertebrates?
3. What distinguishes these two groups?
4. Explain how a closed circulatory system works.

## 4.3.2 Ectotherms

### OBJECTIVES

- Explain why the term *ectotherm* is more accurate than the term *cold-blooded*.
- Describe several behaviors that ectotherms use to regulate their body temperatures.

Imagine leaving your warm house to walk to school on a winter morning. As you step into the cold air, something strange happens. Your body begins to slow down, and your legs become sluggish. What is going on? Can this be normal? This may not be normal for you, but it is normal for ectotherms. Because ectotherms cannot internally regulate their body temperatures, their body temperatures and their body functions are controlled by outside conditions.

Animals that cannot internally control their body temperatures were once called *cold-blooded* animals, but that term can be misleading. An ectotherm can have a higher body temperature than an endotherm. Tuna, for example, generate large amounts of heat from inside their bodies; but since any heat generated by water ectotherms is quickly dissipated in the water, the body temperature of tuna is never much higher or lower than the water they live in.

Because air temperatures change more than water temperatures do, the body temperatures of reptiles and amphibians, which spend part or all of their time on land, vary more than the body temperatures of fish. To exist in their environments, God gave reptiles and amphibians the ability to lower their body temperatures in hot weather and raise their body temperatures in cold weather. Some reptiles and amphibians lift their bodies off the ground to allow air to cool their undersides. Others bury themselves or seek shade to cool off. Some reptiles lighten their skin to reflect heat or darken it to absorb heat. (You may use this trick by wearing light colors in the summer and dark colors in the winter.) Still others lie in the sunshine to warm themselves, a behavior called *basking*.

 **TRY THIS**

**Heat Capacity**

Water has a higher heat capacity than air, so water gains or loses heat more slowly than air. Place a thermometer in a jar of water. Place another thermometer in an empty jar. Record the temperature of the water and the air. Place both jars in a refrigerator and compare the temperatures after 10 minutes. How does this comparison relate to ectotherms that live in the water and those that live on land?

At night many reptiles and amphibians hide from predators. Without the sun's energy, their body temperatures drop, and they become sluggish. Moving slowly can make them easy prey. Other reptiles and amphibians come out only at night. For example, some desert reptiles come out at night so that they do not get too hot. God designed each animal to thrive in the environment in which He placed it.

Tortoises bury themselves to stay cool.

## LESSON REVIEW

**1.** What is the distinguishing feature of ectotherms?
**2.** Why is *cold-blooded* a confusing term to describe ectotherms?
**3.** Which organisms have an easier time regulating their body temperatures—ectotherms that live on land or those that live in water? Why?
**4.** What are some ways that ectotherms regulate their body temperatures?

### TRY THIS

**Colors and Heat**
Cut out squares of fabric in a variety of colors. Place thermometers under the squares. Leave the squares in the sun for 15 minutes. What is the temperature difference? How can changing color help amphibians and reptiles?

Iguanas are ectotherms. They bask in the sun to warm their body temperatures and get up on their feet to cool off.

### 4.3.3 Reptiles

## OBJECTIVES

- Indicate the common characteristics of reptiles.
- Identify various types of reptiles and describe each group.

## VOCABULARY

- **constrictor** a snake that squeezes its prey
- **herpetology** the study of reptiles and amphibians

Scientists have identified over 10,000 species of reptiles, and they are discovering more all the time. The study of reptiles and amphibians is called **herpetology**. The word *herpo* is Greek for "to creep or crawl." When you think about some common reptiles and amphibians, such as snakes, lizards, turtles, and salamanders, creeping and crawling seems like a good description.

But creeping and crawling is not enough to define reptiles. God gave reptiles many common features. All reptiles have lungs. The outer layer of a reptile's skin is dry, thick, and scaly. Most reptiles shed this entire skin all at once instead of one cell at a time like humans do. Reptiles shed their skin as they grow, so they shed more often if they have plenty of food. The scales of their skin form a waterproof barrier that keeps the inner skin from drying out. Reptiles have bone skeletons, and most reptiles have teeth. A reptile's teeth vary from simple blunt teeth for crushing prey to the hollow needlelike fangs of snakes. Many reptiles have a muscular, flexible, sticky tongue for catching insects. Reptiles generally have well-developed organs, although this is not always the case; snakes, for example, are deaf, and some lizards that live underground do not have eyes.

Reptiles reproduce through internal fertilization. Some reptiles lay leathery eggs that hatch into fully developed young. The

Alligators have rough, scaly skin and large, sharp teeth. Although they are similar to crocodiles, alligators belong to a different family. How can you tell the difference between an alligator and a crocodile?

leathery covering keeps the eggs from drying out and breaking easily. A few reptiles, such as certain lizards and snakes, keep the eggs inside their bodies while the embryos are developing. The eggs hatch inside the mother, and the young are born live. But whether they hatch or are born live, young reptiles look like miniature adults. Reptile mothers do not take care of their young after they are born.

Leatherback sea turtle returning to sea after laying its eggs

The most common reptiles can be divided into three major groups: turtles, lizards and snakes, and alligators and crocodiles. Turtles are reptiles whose bodies are protected by a shell formed of plates of bone. Some turtles have hard shells that can support a weight 200 times greater than their own! How much could you support if you could do that? Turtles are toothless, but they have a beaklike structure that they use to eat both animals and plants. Turtles have either legs or flippers, which are very strong. Tortoises are turtles that live completely, or almost completely, on land.

Turtles are an amazing group of creatures. The leatherback sea turtle is the largest turtle. It is about 2 m long and weighs about 900 kg—about as much as a small car. The smallest turtle, the speckled Cape tortoise, is under 10 cm long and weighs about 150 g.

Certain sea turtles head to the ocean after hatching, only to return—sometimes over hundreds of kilometers—to the same beach where they were born to lay their own eggs! They use wave motion and the Earth's magnetic field to make their way back to their birthplace.

Gecko

A boa constrictor swallowing its prey

The second group of reptiles includes lizards and snakes. Lizards have slender bodies, long tails, four legs, and different types of claws. Most lizards eat insects, either by catching the prey on their sticky tongues or by lunging for it. A closer study of lizards reveals God's creativity. Most lizards are less than 30 cm long, but some species of lizards range in size from the recently discovered 3 cm *Brookesia micra* chameleon to the 3 m Komodo dragon. The marine iguana of the Galapagos Islands spends most of its time in the ocean grazing on algae. The frill-necked lizard of the tropical woodlands of northern Australia opens an enormous frill to scare off predators. Chameleons can rapidly change their skin color to match their surroundings, and they have eyes that rotate independently of each other so they can look in two different directions at once. Chameleons flick their long body-length tongues with incredible speed to capture insects. The tree-dwelling flying lizards of Indonesia have five to seven pairs of extended ribs connected by a membrane that act like a parachute, enabling them to glide through the air. The basilisk lizard of Costa Rica can run on water when threatened,

*Brookesia micra* is one of the world's tiniest lizards reaching a length of just 29mm.

The Komodo dragon is the largest lizard in the world. Komodo dragons are excellent swimmers and can run up to speeds of 24 kph. They use their long tails to crush prey. They also have a dangerous bite. Until recently, it was believed that bacteria in a dragon's mouth caused an infection that killed its prey. However, scientists have found that Komodo dragons have venom ducts between their teeth. The venom is what kills their prey.

Snakes do not have limbs. They move by wriggling their long, muscular body. Snakes have more than 400 vertebrae in their backbones; humans have 32. The greater number of vertebrae makes a snake very flexible. A snake can move its entire body over many types of surfaces. Scales on snakes' bellies grip a surface and help pull the snakes forward even across smooth surfaces.

God designed snakes to sense vibrations on the ground because they do not see or hear well. Snakes breathe with their nostrils, but they sense their surroundings by flicking their tongues into the air and collecting air molecules. The odors a snake senses with its tongue tell it if prey or predators are nearby.

Honduran milk snake

One way that snakes help people is by eating small animals like rats and mice. Snakes eat a wide variety of things from eggs and ant larvae to antelopes and kangaroos. Many people hate snakes and want to get rid of them all. What do you think would happen to the rodent population if all snakes were destroyed?

The teeth of a snake are designed for grabbing their prey, but not for chewing. Since they cannot tear their prey into pieces, snakes must swallow their food whole. Imagine trying to swallow something as big as your head whole! You cannot do it, but snakes can because they can unhinge their lower and upper jaws. An African rock python can even swallow a 59 kg antelope.

There are 3,000 species of snakes, but only about 600 are poisonous. Poisonous snakes have hollow fangs that inject venom into their prey. Other large snakes, such as boas and pythons, are constrictors. A **constrictor** is a snake that squeezes its prey.

Different snakes live on land, in water, or both. Snakes live mainly in tropical or temperate climates. They range in size from the 10 cm thread snake to the anaconda, which is 10 m long and about 227 kg.

The third group of reptiles includes alligators and crocodiles, large reptiles with short legs and powerful jaws. They eat water

Chameleon

bugs, fish, turtles, birds, and mammals. They range in size from the Cuvier's dwarf caiman, a crocodilian that grows to 1.5 m, to the saltwater crocodile, which grows to 7 m.

Alligators live in freshwater and saltwater in the southeast United States and in China. The species that lives in the southeast United States grows up to 5 m long, and the species that lives in China grows up to 2.2 m long. Their nostrils and eyes are at the top of their head, so they can breathe and watch their surroundings when most of the body is hidden underwater. Their broad, heavy heads help them catch prey and crash through heavy vegetation in swampy areas. Alligators' teeth fit evenly into their jaws when they close their mouths. Their snouts are wide and U-shaped. Caimans, which live in South and Central America, have features similar to those of alligators.

Crocodile snouts are V-shaped and narrower than alligators. Most crocodiles live in saltwater. Crocodiles have salt glands on their tongues to get rid of the salt from the water that they drink. When crocodiles close their mouths, the fourth tooth in their lower jaw shows. This visible tooth is an easy way to tell a crocodile from an alligator or a caiman. Gharials are crocodilians that eat fish.

## LESSON REVIEW
**1.** What do all reptiles have in common?
**2.** What are the main groups of reptiles?
**3.** How do reptiles reproduce?
**4.** How do reptile eggs protect the young?

The word *amphibian* means "double life." That is a good description for amphibians because most of these vertebrates live the first part of their lives as plant eaters underwater and the second part of their lives as predators on land. Amphibians hatch from eggs laid in water and develop into gill-breathing larvae called *tadpoles*. The tadpoles mature into adults that live on land and breathe through both their skin and their lungs.

Many amphibians reproduce through external fertilization: a female lays eggs that a male fertilizes. The eggs develop into tadpoles, which breathe underwater and feed on algae or plants. The tadpole then undergoes metamorphosis, changing from a tadpole to an adult. In an amazing sequence of changes, it loses its tail (except for salamanders, which retain their tails) and develops two pairs of legs. As its gills disappear, its lungs develop. The tadpole is now an adult that can live on land as long as it stays near moisture.

An amphibian's skin is one of its unique features. It is moist, scaleless, and it contains many glands. These glands secrete mucus, which helps keep the skin moist. Some amphibians can secrete toxic chemicals through their glands to protect themselves from predators. Many South American rain forest frogs, for example, give off a poison that some native people use on darts for hunting. One 5 cm poison dart frog can produce enough poison to kill 10 grown men. However, these natural poisons may help people do more than hunt prey. Scientists are researching the

**A Frog's Life**

Adult frog

Frog eggs

Older tadpole with legs

Newly hatched tadpoles

The skin on this red salamander is smooth and moist.

Caecilians live in the soil and leaf litter of tropical rain forests near streams. Their diet consists of worms, insects, and insect larvae.

use of such potent venoms in medical treatments as painkillers.

Most amphibian species live in moist, humid environments so they do not dry out. Water and gases can pass through amphibian skin easily, allowing amphibians to breathe through their skin. But because their skin is so permeable, air pollution and water pollution easily invade amphibians' bodies and harm them.

Amphibians have skeletons that include cartilage, and they have small teeth for grasping prey. Their flexible, muscular tongues are often coated with a sticky substance that helps them snatch insects. Most amphibians, but especially frogs, have a great sense of hearing.

Scientists have identified over 6,500 species of amphibians, and many more species are discovered every year, particularly in rain forests. Amphibians are found throughout most of the world, but rain forests hold the greatest variety of them. In fact, Brazil has the highest number of identified species of amphibians.

Amphibians can be divided into three main groups: caecilians, salamanders, and toads and frogs. Caecilians are rare amphibians that live underground. They are burrowing animals with small eyes or no eyes at all. Caecilians are the only amphibians without legs. Scientists do not know very much about them because they stay hidden.

Salamanders are the second group of amphibians. The word *salamander* comes from a Greek word that means "lives in fire." For many centuries Europeans believed that the black and yellow salamander could walk

The tropical poison dart frog is one of the many exotic species of amphibians that live in the rain forest.

through fire. The bright colors and poisonous secretions of some species of salamanders may be the reason behind its name.

Toads and frogs make up the third group of amphibians. Toads have plump bodies and rough skin, and they live mostly on land, although most toads must stay near water. Frogs have streamlined bodies with smooth skin, and they usually live in water. Each frog species has a specific call to attract mates. The males make sounds to attract females or to warn other males to stay away. Most female frogs are mute.

Frog populations throughout the world are declining. Several species have become extinct, and the populations of many other species are dropping rapidly. Individual frogs in some regions are deformed. Scientists have discovered that a fungus is killing many species. The loss of habitats and increased pollution may also be to blame. Amphibians are more sensitive to pollution, so some scientists believe the decline of frog populations indicates the biosphere is becoming more unhealthy.

## LESSON REVIEW

1. What features make amphibians different from other vertebrates?
2. What are the types of amphibians?
3. Describe the life cycle of an amphibian.
4. Why are amphibians so sensitive to pollution?

This marbled newt belongs to the group of salamanders. All newts are salamanders, but not all salamanders are newts. How can you tell the difference? The term *newt* is sometimes used to describe salamanders that spend most of the year living on land.

Corroboree frogs are some of the world's most endangered frogs. The frogs' population decline over the past three decades has been linked to amphibian chytrid fungus. Their natural habitat is in Australia.

The oriental fire-bellied toad

You can tell by looking at fish that they are designed for an underwater life. Most fish have streamlined, scale-covered bodies that help them glide through the water. Most fish have fins to keep them upright and to help them navigate. All fish have gill slits for breathing underwater. Oxygen passes from the water into the blood vessels of the gills, and carbon dioxide passes out from the gills into the water.

Fish have well-developed nervous systems to sense their environments. For example, most fish have eyes that can see color, although some fish that live in caves do not have eyes. Fish can also smell and taste. For example, sharks have a keen sense of smell; some sharks would be able to sense a drop of blood in a backyard pool. This ability enables them to locate wounded animals. Most fish cannot hear well, but they do detect vibrations in the water.

Most fish are either male or female, but some species of fish are born as males and develop into females. Others are born as females and develop into males. Most fish reproduce through external fertilization. The female releases jelly-coated eggs

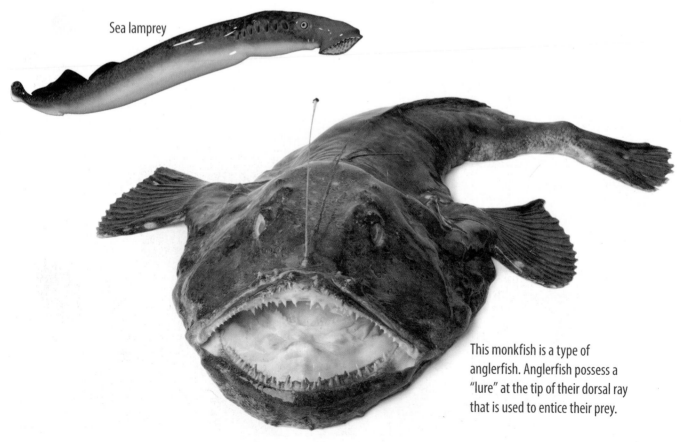

Sea lamprey

This monkfish is a type of anglerfish. Anglerfish possess a "lure" at the tip of their dorsal ray that is used to entice their prey.

into the water, and the male fertilizes them with sperm. The fertilized eggs then hatch. A few species of fish use internal fertilization. In these fish, the fertilized eggs develop inside the mother's body. When the young fish are developed enough, the eggs hatch inside the mother's body, and the young fish are born.

As a group, fish eat almost anything from algae to worms to coral to dead fish. Fish use a variety of different hunting techniques. Swordfish slash through schools of fish and return to capture the wounded prey. Anglerfish dangle wormlike "lures" in front of their prey. Other fish blend in with their prey and then launch a sneak attack.

The known fish species can be divided into three main groups: jawless fish, cartilaginous fish, and bony fish. Jawless fish have round mouths that work like suction cups, but no jaws. Their skeletons are made of cartilage, and they do not have scales. Hagfish and lampreys are jawless fish. Hagfish live in the ocean and devour large dead carcasses from the ocean floor. Lampreys have over 100 teeth, which they use to drill holes through another fish's scales, attach themselves to the host's body, and suck out the blood and body fluids.

## HISTORY

**Other Fish in the Sea**
In 1976, a U.S. Navy research vessel off the Hawaiian Islands found a large shark tangled in its anchor. The shark was from a previously unidentified species. Scientists named this new species *the megamouth shark.*

Manta rays lead solitary lives, but they will gather where food is concentrated and to find mates. Manta rays have filters in their mouths through which they feed on small fish and tiny crustaceans.

The mandarinfish is a brightly colored reef dweller native to the Pacific (from the Japanese Ryukyu Islands south to Australia). They eat small crustaceans and other invertebrates.

Courtesy of Luc Viatour

The second group of fish are the cartilaginous fish—sharks, skates, and rays. Like hagfish, these fish have skeletons made of cartilage. The fish in this group are covered by small, pointed structures called *denticles*, which make the skin coarse like sandpaper. Rays and skates have flat bodies with large, winglike pectoral fins and a long, thin tail. The largest ray is the manta ray. Ocean mantas can measure 7 m from wing tip to wing tip and can weigh over 2,000 kg. The largest shark, which is also the largest fish, is the whale shark. Whale sharks can grow to over 12 m long.

Scientists have identified a few hundred species of sharks, but the oceans probably contain many undiscovered species. Most sharks actively hunt large prey. Many people are afraid of sharks, but sharks generally kill fewer than 10 people each year. In some parts of the world, sharks are overfished for food and sport. Sharks were once killed in large numbers for sport, but because of conservation concerns, this does not happen as often now. Many shark-catching competitions are now catch and release.

The third group of fish are the bony fish. These fish have skeletons of bone and scales of a bony material. Over 28,000 known species of bony fish range in size from the stout

**Fish**

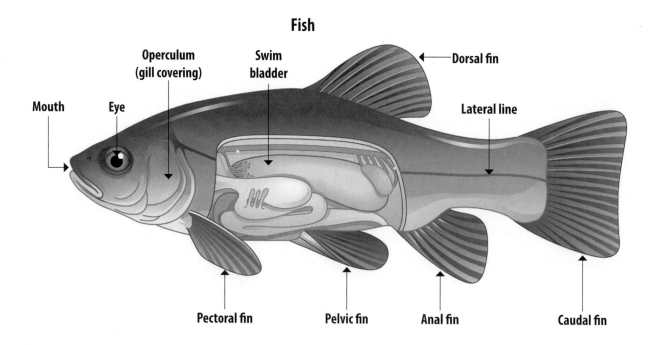

**FYI**

### A Costly Kettle of Fish

People have always eaten fish. Today new technology enables people to catch more fish than ever before. Radar systems provide the ability for boats to go out even in the thickest fog; sonar systems track schools of fish deep in the ocean; and satellite positioning systems map the open seas. Modern fishing boats can use up to 130 km of longlines, fishing lines with thousands of baited hooks. They have trawl nets large enough to hold several jumbo jets and drift nets up to 65 km long.

In 1940, commercial fishing brought in 9 million kg of fish. Today commercial fishing harvests 81 billion kg of fish each year. Populations of commercial fish, such as tuna and snapper, are shrinking. Until the 1990s, drift nets routinely netted not only fish but also accidentally trapped and killed birds, dolphins, whales, sharks, and turtles. Drift nets are now regulated so that not as many animals are accidentally killed, although the problem still exists on a smaller scale.

The oceans' resources will not last forever. Overfishing jeopardizes the balance that God intended for ocean ecosystems. Distinguishing between stewardship and exploitation is a concern for Christians, who read in Genesis 1:22 that "God blessed them [ocean and winged creatures] and said, 'Be fruitful and increase in number and fill the water in the seas.'"

infantfish at under 1 cm to the giant oarfish at 36 m. Most bony fish have two large eyes without eyelids, nostrils (for smelling, not breathing), tongues (for touching, not tasting), and swim bladders. **Swim bladders** are balloonlike organs that bony fish can fill or empty of gases to allow them to ascend or descend in the water. They also have a hard bony flap that covers and protects their gills called the **operculum**.

## LESSON REVIEW

**1.** What are the general characteristics of fish?
**2.** How are fish equipped to live in water?
**3.** What are the three main groups of fish?
**4.** How do swim bladders work?

## OBJECTIVES

- List the advantages that endotherms have over ectotherms.
- Identify the challenges that endotherms face.
- Summarize how endotherms maintain an ideal body temperature.

Kangaroos keep cool by licking their forearms.

If you took a dog on a walk on a hot day, both of your bodies would respond to the heat. You would begin to sweat, and the dog would begin to pant, but your body temperatures would not change. Sweating and panting are processes designed to keep an animal's body from overheating. Endotherms use these and other processes to maintain a consistent body temperature no matter what the temperature around them is. So whether you are walking along a sunny beach in July or at a sledding party in January, your body temperature will remain about the same.

How can this be? All endotherms, which include birds and mammals, have a thermostat that regulates body temperature. God gave endotherms body features and certain behaviors to help them stay warm in cold environments and cool in hot environments. For example, fat, fur, and feathers help endotherms retain heat. Desert mammals have larger exposed body surfaces than animals living in colder environments. The fennec fox that lives in North African deserts has large ears to radiate away heat, whereas the Arctic fox has small ears to conserve heat. Small animals often instinctively huddle together to keep warm. Larger mammals keep cool by sweating; smaller mammals pant, passing air quickly over the saliva on their tongues, which draws heat away from their bodies. A dog may pant at a rate of 300 times a minute when it is hot, compared

Fennec fox

with breathing 30 times a minute when it is cool. Because of their ability to maintain their body temperature in the cold, endotherms can live in colder parts of the world than most ectotherms. They do not become sluggish in colder weather, so they are not easy targets for predators when it is cold. Endotherms can be active at night when the temperature drops. In hot climates many mammals are active only at night when it cools off. They seek food and even mates under the cover of darkness.

Arctic fox

Although endotherms can maintain their internal temperatures, they are affected by the air temperature around them. Endotherms can overheat in hot conditions and they can freeze if unprotected in cold weather.

Aquatic endotherms such as whales, seals, walruses, and penguins have the challenge of staying warm in water. Their body fat and thick skins provide insulation to help them conserve heat. So even though their skin temperatures may drop dramatically, they maintain safe internal temperatures. Other animals, such as bears, store up body fat to maintain a constant body temperature while they hibernate for the winter. Hibernation allows them to conserve energy rather than using more energy to hunt for food in cold temperatures. Other endotherms use burrows or dens to keep warm in cold weather.

The energy that keeps endotherms warm is the chemical energy that comes from food. It takes a lot of energy to maintain a consistent body temperature, so endotherms have to eat a lot

## TRY THIS

**Sweating**
Leave a thermometer on the table for three minutes to determine the room's temperature. Moisten a cotton ball with rubbing alcohol and spread a thin layer of cotton around the thermometer's bulb. Blow on the cotton ball 20 times. What happens to the temperature? Why? How can sweating help regulate the body temperature of endotherms?

Walruses have a blubber layer that insulates them and streamlines their body. The blubber also functions as an energy reserve. The blubber layer can be up to 10 cm thick.

more than ectotherms do. A mammal uses far more energy than a reptile of the same size and weight. Endotherms must also digest nutrients faster to provide this extra energy. Mammals have longer digestive systems, which absorb nutrients from food more quickly than a reptile's digestive system does.

Smaller endotherms generally have a higher metabolism than larger endotherms. Animals with faster metabolisms convert food into heat and other forms of energy quicker than animals with slower metabolisms. It takes more energy for a mouse to sustain one gram of tissue than it does for an elephant to sustain one gram of tissue because smaller animals have a higher surface-to-volume ratio. In other words, small animals have more skin exposed to the air compared to the volume inside their bodies. Because they have more surface area exposed to the air, small animals lose heat faster than large animals do.

## LESSON REVIEW

**1.** What advantages are there to being an endotherm?
**2.** What challenges do endotherms face?
**3.** How do endotherms maintain a constant body temperature?
**4.** Suppose someone brought a "new" mammal to class that was nocturnal and had large ears. Where would you guess that this animal had lived? Why?

Opossums forage at night when competition and predation are lowest.

Birds are among some of God's most fascinating creatures. They range in size from the 1.6 g bee hummingbird to the 160 kg African ostrich. Many birds are brightly colored, and many of them sing beautifully. A bird in flight is an amazing thing to watch!

Birds are diverse creatures, but they all have several features in common. All birds have feathers, which are made of a protein called *keratin*. The hair, claws, and horns of other animals are also made of keratin. Feathers protect the bird's skin and insulate it to keep the bird warm. Feathers also make it possible for birds to fly. The number of feathers a bird has depends on its species. Hummingbirds have approximately 1,000 feathers, and tundra swans have approximately 25,000.

Feathers cover the bird's entire body except the beak, scaled leg parts, and feet. Birds have three basic types of feathers, and each type has a separate function. Contour feathers streamline the body to help the bird fly. They also protect the bird's skin from dirt and the sun's harmful rays. Flight feathers, which are stiffer and longer than contour feathers, provide motion and lift for flying. Down feathers, which are soft and fluffy, insulate the bird against the cold.

All birds have wings, and most birds can fly. Flying birds have one of three wing shapes: broad, rounded wings; slender, curved wings; and straight, narrow wings. Broad, rounded wings are best for accelerating and maneuvering over short distances. This type

## OBJECTIVES

- Distinguish birds from other endotherms.
- Summarize how the various features of birds help them thrive.
- Describe the functions of different types of bird features.

## VOCABULARY

- **keel** the breastbone of a bird
- **migrate** to travel from one place to another in response to seasons or environmental conditions

A flight feather        A contour feather        A down feather

Notice the shape of the wings on this dove.

The wings on snow goose are shaped perfectly for long distance flights.

The wings on this hawk enable it to soar through the air with ease.

of wing is common among woodland birds, such as grouse and woodpeckers, and among birds that spend much of their time on the ground, such as quail. Pigeons and doves use their wings to quickly accelerate to speeds of 80 kph.

Birds that fly long distances have slender, curved wings. Ducks, geese, and other long-distance travelers, use their wings to cruise at speeds of around 50 kph for many hours at a time. Snow geese, for example, can cover over 3,800 km in a week or less. The alpine swift can stay in the air continuously for six months.

Straight, narrow wings allow birds to soar and glide. Flying takes a lot of energy—a bird uses about 15 times more energy to flap its wings than it does when sitting still. Large birds of prey, such as eagles, hawks, and vultures, conserve energy by soaring on updrafts, which are warm air currents pushed up from the ground. Tiny hummingbirds beat their straight, narrow wings 80–120 beats per second to hover in midair.

Feathers and wings are not the only features that help birds fly. Birds' bones are hollow and light. For example, the frigate bird, a large sea bird, has a wingspan of over 2 m but a skeleton of only 113 g. Birds' lungs are more efficient than mammals' lungs. They have multiple air sacs that allow them to get more oxygen with each breath. The air sacs in birds' lungs also decrease their body density and keep birds light. Flying birds have a large breastbone called the **keel**, which supports their chest muscles. Birds' chest muscles power their flight. Birds also have higher body temperatures than mammals.

Not all birds can fly, however. Ostriches, emus, and penguins are all flightless birds. Ornithologists, the scientists who study birds, calculate that no bird heavier than 18 kg can fly.

Birds have different types of feet that help them thrive in their environments. Perching feet have a single hind toe and three toes in front. Half of bird species, including common birds such as sparrows, starlings, and crows, have perching feet. Climbing feet have two toes in the front and two in the back. Woodpeckers have this kind of foot. Other types of feet include talons, webbed feet, and wading feet. Birds of prey, such as eagles, use talons for tearing flesh. Ducks and geese use webbed feet for swimming. Wading feet, such as those of the blue heron, have long toes to prevent the bird from sinking into the mud near the banks of rivers, ponds, or the ocean.

In a similar way, birds have beaks designed for the kind of food they eat. Most birds have one of four different types. Seed eaters like finches and sparrows have cone-shaped beaks. Ducks and geese have flat or shovel-shaped beaks to strain food out of the water. Birds of prey have beaks that are hook shaped for tearing into their prey. Insect-eating birds, such as robins, have beaks that resemble tweezers, which help them grasp bugs.

All birds reproduce by laying eggs. An egg consists of a yolk, albumen (egg white), and shell. The yolk, a fluid that is rich in fats and protein, provides food for the developing embryo. The albumen absorbs shocks to protect the embryo if the egg falls or gets knocked around. The hard shell also protects the embryo as it develops. God created bird eggs in a variety of colors, patterns, and sizes, ranging from the pea-sized bee hummingbird egg to the 1.3 kg ostrich egg. The heaviest known bird, the extinct elephant bird, laid an 11 kg egg. This egg was over 30 cm long!

Webbed feet for swimming

Talons for grasping and tearing

Perching feet

**Bird**

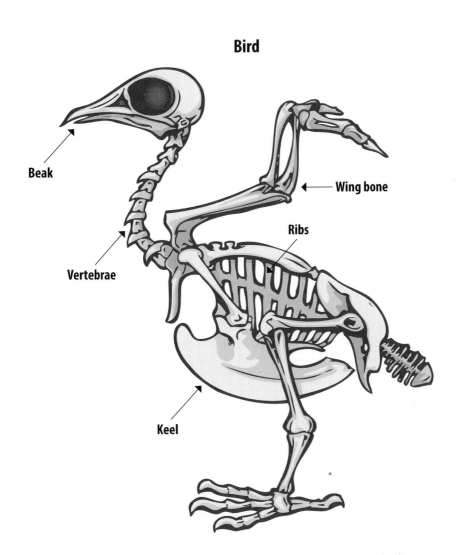

Beak

Wing bone

Ribs

Vertebrae

Keel

An eagle's hook-shaped beak

A cardinal's cone-shaped beak

A mallard's flat beak

A barn swallow's tweezer-shaped beak

 **FYI**

**It's a Bird**
What is the fastest animal? Cheetahs are the fastest mammal, but the fastest animal is a bird. Falcons can dive at speeds over 350 kph, and swifts can glide at speeds of 170 kph.

After they breed, most birds build nests in which to lay their eggs and care for their young. Nests come in a variety of shapes and sizes with the most elaborate designs being made by tropical birds. Birds use a wide variety of materials for building nests, including mud, twigs, seed heads, feathers, fur, hair, aluminum foil, string, lichen, grass, wool, and paper.

Most birds produce sounds or songs. Each species of bird has its own song. A bird's song is one way ornithologists and amateur bird-watchers identify birds. Songs are used to court mates and warn other birds.

Many birds **migrate**—they travel from one place to another in response to the seasons or environmental conditions. Birds in the Northern Hemisphere migrate south, leaving their northern breeding sites before cold weather to find food in a warmer area. Many waterfowl, such as ducks, migrate south in the autumn because the plants they eat become trapped beneath frozen ponds and lakes. The Arctic tern's migration is 40,000 km each year—that is the distance around Earth!

## LESSON REVIEW
**1.** How are birds different from other endotherms?
**2.** What are the functions of the three main kinds of feathers?
**3.** What kind of feet do birds of prey have? Why?
**4.** How does the shape of a bird's beak help it survive?

Think of all the different mammals in the world: dogs, cats, bats, whales, elephants, duck-billed platypuses, kangaroos, rats, cows, zebras, giraffes, otters, and more. There are over 5,000 different species of mammals living on Earth today. God created a vastly diverse group of mammals that range in size from the Kitti's hog-nosed bat to the blue whale. A wolf may journey in an area as large as 130 km², but a naked mole rat never leaves its burrow. The Virginia opossum can give birth to a litter of 20, and the orangutan has only one offspring at a time. Elephants can live 60 years or more, but the male brown antechinus, a shrewlike animal, does not live for even a year in the wild.

## OBJECTIVES

- Identify the basic characteristics of mammals.
- Describe what sets humans apart from other animals.
- Explain the three basic groups of mammals.
- Observe and describe the structure and behavior of mammals.

## VOCABULARY

- **echolocation** a method of locating objects by interpreting reflected sound waves
- **mammary gland** a structure in female mammals that produces and secretes milk
- **marsupial** a mammal whose young are born and then develop further in the mother's pouch
- **placenta** a tissue in the womb of certain mammals that nourishes the developing young

An orangutan and her baby in Borneo

Female mammals produce milk for their young.

What do all mammals have in common? Mammals are endotherms that have hair or fur. The hair or fur of a mammal protects the animal and keeps it warm. Not all mammals are completely covered with hair, but even whales, dolphins, and porpoises have hair bristles. Female mammals have **mammary glands**, which are structures that produce and secrete milk. Mammal mothers stay with their young after birth because their young rely on them for food. Depending on the species, a mammal may have from one to eleven pairs of mammary glands. Species that have larger litters usually have more glands.

Mammals have many other features in common. All mammals, even those that live in the water, breathe with lungs. Mammals also have highly developed senses. For example, chimpanzees can see colors, bats depend on keen hearing, and dogs recognize people by their smell. Mammals also have the most developed brains of all animals. A mammal's brain coordinates movement and makes learning and understanding possible.

Mammals reproduce through internal fertilization, but not all mammals reproduce in exactly the same way. Mammals can be divided into three groups according to the way that they reproduce—egg-laying mammals, marsupials, and placental mammals. Egg-laying mammals do not bear live young. They lay eggs with tough, leathery, reptilelike shells. Egg-laying mammals include duck-billed platypus and two species of spiny

## FYI

**They Said It Wasn't So!**
In the 1700s when European scientists heard reports of the platypus, they thought it was a hoax. An egg-laying mammal? Impossible! Naturalists in Australia finally sent a live specimen back to Europe, and the European scientific community finally recognized the platypus species.

Duck-billed platypus

anteaters. The platypus has a snout that looks like a duck's bill, a flat tail like a beaver, and webbed feet. All egg-laying mammals live in Australia and New Guinea.

**Marsupials**, the second group of mammals, have pouches for their young. As soon as they are born, young marsupials, which are not fully developed, crawl to the mother's mammary glands inside the pouch and remain attached to them as they continue to develop. Even after the suckling stage, the young return to the mother's pouch for shelter and transportation. The approximately 300 species of marsupials include the koala, kangaroo, bandicoot, mouse opossum, wombat, Tasmanian devil, and Tasmanian tiger.

Most mammals belong to the third group, which is placental mammals. Placental mammals give birth to well-developed young. Inside the mother's womb is a **placenta**, a tissue that nourishes the developing young and helps the unborn mammal thrive. The placenta delivers food and oxygen from the mother to the young and exchanges wastes as well.

Kangaroo with joey

Because various placental mammals have similar features, they are further divided into many different orders. The major orders are the following:

**Insect Eaters (Insectivora)** Several hundred species of mammals are insect eaters, including moles, hedgehogs, and

**TRY THIS**

**Mammary Glands**
Count the number of mammary glands on different species of animals. Relate the number of glands to the average size of the litter that the mammal produces. Graph the relationship.

Contrary to what most people believe, moles are not actually blind. They have very small eyes that are covered by fur, which is the reason for their poor eyesight. Moles cannot see colors, but they can detect motion and light from dark.

## HISTORY

**Rats . . . Black Death!**
The bubonic plague begins with rats that carry the bacterium *Yersinia pestis*. Fleas drink rat blood and pass the bacteria on to humans when they bite them. The plague has killed people throughout history, but three major outbreaks stand out. The Plague of Justinian (542–543) killed 70,000 people in the city of Constantinople. The Black Death (named after the black splotches on people's skin), killed 25 million people—one-third of Europe's population—from 1346 to 1350. The third major outbreak, from 1890 to 1910, killed 12.6 million people, mostly in India and Asia but also in North America.

shrews. Insect eaters have long, narrow snouts; powerful digging claws for living underground; and high metabolic rates because they use a lot of energy. Shrews, for example, must eat up to twice their body weight daily to get enough energy to live. How much food would you have to eat if you did that?

**Flying Mammals (Chiroptera)** There are over 1,300 species of flying mammals, all of which are bats. In fact, about 20% of all mammal species are bats. Bats have membranes between their "finger bones," enabling them to use their arms as wings. Bats are active at night. They have poor eyesight, but they can navigate and find insects by echolocation. **Echolocation** is a method of locating objects by interpreting reflected sound waves. A bat screeches a high-pitched sound into the night, and the sound bounces off an insect and comes back to the bat. From this reflected sound the bat can determine an insect's location, distance, and size—or where the wall of a cave is. Bats are important for controlling insect populations. One brown bat can eat 1,200 mosquitoes in an hour. The 100 million Mexican free-tailed bats that live in Central Texas eat over 900,000 kg of insects each night.

**Rodents (Rodentia)** Rodents make up almost 40% of all mammal species. The many types of rodents include mice, rats, guinea pigs, lemmings, muskrats, gerbils, gophers, squirrels, porcupines, pacas, and beavers. Rodents thrive in every habitat. Lemmings, for example, live in the snow-covered tundra, and gerbils live in the hottest, driest deserts. Many rodents live near

The dwarf epauletted fruit bat lives in Africa and is about 10 cm long!

**Rodent Observation**

Set up an aquarium or cage to house small rodents such as mice or gerbils. Be sure to supply plenty of materials for the rodents to chew on. Observe the rodents for several five-minute intervals over the course of several days. Record the amount of time the rodents spend doing various activities such as chewing, grooming, sleeping, huddling, or exploring.

Place the rodents' food into different types of containers, presenting problem-solving opportunities to be observed and timed. Give the rodents the same food problem-solving dilemma for three days; observe and time how quickly they learn to get the food. Present a second food dilemma for the next three-day period.

Using a triple-beam balance or scale, determine the mass of the rodents' food and calculate the average amount of food one rodent consumes daily. After several observation sessions, use the collected data to draw conclusions about the rodents' metabolic rate, preferred activities, and ability to learn new things about their environment.

The spotted paca has a wide distribution throughout Latin and South America.

The snowshoe hare is white in the winter and grayish-brown in the summer. These color changes help it blend in with its surroundings.

human populations, eating stored food and spreading disease. Around the world, over 35 diseases are spread by mice and rats, and many of these can be spread to humans. Even so, rodents are important sources of food for predators in most ecosystems. Rodents have sharp teeth that never stop growing, which is why gerbils, hamsters, and mice are always chewing on things. They must continually wear down their teeth. Otherwise, their teeth will become too long, and they will starve because they will not be able to open and close their mouths.

**Rabbits and Hares (Lagomorpha)** Rabbits and hares have long, soft fur, and large ears. Their eyes are set high on the sides of their heads, giving them a wide field of vision. These mammals are a favorite prey of many larger animals, but their high death rate is balanced out by their high birthrate—the phrase "multiply like rabbits" is well-founded!

**Toothless Mammals (Xenarthra)** Twenty-eight species of mammals have no teeth or very simple teeth. Anteaters do not have any teeth, and sloths and armadillos have only rootless molars that grow throughout their lives. These animals have long, sticky tongues that help them catch ants, termites, and other insects.

The giant anteater lives a solitary life in South America. When it sleeps, it curls its bushy tail (which is as long as its body) around itself to keep warm.

**Carnivores (Carnivora)** The over 270 species of meat-eating mammals can be divided into nine families that live on land.

The palm civet is native to Asia. Its body length is 43–71 cm, and it has a tail length of 40–66 cm.

The aardvark is a nocturnal mammal native to Africa.

These families are the cats, dogs, bears, raccoons, weasels, skunks, mongooses, civets, and hyenas. There are also three aquatic families that include sea lions, seals, and walruses. Carnivores have sharp teeth for holding prey and cutting flesh, eyes in the front of their heads for depth perception, and bodies built for speed.

**Hoofed Mammals (Perissodactyla, Artiodactyla, Tubulidentata, and Proboscidea)** About 250 species of hoofed animals live on Earth. These animals, which are also called *ungulates*, include elephants, aardvarks, horses, zebras, rhinos, camels, sheep, deer, and goats. Because hoofed mammals are prey for large carnivores, their hard hooves are designed so they can run quickly, and their eyes are on the sides of their heads so they can see predators coming from any direction.

**Sea Mammals (Cetacea and Sirenia)** There are over 100 species of sea mammals, which include whales, porpoises, dolphins, and manatees. Most sea mammals spend their entire lives at sea. They have streamlined bodies, flippers for speed in the water, and thick layers of blubber, or fat, to stay warm in the cold water. Like bats, some sea mammals also use echolocation to find food and each other.

**Primates (Primates)** Primates have flexible fingers and toes, flat nails instead of claws, and the most highly developed

Manatees swim by moving their large paddlelike tails up and down.

# 🔖 BIOGRAPHY

### Jane Goodall

Jane Goodall (born 1934), a British ethologist (someone who studies animal behavior), is famous for her pioneering research on chimpanzees. As a young girl, Jane was fascinated by animals. She read stories about Tarzan and the book *The Story of Dr. Doolittle*, and she dreamed of going to Africa. In her spare time she studied local animals and read books on zoology (the study of animals).

In 1960, she set off for Africa to begin research on chimpanzees. At that time people thought that chimps might be vicious and that she might not survive. Goodall's research, which was the longest field study ever in the Gombe National Park in Tanzania, proved otherwise. Over 40 years and three generations of chimpanzees later, people know far more about chimpanzees, thanks to her observations of behaviors such as chimps eating meat, making tools, planning strategies, using human-made objects, mounting warfare against other groups of chimps, adopting orphaned chimps, and forming coalitions to maintain their power in the group.

Goodall has used the information she gathered to protect chimpanzees. She founded the Jane Goodall Institute, which supports the research at Gombe as well as chimp orphanages and rehabilitation centers. In 1991, she started a program for schoolchildren in Tanzania called Roots and Shoots. The program encourages students to improve the environment around them, to care for their human community, and to make the world a better place for the animals in their area. This program for students from preschool through university is now in more than 130 countries. "Dr. Jane" now travels all over the world, visiting clubs, zoos, and chimp rehabilitation clinics and giving talks to increase awareness of the impact of people's everyday decisions on the environment.

Jane Goodall entered Cambridge University as a Ph.D. candidate in 1961. She is one of very few people to be admitted to the university without a college degree. She earned her Ph.D. in ethology in 1966. Dr. Goodall has written many books and articles and received many scientific awards and honorary degrees. She is the only non-Tanzanian to have received the Medal of Tanzania, and Queen Elizabeth II awarded her the Commander of the Order of the British Empire.

In an interview, Goodall was asked what she would like young people to know. She replied, "Every individual matters. Every individual has a role to play. Every individual makes a difference."

brains of all mammals. Primates include lemurs, monkeys, baboons, orangutans, chimpanzees, and gorillas. Humans are also classified as primates because they have the physical characteristics that define this order. Of course, humans are more than primates. God formed humans, unlike animals, in His image, and humans alone have souls. God set people apart from the animals. He commanded people to care for the rest of creation. The psalmist wrote in Psalm 8:3–5, "When I consider Your heavens, the work of Your fingers, the moon and the stars, which You have set in place, what is mankind that You are mindful of them, human beings that You care for them? You have made them a little lower than the angels and crowned them with glory and honor." So even though physically you may be considered a primate, remember that you are created in His image!

## LESSON REVIEW

1. What features do mammals have in common?
2. What are the three basic groups of mammals?
3. Choose two of the orders of placental mammals and describe their structures and behaviors.
4. Explain what sets humans apart from other animals.

Gorillas eat mostly plants, such as celery, nettles, bamboo, and thistles. This one is eating a nectarine.

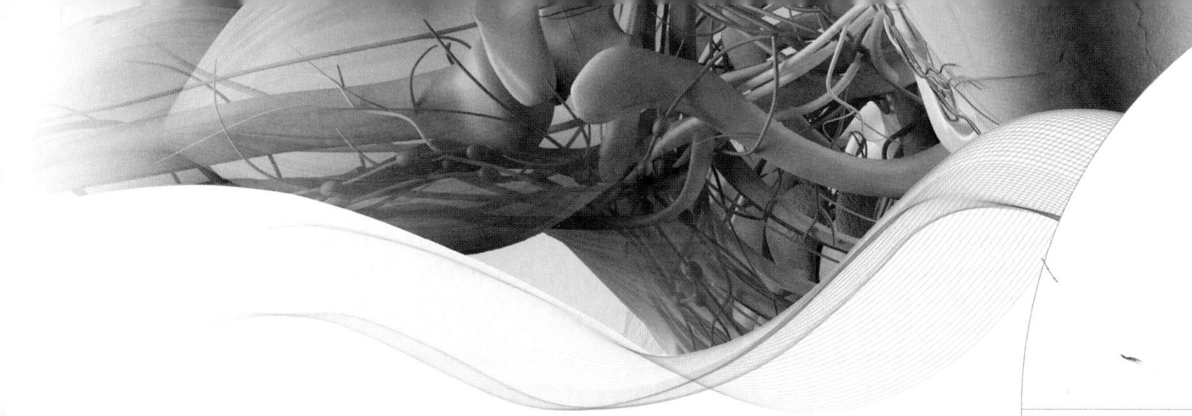

# UNIT 5

# Human Body

## Chapter 1: *Cardiovascular, Digestive, and Immune Systems*
## Chapter 2: *Respiratory and Excretory Systems*
## Chapter 3: *Movement and the Senses*
## Chapter 4: *Human Development*

*Vocabulary*

addiction
adrenal gland
allergen
alveoli
antibody
antigen
atrium
axon
biological clock
bronchi
central nervous system
circadian rhythm
colon
cone
cranium
dendrite
dependence
dermis
element
epidermis
esophagus
fallopian tube
gland
ligament
marrow
melanin
molecule
nephron
ovulation
pelvis
peripheral nervous system
peristalsis
pituitary gland
plasma
platelet
rod
sebaceous gland
semen
spleen
striated muscle
sweat gland
synapse
tendon
trachea
tympanic membrane
ureter
urethra
uterus
vagina
ventricle
withdrawal

*Key Ideas*

- Evidence, models, and explanation
- Change, constancy, and measurement
- Equilibrium
- Form and function
- Abilities necessary to do scientific inquiry
- Understandings about science and technology
- Structure and function in living systems
- Reproduction and heredity
- Regulation and behavior
- Personal health
- Risks and benefits
- Science and technology in society
- Science as a human endeavor

**SCRIPTURE**

May God himself, the God of peace, sanctify you through and through. May your whole spirit, soul and body be kept blameless at the coming of our Lord Jesus Christ.

1 Thessalonians 5:23

### OBJECTIVES

- Distinguish between cells, tissues, organs, and body systems.
- Name the four elements that form most biochemicals.

### VOCABULARY

- **element** a pure substance that cannot be broken down into simpler substances by chemical or physical means
- **molecule** a particle consisting of two or more atoms chemically bonded together

Pliny the Elder (23–79 AD) was a Roman writer who wrote the following statement in his 37-volume work called *Natural History*: "The arteries … are without blood, nor do they all contain the breath of life…. The veins spread underneath the whole skin, finally ending in very thin threads, and they narrow down into such an extremely minute size that the blood cannot pass through them nor can anything else but the moisture passing out from the blood in innumerable small drops, which is called sweat."

Even if you do not know much about body systems, you probably know enough to recognize that Pliny the Elder had the wrong idea about the nature of the systems that carry blood and oxygen throughout the body and that protect the body from infection. Misconceptions about these systems continued for centuries. Some people thought that blood was constantly made by the liver and used up by the heart. Others thought that there were two kinds of blood—a blood that carried nutrients and a blood that carried breath and "vital spirits."

Today more is known about the amazing body systems that God designed. Humans have self-regulating systems that carry blood throughout the body, supply nutrients and oxygen to every cell, and provide protection from disease. The circulatory, respiratory, and immune systems are examples of God's amazing work.

### Human Body

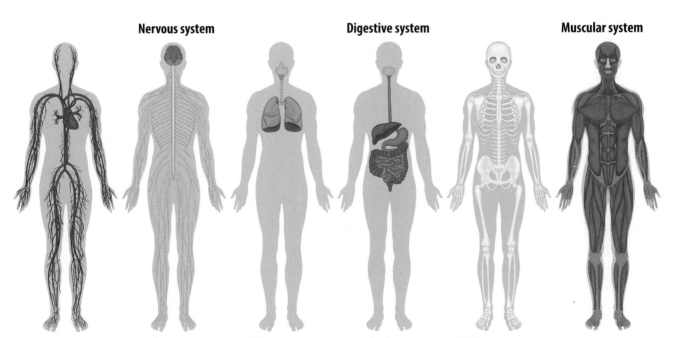

Nervous system     Digestive system     Muscular system

Circulatory system     Respiratory system     Skeletal system

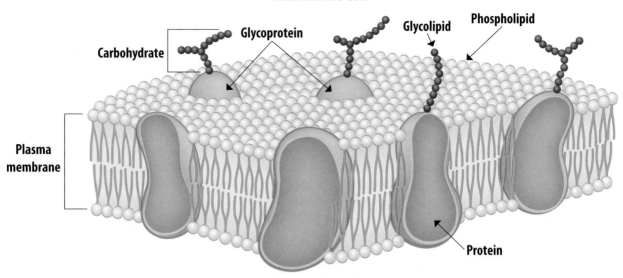

**Outside the Cell**

Carbohydrate

Glycoprotein

Glycolipid

Phospholipid

Plasma membrane

Protein

**Inside the Cell**

Your body is a lot like a city. A city is a network of systems working together in harmony. Your bones and muscles are a framework like the streets of a city. Your nervous system carries messages like electric power lines do. Your respiratory and circulatory systems are busy each minute of the day with heavy traffic, delivering food and oxygen to your body's cells and tissues. Your digestive system is like the grocery trucks that deliver food to the store. Your excretory system is like a garbage truck, removing waste from your body. Your endocrine system is like the traffic lights, determining when and how hormones flow. Your immune system is like the police, rushing to your assistance when invaders try to harm you. Your body systems work together to keep everything functioning in harmony.

Think of all the different materials used to build a city—glass, wood, concrete, steel, aluminum, copper, brick, and plastic. The basic materials from which your body is woven are called *elements*. An **element** is a pure substance that cannot be broken down into simpler substances by physical or chemical means.

The smallest particle into which an element can be divided and still be the same substance is an atom. Elements sometimes exist in pure form, but usually atoms from different elements join to make molecules. A **molecule** is a particle consisting of two or more atoms chemically bonded together.

All living things contain combinations of elements called *biochemicals.* Amazingly, just four elements combine in different configurations to form 98% of the biochemicals that make up living things. These four elements are carbon, hydrogen, nitrogen, and oxygen. Think of it—the vast majority of your complex body contains combinations of just four elements!

The cell is the basic unit of structure of all living things, and it is the smallest unit that can carry out life processes. Each cell is formed from biochemicals. The biochemical molecules perform specific jobs in the cells. Although by itself a biochemical molecule is not alive, a cell—a careful arrangement of many different biochemicals—is alive because it can carry out the processes of life.

Although each of the body's trillions of cells share some common features (a nucleus, a membrane, and various organelles), cells develop differently and function differently in your body. For example, bone and teeth cells are structured to attach to mineral matter. Skin cells are formed largely from keratin, a protein that does not dissolve in water. Individual heart cells have a beat. Cells also vary widely in shape. Muscle cells are long and smooth. Nerve cells are long with branching extensions. Red blood cells are round with a depression in the middle.

**Levels of Organization**

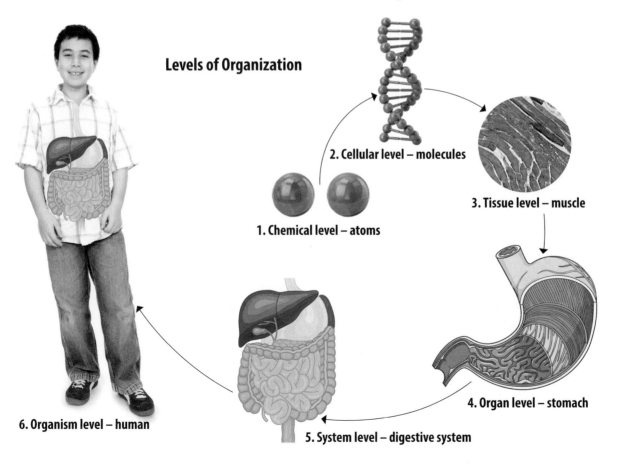

2. Cellular level – molecules

1. Chemical level – atoms

3. Tissue level – muscle

4. Organ level – stomach

5. System level – digestive system

6. Organism level – human

### Physicians

Only God fully understands the way that people are put together, but over the past 200 years much has been learned about how the human body functions. There are still mysteries to unravel. The volume of knowledge about the human body has led to more and more specialization among doctors because no one person can be an expert on everything there is to know about the human body.

A family physician, sometimes known as a general practitioner (GP), often cares for a whole family and can give advice regarding a healthy lifestyle (eating and exercise) to prevent medical problems. Family physicians treat the common problems that arise and, if necessary, recommend treatment by a specialist. When a specialist's skills are needed, the family physician usually remains part of the medical team that cares for the patient.

Doctors may choose to specialize in different body organs or systems: cardiologists, heart and circulatory system; neurologists, nervous system and brain function; dermatologists, the skin; nephrologist, kidneys; urologists, urinary tract; gastroenterologists, stomach and digestive tract; internal medicine specialists, internal organs; endocrinologists, glands and hormones; and gynecologists, women's reproductive systems.

Some physicians specialize in treating certain groups of people. Pediatricians treat children, and gerontologists treat the elderly. Sports medicine specialists care for athletes. Allergists see people with allergies. Anesthesiologists put people to sleep and manage pain during surgery.

Surgeons specialize in performing surgery, cutting bodies open in order to remove or repair tissues or organs. General surgeons perform a wide variety of surgeries, but many surgeons specialize. Eye surgeons operate on eyes, and neurosurgeons perform brain surgery. Plastic surgeons may reconstruct people's skin or features after an accident. Orthopedic surgeons may treat bone and joint disorders or perform hip and knee replacement surgeries.

With all that there is to learn about the human body, it is not surprising that it takes many years of study to become a physician. After medical students complete a college or university degree, they attend medical school for another four years. After that they serve as an intern for one or two years, working in a hospital to get more practical experience. Then they must study additional years in a residency program as they learn more about their area of specialization. That all adds up to 10 to 12 years of school after high school! Becoming a doctor is not easy, but physicians use what they learn to explore the majesty of the human body and to heal their patients.

## TRY THIS

### Cheek Cells

With a toothpick, scrape some cells from the inside of your cheek. Use methylene blue to prepare a wet-mount slide. Examine the cells under a microscope and sketch your observations. Carefully follow your teacher's safety instructions during the activity and during cleanup.

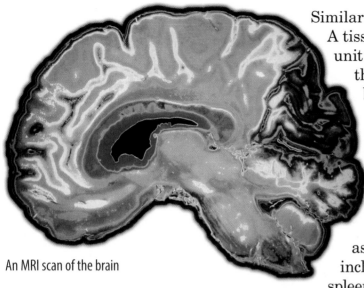
An MRI scan of the brain

Similar types of cells are grouped into tissues. A tissue is a group of cells organized into a unit in the body. Tissues are distributed throughout the body. Tissues include blood, skin, fat, nerves, muscle, bone, and connective tissue such as cartilage.

Organs have a distinct structure and are made of two or more tissues that work together to perform a specific job in the body. Some organs, such as the brain, occur individually; others, such as the lungs, come in pairs. Other organs include the heart, liver, brain, kidneys, spleen, and stomach.

Tissues and organs are organized into body systems such as the circulatory, respiratory, and nervous systems. Each system plays an important role in your health and well-being and depends on other systems. If one body system is weak, the whole body usually suffers.

Amazingly, human bodies work far more efficiently than any city. Every day you draw thousands of breaths and your heart beats thousands of times. The glands of your endocrine system send messages to your bones to grow. Your immune system automatically rushes to your defense to heal cuts and diseases. God created a delicate balance among your many body systems and cells to sustain your body. Each system works in harmony to allow you to live and thrive!

Being made in God's image, you are far more valuable than the sum of the biochemicals in your body. You can honor God by caring for the amazing body He gave you. You care for your body by making good choices about food, exercise, and rest as well as by avoiding risks and stress.

## LESSON REVIEW

**1.** What are some features that all cells have in common?
**2.** What is a tissue? Give two examples.
**3.** What is an organ? Give two examples.
**4.** What is the difference between tissues and organs?
**5.** Give two examples of a human body system.
**6.** What are the four elements that form most biochemicals?

Cells die without nutrients. Since the cells of your toes and nose cannot wander over to your stomach or intestines to pick up some nutrients for lunch, God created the cardiovascular system—a system of vessels and organs that transports materials to and from body cells. The word *cardio* comes from the Greek word for "heart," and the word *vascular* comes from the Latin word for "vessel." The cardiovascular system is also known as *the circulatory system*. This system has two functions: to carry oxygen and nutrients to different parts of the body and to pick up waste products from the cells.

Most states and provinces have thousands of kilometers of roads, ranging in size from superhighways with four or more lanes of traffic to narrow alleys. In the same way, the cardiovascular system has thousands of meters of blood vessels, ranging in size from the large veins and arteries that enter and exit the heart to capillaries, the tiniest blood vessels. Capillaries are so narrow that blood cells must pass through them single file. Blood moves continuously through arteries and veins in a one-way circuit. In general, arteries are blood vessels that carry blood away from the heart to the body. Veins are the vessels that take blood from the body to the heart.

The cardiovascular system has two main circuits: pulmonary and systemic. Pulmonary circulation is the circulation of blood between the heart and the lungs. The word *pulmonary* comes from the Greek word for "lung." Systemic circulation is the circulation of blood between the heart and the rest of the body (excluding the lungs). Both of these circuits start and end at the heart, the major organ of the cardiovascular system.

Your heart is located between the lungs and is about the size

## OBJECTIVES

- Explain the functions of the cardiovascular system.
- Identify the parts of a heart and trace the path of blood through the heart.
- Summarize several factors that can lead to heart disease.

## VOCABULARY

- **atrium** an upper chamber of the heart
- **plasma** the thin, yellow fluid part of the blood
- **platelet** a cell fragment that helps in clotting blood
- **ventricle** a lower chamber of the heart

**Cardiovascular System**

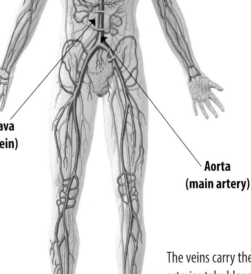

Heart

Vena cava
(main vein)

Aorta
(main artery)

The veins carry the blood to the heart. The arteries take blood away from the heart. All the blood vessels in the body are linked to the main artery or the main vein. *Veins are shown in blue; arteries are shown in red. The vessels in the hands and feet are shown separately for easier visibility.*

## Heart

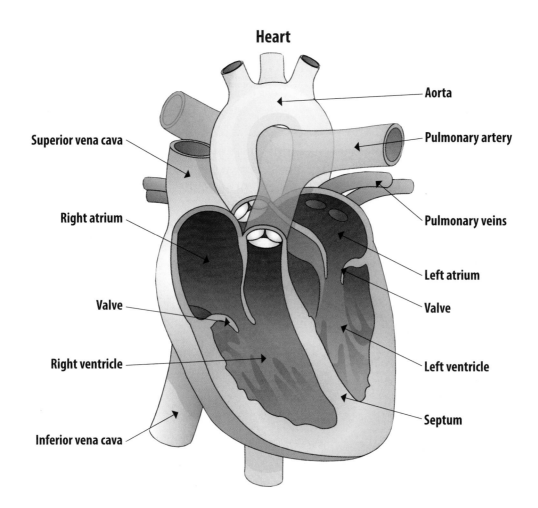

Aorta

Pulmonary artery

Superior vena cava

Pulmonary veins

Right atrium

Left atrium

Valve

Valve

Right ventricle

Left ventricle

Inferior vena cava

Septum

## TRY THIS

**Every Heartbeat**
Your heart beats nonstop every minute of every day of your life. It beats about 100,000 times each day. Calculate how many times it beats in a year. Calculate about how many times your heart has beat so far in your life.

of your closed fist. This thick, hollow muscle has a right side and a left side that are separated by a sheet of muscle called *the septum*. Each side is divided into two hollow chambers. Each upper chamber of the heart is called an **atrium**, from the Latin for "entrance chamber." Each lower chamber of the heart is called a **ventricle**. Between the chambers and each of the arteries and veins that connect to the heart, one-way valves prevent blood from flowing in the wrong direction. With each beat, the heart acts as a powerful pump, continually pushing the blood along.

The blood travels in a circuit throughout the body. It enters the heart through the right atrium. The right ventricle relaxes between heartbeats, and the blood in the atrium flows through a valve and down into the right ventricle. During the next beat, the powerful muscle wall around the ventricle squeezes, pushing the blood through a valve that opens into the pulmonary artery. The blood travels to the lungs through blood vessels that branch out into tiny capillaries. In the lungs, the blood gives up the carbon dioxide waste it collected from the body cells and picks up oxygen to deliver to the cells.

232

The oxygen-stocked blood travels back from the lungs through veins to the heart. The blood enters the left atrium through the pulmonary veins. As the left ventricle relaxes between beats, the blood flows through a valve and into the left ventricle. Finally, as the left ventricle contracts during the next heartbeat, the blood is pushed through a valve and enters the aorta, the blood vessel that carries the blood directly from the heart.

The aorta branches out into arteries that lead in a number of directions. Two branches take blood up each side of the neck into the head. Two others take the blood into the arms. The main branch travels down along the backbone. That branch, called *the abdominal aorta*, has branches that surround each of the organs in the abdomen (stomach, intestines, liver, and kidneys) as well as branches that go down each leg. The branching continues to the point of forming about 10 billion capillaries in an adult body. The walls of capillaries are so thin that oxygen and nutrients can diffuse into the adjacent cells, and waste products can diffuse from the cells into the blood.

 **BIOGRAPHY**

### William Harvey

William Harvey (1578–1657) was an English physician who laid the foundation of modern medicine by discovering that blood circulates through the body. In 1628, he published *De Motu Cordis—On the Motion of the Heart and the Blood in Animals*. Harvey's discovery ended the teachings of Greek physician Galen, which had been accepted for 1,400 years. According to Galen's proposed circulation system, the liver produced the natural spirit, the heart the vital spirit, and the brain the animal spirit.

Harvey worked at Saint Bartholomew's Hospital in London and was a professor there from 1615–1643. In 1618, he became the court physician for King James I and then later for King Charles I. Harvey examined the heart and blood vessels of mammals and deduced that the blood in the veins flows toward the heart. He calculated the amount of blood that left the heart at each beat and realized that the same blood circulates continuously around the body. He also reasoned that blood passes from the right side of the heart to the left through the lungs.

Harvey also speculated that mammals reproduce through the union of an egg and sperm. It was 200 years before fertilization of a mammalian egg was finally observed, but Harvey's theory made so much sense that the world assumed he was right long before the discovery was made.

## FYI

**Carbon Monoxide**
Carbon monoxide is a colorless, odorless gas that results from incompletely burned fossil fuels. It acts as a poison when it attaches to hemoglobin. When this happens, the hemoglobin cannot transport oxygen to the different parts of the body because it is carrying carbon monoxide instead. The person may get sick or even die. You might have a carbon monoxide detector at home to alert you if your furnace begins to release carbon monoxide. Cigarette smoke also produces carbon monoxide. As much as 15% of the hemoglobin in a smoker's blood may be attached to carbon monoxide. A pregnant woman who smokes deprives her growing baby of oxygen. Such babies are often smaller at birth and have more health problems.

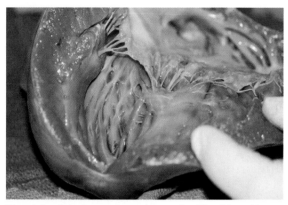

The walls of the heart are very muscular. The thin, white strings are called *chordae tendineae.* They connect the heart muscle to the valves.

On its way back to the heart, the blood in the capillaries flows together into veins. Because most of the blood in veins has to move against gravity to get back to the heart, the veins have valves that do not allow blood to drain back. The blood flows over and around the digestive system, picking up nutrients and waste. This blood then travels through the liver, where toxins from the digestive system are chemically changed, made water soluble, and sent to the kidneys to be excreted in the urine.

Heartbeats create pressure within the heart and blood vessels. This pressure keeps the blood flowing. The body has many remarkable means of regulating blood flow and blood pressure so that tissues receive what they need when they need it. Nerves, hormones, and chemicals produced by the tissues themselves all send messages to regulate blood flow.

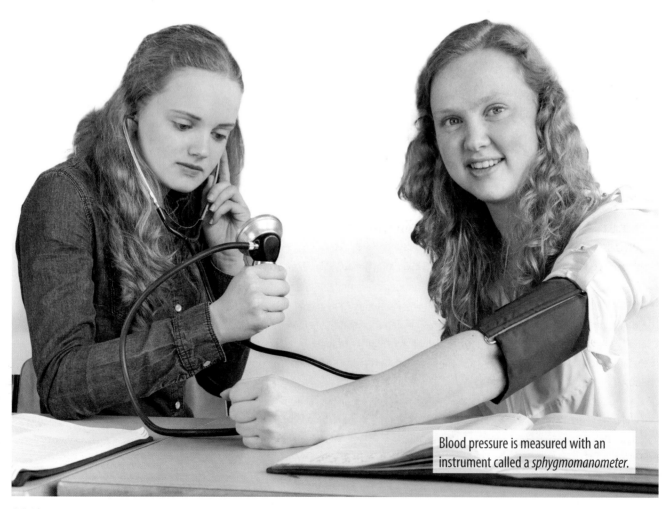

Blood pressure is measured with an instrument called a *sphygmomanometer.*

Have you ever had your blood pressure measured as part of a routine medical checkup? Two numbers are given in a blood pressure reading; the first number is always higher than the second. The higher number, the systolic pressure, indicates the pressure in the blood vessels just after the heart beats, as a fresh wave of pressure pulses through the aorta and arteries. The pressure eases up between heartbeats, resulting in the lower number, the diastolic pressure. A person's blood pressure is affected by many factors: the flexibility of the blood vessel walls, the blood's thickness (called *viscosity*), and the person's weight, diet, and hereditary factors. Hypertension, blood pressure that is too high, may cause cardiovascular damage or disease.

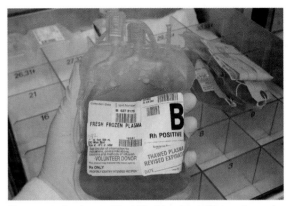

Plasma can be donated just like blood can. Plasma is used to treat people with rare diseases and disorders such as immunodeficiency and hemophilia.

God designed the elaborate cardiovascular system so blood can flow to every cell in the body. Blood is truly a remarkable tissue. Because it holds heat, it helps keep the body's temperature constant by taking heat from the inside of the body to the skin, where it can be released. Blood contains biochemicals, such as antibodies that fight disease. Hormones and enzymes in the blood help keep the blood's pH at the proper level. Other specialized blood cells protect the body against invaders.

Blood taken from the body looks like a thick red liquid. When a test tube full of blood is spun in a centrifuge, the blood separates into two parts: the thin, yellow fluid part of blood called **plasma** on the top (about 55% of the total) and a thick red mass (the other 45%) on the bottom. Although plasma is about 92% water, it contains some very important molecules including enzymes, hormones, nutrients, waste products, gases, and various proteins that move lipids, carbohydrates, and hormones through the body.

Red blood cells, white blood cells, and platelets float in the plasma. Most of the floating particles in plasma are red blood cells. A critical component of red blood cells is hemoglobin, the protein in red blood cells that attaches to oxygen so it can be transported. Hemoglobin includes iron atoms and makes the cells red. White blood cells can slip through the walls of capillaries. They are attracted to foreign material, such as dirt or bacteria, and dead cells. They surround the invading dirt or bacteria, and then they die. Pus is your body's natural response to an infection. It is made up of many dead white blood cells

## FYI

**Open Those Capillaries**
Blood flow through skeletal muscles greatly increases during exercise. When skeletal muscles are resting, only about 25% of the capillaries in the muscles are open. During exercise, most of the capillaries in the muscles are open, which allows more oxygen and nutrients to reach those muscles.

**A antigen**     **B antigen**

**Blood type A**

**Blood type B**

**Blood type AB**

**Blood type O**

and some cell debris. **Platelets** are cell fragments that clump together at the site of a wound to help form a blood clot.

You cannot live without blood. Sometimes an accident victim or someone having surgery needs a blood transfusion to replace the blood he or she has lost. When doctors first began performing transfusions, they noticed that not all blood works in all bodies. Sometimes the transfused blood clotted together in the veins because the recipient's body was attacking the blood as if it were a foreign invader. Since those early transfusions, scientists have identified particular molecules called *antigens* on the surface of red blood cells. The presence or lack of specific antigens determines your blood type. Four blood types are very common: type A, type B, type AB, and type O. The blood of people who share a blood type is compatible. People with any of these blood types can receive type O blood cells in a transfusion. People with type AB can receive blood cells from any blood type.

Another important blood grouping is the Rh factor. People are Rh positive if the Rh antigen exists on the surface of their red blood cells. People without that antigen are Rh negative. It would be dangerous to give an Rh-negative person Rh-positive blood in transfusion. No harmful effects occur the first time incompatible blood is given, but the immune system responds to the foreign Rh antigen by producing anti-Rh antibodies. If an Rh-negative person had Rh-positive blood transfused a second time, after these antibodies had formed, the antibodies would attack the

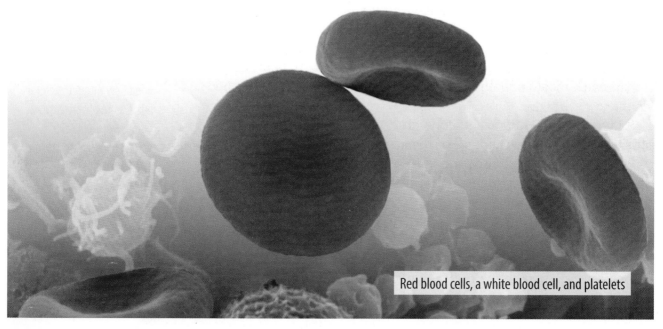

Red blood cells, a white blood cell, and platelets

foreign red blood cells, causing them to clump together. This reaction would destroy the red blood cells, causing illness or death.

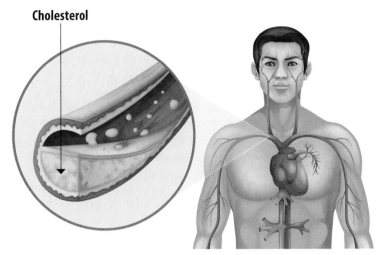

Cholesterol blocking an artery

Occasionally harmful blood conditions develop. When there is not enough hemoglobin in the blood, a person has a medical condition called *anemia*. Because hemoglobin carries oxygen, the cells of people with anemia do not get enough oxygen, so these people lack energy and tire easily. Some types of anemia result from nutritional deficiencies of vitamin B$_{12}$, folic acid, or, more commonly, from iron deficiency. Iron is the one nutrient that menstruating girls and women cannot get enough of in their diets without overeating, which is why doctors often advise women to take an iron supplement to prevent anemia.

Leukemia is a cancer in which abnormal white blood cells are produced. These abnormal cells are produced so rapidly and use so many nutrients that the body has trouble supplying the body's tissues with nutrients and making enough red blood cells. The white blood cells that are formed do not function normally, leaving the body at risk of infection.

Heart attacks are still a leading cause of death throughout the world. Atherosclerosis, also called *hardening of the arteries,* is a condition that occurs when fatty material builds up on the inside walls of arteries. Later, connective tissue and even calcium deposits can form on top of the fatty substances. This buildup restricts the flow of blood through the arteries and makes the heart work harder to pump blood. Atherosclerosis contributes to heart disease, and it develops faster in people who are overweight, who eat a lot of saturated fats, who do not exercise, or who smoke. In fact, people who smoke are much more likely to get heart disease than people who do not smoke.

High levels of cholesterol in the blood also increase the risk of heart disease. Cholesterol is a fatlike substance found in all cells of the body. Low density lipoprotein (LDL) cholesterol is a major component of the fatty material deposited in the arteries in atherosclerosis. High density lipoprotein (HDL) cholesterol helps carry LDL cholesterol away from the arteries. Regular exercise can raise HDL cholesterol and lower LDL cholesterol.

High blood pressure also contributes to heart attacks. High blood pressure forces the heart to work harder to pump the blood and can injure the heart muscle itself. It contributes to atherosclerosis and increases the risk that blood vessels may rupture, causing internal bleeding. The biggest cause of high blood pressure is obesity. People can often lower their blood pressure by losing weight and eating a diet low in salt. Regular exercise can also help lower high blood pressure.

## LESSON REVIEW

**1.** Name two functions of the cardiovascular system.
**2.** Name the five parts of the heart.
**3.** Sketch the path that blood takes from the time it enters the heart through the right atrium until it enters the heart again.
**4.** Explain what blood pressure is and what medical risks high blood pressure poses.
**5.** List five factors that can contribute to heart disease.

### FYI

**How to Be Heart Smart**

You probably do not have heart problems, but you still need to take care of your heart. What you do now will affect you from now on. Here are some ways that you can be "heart smart."

• Eat fruits and vegetables regularly and eat junk food only once in a while.

• Exercise several times a week. If you do not enjoy one kind of exercise, try different kinds until you find something that you enjoy. You will probably have more fun exercising with a friend or family member than exercising alone.

• Relax! Laugh a lot and take it easy, especially when you are under stress. Tension, frustration, and sadness have been medically proven to put a strain on the heart. Learn how to deal with your stress. First, ask God for help. Then find out what additional methods can help you relax. Many people relax by writing, playing sports, drawing, talking to a friend or family member, or listening to music.

You know that food is the fuel that your body needs to function, grow, and be healthy. But how does the food that you eat strengthen muscles and bones and nourish blood? How does yogurt strengthen your bones, and how does peanut butter renew your red blood cells?

Food in its original form is too large to be absorbed by the cells in your body. It needs to be broken down into its microscopic nutrients. God designed the complementary processes of digestion and absorption to meet that need. Digestion is the process by which food is broken down into molecules small enough to be absorbed by the body. This process happens in two ways. First, food is broken down physically by chewing. Then, food is broken down chemically through the use of enzymes. Enzymes are protein molecules that make biochemical reactions happen much faster. Enzymes break food into tiny nutrient molecules. Absorption is the process by which nutrient molecules are moved into the bloodstream. The bloodstream acts like a highway that brings nutrients to every cell. Most absorption occurs in the walls of the small intestine.

The major component of a person's digestive system is the gastrointestinal tract, a series of tubes and body organs. The gastrointestinal tract includes the mouth, throat, esophagus, stomach, small intestine, large intestine, rectum, and anus. The digestive system works with the nervous system and the muscle system. The gastrointestinal tract is lined with long smooth muscles and with nerves that send messages to the muscles. The wavelike muscle contractions that push food through the digestive system is called **peristalsis**. A second type of contraction called *segmentation contractions* function mainly in the small intestine where segmented rings contract. These contractions mix the food and the digestive enzymes and aid in absorption. Glands lining the gastrointestinal tract produce two types of secretions to help digestion: digestive enzymes, which break the food molecules apart, and mucus, which lubricates and protects the lining of the gastrointestinal tract.

How does all this affect the food that you eat? You can follow some food through the gastrointestinal tract to find out. Digestion begins in your mouth. As you

## OBJECTIVES

- Describe the path food takes as it travels through the gastrointestinal tract.
- Identify the parts of the digestive system and explain their functions.

## VOCABULARY

- **colon** the large intestine
- **esophagus** the long tube that connects the throat to the stomach
- **peristalsis** the wavelike muscle contractions that push food through the digestive system

## FYI

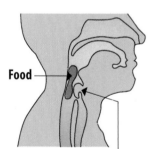

**Epiglottis covers entrance to trachea.**

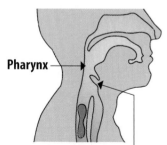

**Epiglottis reopens as the food passes.**

chew the food into smaller pieces, the food mixes with saliva from the salivary glands. The enzymes in saliva begin to chemically break the long chains of complex carbohydrates into shorter chains.

When the food has been chewed into a soft mass called *a bolus*, it is swallowed. Swallowing is more complicated than it seems because both inhaled air and ingested food travel through the back of the mouth. God provided a way to ensure that the air gets to the lungs and the food gets to the stomach by designing a flap called *the epiglottis*. This little flap of cartilage folds over the windpipe during swallowing and prevents food from going down into the lungs.

Once the bolus passes the epiglottis, it travels by peristalsis through the **esophagus**, the long tube that connects the throat to the stomach. Just before the bolus moves into the stomach, it passes through a thick, narrow muscle called *a sphincter*. Various sphincters are strategically placed throughout the gastrointestinal tract to keep the food mass from moving too quickly or backing up in the wrong direction.

Significant mixing and digestion occur in the stomach. Glands cover most of the stomach wall. Some glands secrete a very strong acid called *hydrochloric acid*. Stomach acid has a pH of 1 and kills most of the bacteria that you eat. Other glands secrete enzymes, such as pepsin, which split proteins into

**Gastrointestinal Tract**

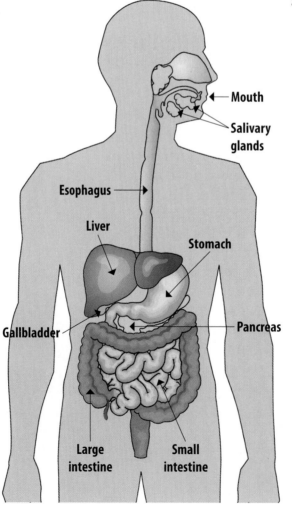

240

**Peristalsis**

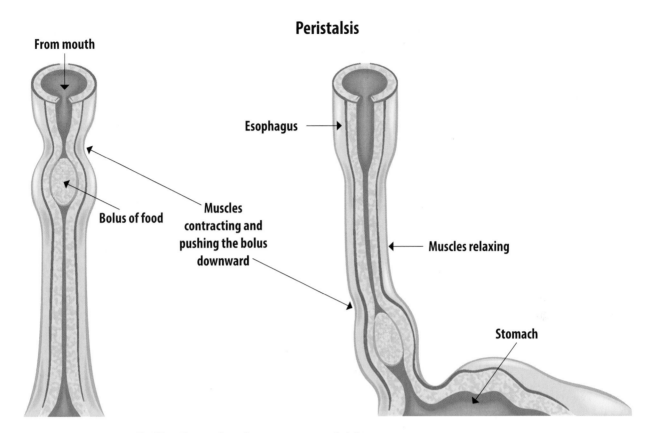

From mouth

Esophagus

Bolus of food

Muscles
contracting and
pushing the bolus
downward

Muscles relaxing

Stomach

their amino acids. Still other glands secrete a thick mucus to coat the stomach walls so the powerful acid and enzymes do not digest the stomach itself. The stomach walls contract to blend the secretions and food into a soupy mixture called *chyme*. The chyme must reach a pH of 1.5, which is a very acidic level, before it moves from the stomach to the small intestine.

The duodenum is the first section of the small intestine, and it controls how fast the stomach empties in two ways. When the duodenum senses that the chyme needs more digestion time in the stomach, it sends a message through the nerves to slow down the peristalsis. The duodenum also senses when fatty acids are present in the chyme. The detected fat triggers a hormone that slows down the rate at which the stomach empties. This response is why you feel full longer after eating fatty foods.

Releasing chyme slowly from the stomach also gives the small intestine more time to mix the chyme with fluids from the liver and pancreas. The liver constantly secretes bile, a bitter liquid that helps the small intestine digest and absorb fats. Bile is stored in the gallbladder. When lipids, or fats, enter the small intestine, the gallbladder releases bile into the small intestine. The bile helps make the lipids digestible so that they can be transported into the bloodstream through the intestine wall.

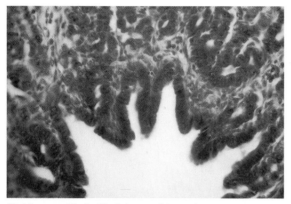
Highly magnified villi of the small intestine

The pancreas is a glandular organ that produces insulin and a juice of powerful enzymes that helps the small intestine digest proteins, lipids, and carbohydrates. The release of this pancreatic juice into the small intestine is regulated both by hormones and by nerve messages that are triggered when food enters the small intestine.

By the time these enzymes are finished with the chyme, most of the proteins have been broken down into amino acids, most of the carbohydrates have been broken down into single-molecule sugars, and most of the lipids have been broken down into fatty acids. Vitamins and minerals, including electrolytes, are released as ions, or free molecules, as the food is digested. Now the nutrients are ready to be absorbed into the body.

Every day the gastrointestinal tract must absorb about 10 L of fluid. About 1.5 L of the fluid come from what is eaten and 8.5 L come from digestive secretions. Most of the fluid is absorbed through the small intestine. This process requires a very large surface area. The small intestines are designed with folds that increase the surface area. Villi are tiny fingerlike projections that line these folds to increase the surface area even more. Even the cells of the 1 mm villi are lined with about 600 microvilli, which are extremely tiny folds of tissue. The villi and microvilli greatly increase the absorption of nutrients from the small intestine. Without these features, the intestines would have to be about 600 times longer than they are to maintain the same level of absorption.

After the food's nutrients are absorbed into the bloodstream, the blood transports them throughout the body so that every cell can receive the energy it needs for growth and repair. When most of the nutrients have been absorbed, the remaining chyme moves into the large intestine, which is also called the **colon**. The first half of the large intestine is called *the absorbing colon*. Approximately 1.5 L of chyme pass into the large intestine each day. Most of the chyme is now water and electrolytes, which are absorbed through the absorbing colon's walls. A number of beneficial bacteria live in the absorbing colon and make small amounts of vitamin K, vitamin $B_{12}$, thiamine, and riboflavin. The second half of the large intestine is called *the storage colon*. It is here solid waste, or feces, is stored until it is eliminated through the anus.

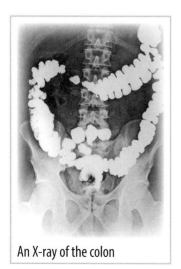
An X-ray of the colon

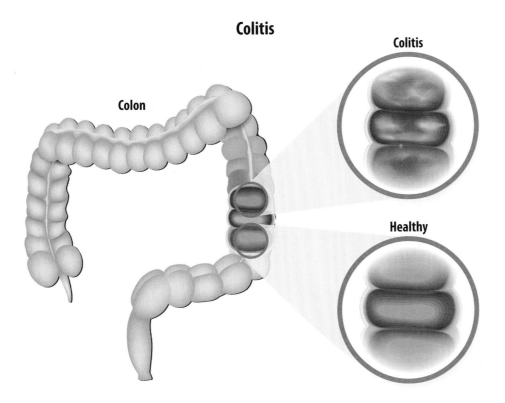

**Colitis**

**Colon**

**Colitis**

**Healthy**

Various things can go wrong with the complex digestion system, but there are measures you can take to keep the system healthy. Abnormal conditions of the gastrointestinal tract include enteritis, which is any inflammation of the intestines, and colitis, which is inflammation of the colon. Both illnesses can include diarrhea and dehydration. Because it contains live bacteria, eating yogurt is one way to help reestablish beneficial bacteria in the colon after a bout of diarrhea.

In contrast, constipation is the slow movement of feces through the large intestine, resulting in hard feces because of the longer time for water absorption in the colon. Eating more dietary fiber such as whole grains, fruits, and vegetables, drinking plenty of water, and exercising regularly help counteract constipation.

## LESSON REVIEW

**1.** Trace the path of food through the gastrointestinal tract. Name the parts and their functions.
**2.** How has God protected a person's stomach from being damaged by the strong acid needed for digestion?
**3.** What are villi and why are they important?

Although at times you may get injured or become sick, God designed your body with an immune system that works continuously to keep you healthy. The immune system as a whole is not located in any one part of your body. It is an army of cells, tissues, and organs that band together to fight any pathogens, the microorganisms that cause disease, that try to invade.

The body's first line of defense includes the skin, mucus, and platelets. The skin acts as a physical barrier to invaders. The top layer of your skin is dead, so most invaders cannot find a live cell to infect. As old, dead cells flake off your skin, they carry along the viruses, bacteria, and other microorganisms that have landed on it. The skin also produces chemicals that protect the body. For example, lysozyme is an enzyme found in tears, sweat, saliva, and nasal secretions that kills bacteria by breaking their cell membranes apart. The skin also produces sebum, which contains an acid that kills microorganisms or keeps them from growing. Mucus is a sticky substance that lines the membranes inside the mouth, nose, and other channels that are open to outside of the body. Mucus traps microorganisms as they enter the body, surrounding the pathogens so they can be digested without harming the body. The body acts to keep out as many pathogens as possible. When you get a cut, blood flows to the injured area, and blood cell fragments called *platelets* help seal the wound to prevent pathogens from entering the body.

If pathogens do get past the first line of defense and enter the body, another line of defense—the lymphatic system—goes into action. The job of the lymphatic system is to carry bacteria and other foreign invaders from the body's tissues and blood. The

lymphatic system includes lymph vessels and spaces between the tissues that are filled with lymph, a clear or yellowish fluid that contains special white blood cells and collects in the lymph vessels. The lymphatic system also includes lymph nodes, which are small organs that remove dead cells and pathogens from the body. Each side of the body has three main groups of lymph nodes—one at the groin, one at the armpit, and one on the side of the neck. Lymph nodes contain many white blood cells, which fight disease. The white blood cells engulf the pathogens or produce chemicals that help destroy them. When pathogens infect the body, the white blood cells multiply. This swift multiplication of white blood cells can cause the lymph nodes to swell and become painful.

Other lymphatic organs include the thymus gland, the spleen, and the tonsils. The thymus gland processes white blood cells called *T cells.* T cells are made in the bone marrow and are designed to determine what immune response is needed when an invader is detected. The thymus sends these cells out through the blood to other lymphatic tissues where they can react to foreign substances. The **spleen** is a lymph organ located behind and above the stomach that acts as a large blood filter. As the blood passes through the spleen, the white blood cells there attack foreign substances and destroy old red blood cells. When red blood cells squeeze through capillaries in the spleen, the older red blood cells rupture. These cells are then broken down and some of their parts are recycled. Tonsils are masses of lymphatic tissue located at the back of the throat. White blood cells in the tonsils defend the body against infection.

The immune system includes various types of cells that target an invader. Macrophages are large white blood cells that engulf pathogens. Macrophages can work alone to stop a few invading

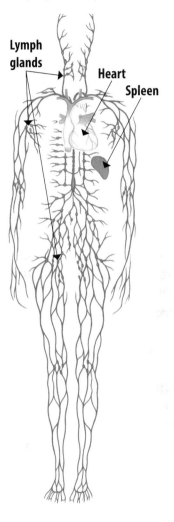

**Lymphatic System**

Lymph glands

Heart

Spleen

Macrophage, lymphocyte

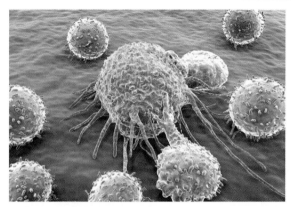

T cells and B cells attacking a cancer cell

microorganisms or viruses, but if millions of pathogens invade your body, the macrophages need reinforcements: T cells and B cells. The macrophages call for these reinforcements by digesting only part of the pathogen and sticking the rest of the pieces to their surfaces. This action signals the rest of the immune system defense troops that help is needed. The substances that generate a response from the immune system are called **antigens**.

The antigens on the surface of the macrophages activate T cells. These white blood cells coordinate the immune system's response by releasing certain chemicals. The T cells activated by the antigens on the macrophages are called *helper T cells*. The helper T cells divide quickly to increase their number. The helper T cells send word to other T cells called *killer T cells*. Killer T cells are T cells that kill any cell that is infected with pathogens. Suppressor T cells shut down the immune system in response to an invader so that the immune system does not overreact and destroy good cells. These T cells suppress the actions of other immune cells. Without the activity of suppressor T cells, immunity could easily get out of hand, resulting in allergic or autoimmune reactions.

The helper T cells also activate B cells, white blood cells that mature in the bones and make antibodies. **Antibodies** are proteins that attach to specific pathogens. Each type of antibody

## Immune System Response

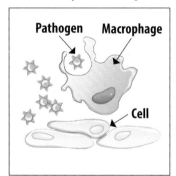

① A macrophage engulfs the invading pathogen.

② The macrophage sticks pieces of the pathogen called *antigens* to its surface.

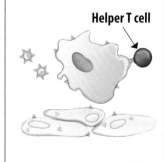

③ The antigens activate helper T cells.

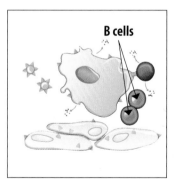

④ The helper T cells release chemicals that fight infection and help T cells and B cells multiply. The chemicals also trigger a response by killer T cells.

246

### Smallpox

Smallpox—one of the deadliest diseases known to mankind—is also the only disease that has been eradicated by the use of a vaccine. Evidence of smallpox lesions date back to Egyptian mummies and has been plaguing man ever since. Thanks to the efforts of Edward Jenner, smallpox is no longer a threat. In 1796, Jenner took pathogens from an infected person's skin and used it to inoculate an eight-year-old boy. The child became mildly ill several days later but then began to feel better. After a second inoculation, no disease developed. Jenner's work was the first scientific attempt to control an infectious disease with the use of a vaccine. Today, many vaccines exist that help prevent a variety of infectious diseases.

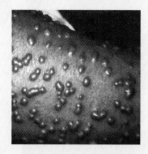

usually attaches to only one type of pathogen because each type of antibody was carefully shaped to match the pathogen like a puzzle piece. Because your body can make billions of types of antibodies, you can be protected from billions of types of pathogens. The antibodies attach themselves to many of the invading pathogens, marking them. Once the pathogens are marked by the attached antibodies, they are attacked by all kinds of immune cells and proteins. Certain proteins attach themselves to the antibodies and punch holes in the pathogens to destroy them. Macrophages then engulf the pathogens that are marked with the antibodies. Through this amazing and organized group effort, the body is freed of invaders.

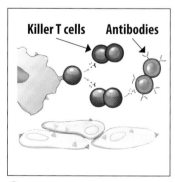

⑤ Killer T cells multiply to attack the pathogen, and B cells make antibodies.

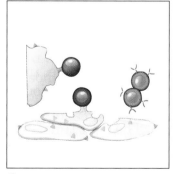

⑥ Killer T cells begin to destroy the pathogen-infected cells.

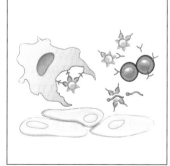

⑦ The antibodies attach to the pathogens, marking them for the other immune system cells to destroy.

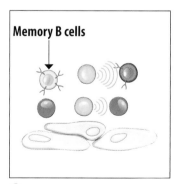

⑧ When the pathogens are gone, suppressor T cells turn off all the T cells and B cells. Memory cells remain to respond if the same pathogen invades again.

People often grumble about their bodies. Perhaps they do not like their hair color, their height, their weight, or some other physical feature. They fail to see the amazing ways their bodies are designed to function. In Psalm 139:13–14, the psalmist praises God for his body: "For You created my inmost being; You knit me together in my mother's womb. I praise You because I am fearfully and wonderfully made; Your works are wonderful, I know that full well."

You eat lunch without thinking of the complex system that helps you gain energy from the food and get rid of the parts that you do not need. When you get a cut or a bruise, you take it for granted that within a few days it will be gone without a trace. You may complain when you catch a cold, but a week or two later you have forgotten that you ever had it. You stretch your muscles and breathe and move without a second thought. But if you stop to consider all of the elaborate and creative systems that work together to keep your body going, you will recognize with the psalmist that you are fearfully and wonderfully made!

How long can you hold your breath? No matter how long you can hold your breath, sooner or later you have to take another one. In addition to needing nutrients, each cell in your body needs oxygen to continue the biochemical reactions that keep it alive. God created a remarkable respiratory system to get oxygen from the atmosphere and to deliver it to each cell in your body.

Respiration is the exchange of gases between living things and their environment. There are two types of respiration. One type is the process of breathing in oxygen and breathing out carbon dioxide, which happens in your lungs every time you take a breath. The other type of respiration, cellular respiration, is the process that occurs in each cell as it uses oxygen to break down biochemicals and release energy from them. Your respiratory system uses both types of respiration.

When you exercise, you inhale and exhale more often and your heart beats faster than when you are at rest. During exercise, you also need additional muscular energy, so your cells have to work faster to release the energy obtained from the food you eat. In order to release that energy, the cells need more oxygen. You breathe more often to take in more oxygen. Your heart pumps faster to deliver oxygen-rich blood to your cells.

The cells use oxygen in a series of chemical reactions, and they form carbon dioxide as a waste product. The cells get rid of the carbon dioxide by diffusing it into the nearby capillaries. The carbon dioxide then starts its trip back to the lungs through the cardiovascular system.

You can follow the path of air through the respiratory system. First, air moves into the nasal cavity through the nose. The interior surface of the nose is lined with cilia and cells that produce a thick mucus, which traps bits of dust and debris. The air is warmed and humidified in the nasal cavity to protect the rest of the respiratory system from being damaged by cold, dry air.

The air passes from the nose to the pharynx, a tubular passage that connects the mouth and nasal cavities with the

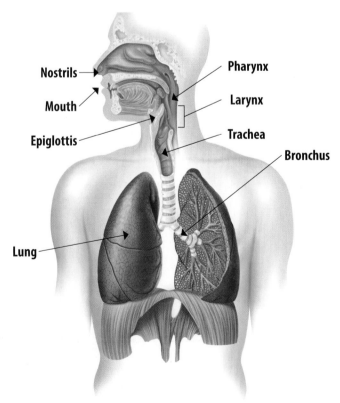

**Respiratory System**

Nostrils

Mouth

Epiglottis

Lung

Pharynx

Larynx

Trachea

Bronchus

A pair of human lungs contains about 700 million alveoli. The wall of each alveoli is one cell thick. Capillary walls are also one cell thick. What a perfect design for quick gas exchange!

esophagus. The epiglottis covers the opening over the larynx when you swallow, directing food and water into the esophagus. The larynx is a muscular structure in the throat that contains the vocal cords. Once the food and water has passed by, the epiglottis opens to allow air to travel into the larynx.

Next the air enters the **trachea**, which carries air to and from the lungs. The trachea is sometimes called *the windpipe*. It is a tube of smooth muscle reinforced with C-shaped pieces of cartilage that keep the air passage open. The base of the trachea lies between the lung's upper lobes, where it splits into two branches. Each of these branches enters a lung. The branches of the trachea that extend into the lungs are called **bronchi**. Bronchi branch several times into smaller and smaller passageways that ultimately lead to **alveoli**, which are clusters of microscopic air sacs deep in the lungs.

The alveoli are surrounded by tiny capillaries that form the intricate network of spongy lung tissue. The walls of the alveoli are very thin tissue, so oxygen molecules and carbon dioxide molecules can pass through quite easily. In the alveoli, the blood releases carbon dioxide and picks up oxygen to deliver to the cells. This exchange of gases all happens in the short time it takes to inhale and exhale.

### Asthma-Inflamed Bronchial Tube

Asthma causes a person's airways to become inflamed and swollen, making breathing difficult.

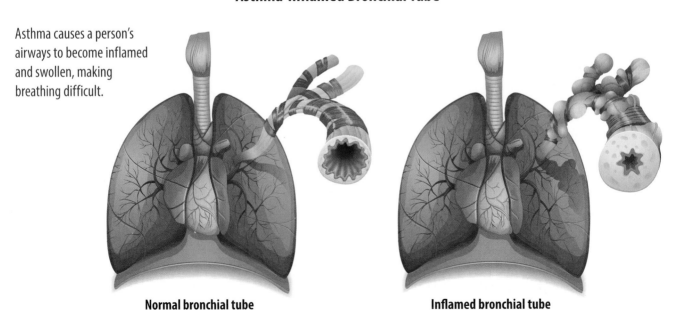

**Normal bronchial tube**          **Inflamed bronchial tube**

## Pneumonia

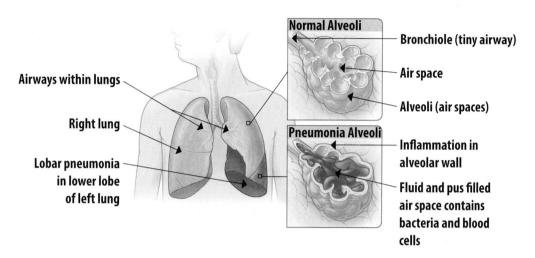

The lungs are extremely flexible. A large, flat muscle called the diaphragm lies underneath the lungs and helps move air into the lungs as it contracts. As the diaphragm contracts, it pulls the chest cavity down, and the lungs naturally fill with air. The muscles that lie beneath the ribs expand the rib cage outward, causing the lungs to fill. The procedure is reversed when you exhale.

Millions of people suffer from respiratory disorders or diseases. Asthma is a respiratory disorder that is often caused by allergies Asthma is characterized by sudden attacks of labored breathing. During an asthma attack, the smooth muscle that surrounds the bronchial tubes contracts, squeezing off the air supply to the alveoli. Bronchitis, the inflammation of the bronchi, is a condition caused by infection, smoking, or air pollution. Bronchitis causes the bronchi to produce more mucus

## FYI

### Voice Box

The larynx is sometimes called *the voice box* because the vocal cords are found here. The vocal cords vibrate when air passes over them. Muscles in the larynx can be used to change the pitch of the sounds they make. The longer the vocal cords, the lower the pitch. Boys' vocal cords lengthen significantly during puberty, which is why their voices become deeper.

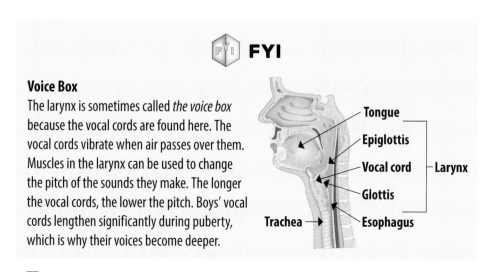

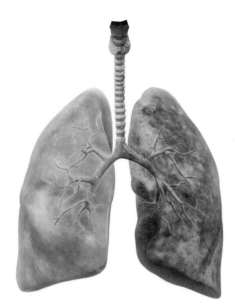

A healthy lung and a smoker's lung. Smoking injects tiny particles of soot into the lungs.

than usual so that gases cannot pass into and out of the lungs as easily. Even more serious is pneumonia, a disease of the lungs that results in fluid buildup in the alveoli. Pneumonia is caused by pathogens such as bacteria, viruses, and fungi. The fluid that builds up in the alveoli makes it difficult for gases to get through to the capillaries and for the exchange of oxygen and carbon dioxide to take place. Asthma, bronchitis, and pneumonia make it hard for a person to breathe.

Smoking injects tiny particles of soot into the lungs as well as hundreds of chemicals that God did not intend for lungs to handle on a regular basis. Smokers often end up with respiratory diseases such as emphysema and lung cancer. Emphysema is a lung disease in which the walls of the alveoli are destroyed so that less and less of the lung tissue can exchange oxygen and carbon dioxide. This disease develops from continual inflammation of the bronchi. People with emphysema cough a lot to get mucus out of the lungs. The continual coughing puts so much pressure on the alveoli that they burst. The lungs become less and less elastic and end up with less and less tissue that can perform the exchange of oxygen and carbon dioxide. Currently there is no cure for emphysema.

Lung cancer causes the most deaths from cancer worldwide. Most people with lung cancer are smokers. Because of the continuous blood flow to and from the lungs, stray lung cancer cells can easily travel to other parts of the body and begin to grow tumors in those locations also.

The most important factor that determines respiratory health is the quality of the air that is breathed. To ensure good respiratory health, people can choose not to smoke, avoid secondhand smoke, and avoid air pollution.

## LESSON REVIEW
1. Describe the two types of respiration.
2. Trace the path of air through the respiratory system and into the blood.
3. List five respiratory diseases or disorders and explain their causes.
4. What are some ways to keep your respiratory system healthy?

## TRY THIS

**Breathing with Emphysema**

Do you want to know what it is like to breath when you have emphysema? Try breathing through a small cocktail straw or a coffee stirrer. People with emphysema struggle like this with every breath they take!

254

Imagine what would happen if the local sanitation department shut down. Garbage would pile up, causing unhealthy living conditions. In the same way, the waste produced daily by your cells and tissues needs to be disposed of in a timely way to keep you healthy. How does the body get rid of the waste?

You already know that the blood carries carbon dioxide from the cells to the lungs and it is breathed out when you exhale. Carbon dioxide is just one chemical waste product that your body's cells produce. Another waste product that cells produce is urea, a toxic chemical produced in all cells as protein is metabolized. As urea is produced, it flows through the cell membrane and into the blood. Unlike carbon dioxide, urea cannot be breathed out through the lungs. The body also needs to keep electrolytes such as sodium, potassium, and chloride in proper balance. If levels of these electrolytes are too high, the body must get rid of the extra. God created the urinary system to deal with these waste products. The urinary system includes remarkable filtering organs called *the kidneys.* The renal arteries carry blood from the abdominal aorta to the kidneys, where the waste is filtered out. The purified blood then travels through blood vessels to the renal veins, which return the blood to circulation.

Your body's two kidneys are bean-shaped organs about the size of your fist. They are positioned at the back of the body cavity, on each side of the backbone just above the waist. The kidney is not

### OBJECTIVES

- Paraphrase how the body filters out waste products through the urinary system.
- Name the major parts of the urinary system and describe their functions.
- Describe three medical problems that can arise in the urinary system.

### VOCABULARY

- **nephron** a tiny filtering device in the kidney
- **ureter** the tube through which urine flows from the kidney to the bladder
- **urethra** the tube that carries urine from the bladder to the outside of the body

## Kidney Structure

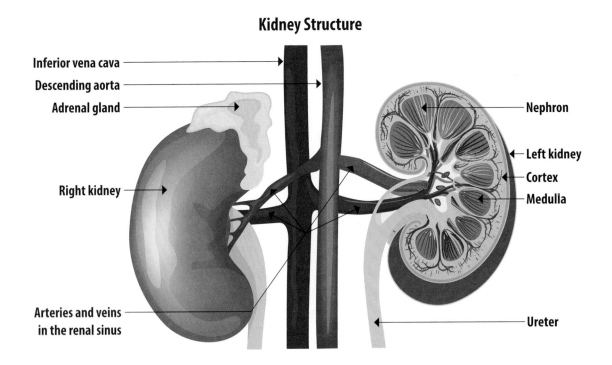

Inferior vena cava

Descending aorta

Adrenal gland

Right kidney

Arteries and veins
in the renal sinus

Nephron

Left kidney

Cortex

Medulla

Ureter

one big filter—it is made up of about a million tiny filters called **nephrons** that remove harmful substances from the blood.

Blood flows into the kidneys through the renal artery. Inside the kidney, smaller arteries branch off until the blood reaches the nephrons. The blood enters the nephron at the renal corpuscle. The nephron  has a filter, called *the glomerulus,* and a tubule. Fluid and waste products pass through the glomerulus, but it prevents large molecules and blood from passing. The filtered blood flows out of the kidney through the renal vein. The liquid that passes through the filtration membrane in the glomerulus is called *filtrate*. The filtrate flows through the  tubule and the loop of Henle. The body resorbs much of the liquid as the filtrate travels through the long, narrow loop of Henle and the tubules inside the nephron. The waste products remain in the filtrate, but some of the smaller molecules that are useful to the body are resorbed, along with most of the water. This process concentrates the waste products. The liquid comes out the other end of the nephron concentrated as urine, which contains dissolved waste materials.

Each day about 180 liters of filtrate pass from the blood through the kidneys, but only 1 to 1.5 liters of urine is actually produced. The urine empties into collecting ducts and travels from the kidney to the bladder through a tube called the **ureter**. The urine is stored in a sac called *the bladder* until it can be

## Nephron Structure

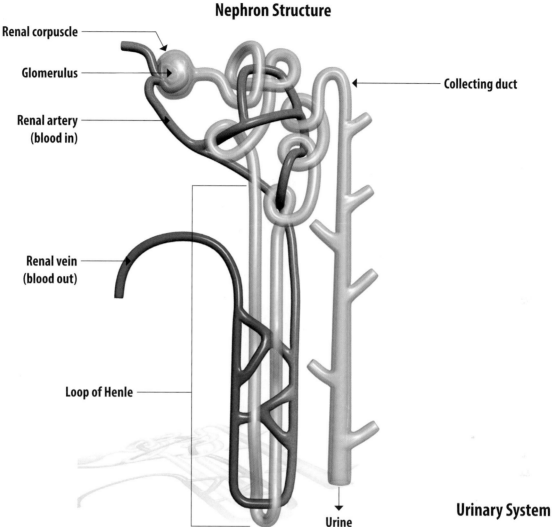

Renal corpuscle

Glomerulus

Renal artery
(blood in)

Renal vein
(blood out)

Loop of Henle

Collecting duct

Urine

## Urinary System

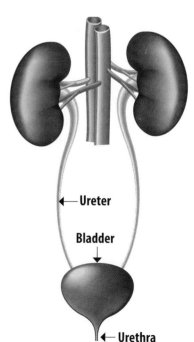

Ureter

Bladder

Urethra

excreted. The tube that carries urine from the bladder to outside of the body is the **urethra**. Two muscles—one where the bladder and the urethra meet and the other where the urethra exits the abdomen—control the release of urine.

Problems sometimes develop in the urinary system. Kidney stones are small (usually 2–3 mm), hard objects that develop in the kidney. The intense pain associated with kidney stones usually occurs when a stone moves into the ureter, irritating the lining of the ureter and causing bleeding, which shows up as blood in the urine. Not all kidney stones are composed of the same chemicals, so the cause of kidney stones is not clear. People who have very concentrated urine often develop more kidney stones than other people. One way to reduce the risk of kidney stones is to drink more water, which dilutes the chemical concentration in the urine.

**Nephrologist**
A medical doctor who specializes in kidney care and treating diseases of the kidneys is called *a nephrologist*. Nephrologists begin their studies in internal medicine and then pursue specialized training in treating patients with kidney diseases. Some of the diseases they treat are chronic kidney disease, polycystic kidney disease, acute renal failure, and kidney stones. Nephrologists are also educated on all aspects of kidney transplantation and dialysis.

Nephritis is an inflammation of the kidney. The various types of nephritis affect the filtration membrane or the ability of the kidney to concentrate urine. Certain types of nephritis can progress to the point that the kidney cannot function.

People can survive with only one-third of their nephrons, but some conditions cause the kidneys to fail completely, which is called *kidney failure*. If the nephrons do not filter out waste products, toxic levels of waste build up. This buildup of wastes can be fatal. Since the 1940s, artificial kidney machines have been used to filter the blood of people who have kidney failure. This process is called *dialysis*. People on dialysis spend about four hours a day three days a week having their blood filtered. In this way a person whose kidneys do not function can be kept alive until a donated kidney is available for transplant.

## LESSON REVIEW
1. Name two waste products the body excretes through the urinary system.
2. Name the major parts of the urinary system and describe their functions.
3. Explain how the body filters out waste products.
4. Describe three medical conditions that can affect the urinary system.

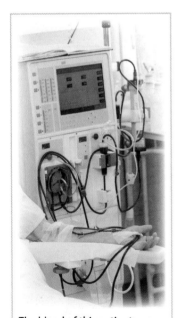

The blood of this patient is being filtered through a dialysis machine.

People think of skin as the organ that protects their internal tissues. That is true, but the skin also plays other roles. Your skin protects inner tissues, helps regulate body temperature, produces vitamin D, and senses pain and touch. Skin also helps the body get rid of waste products, although in smaller amounts than what the kidneys filter out.

Skin has two main layers: the epidermis and the dermis. The **epidermis** is the thin outer layer of the skin. The epidermis, which has no blood cells, has five distinct layers. New epidermal cells are produced at the deepest of the layers. Over time these cells fill with a protein called *keratin* and move to the surface. Hair and fingernails are made of a harder form of keratin. The top layer of the epidermis is composed of dead skin that is 16 or more cells thick. The layer of dead cells forms a barrier for the body that resists abrasion. The dead cells are continually shed, and younger cells then move to become the outer layer.

The pigment that determines skin and hair color is called **melanin**. This pigment is found in a single layer of epidermal cells. All people have about the same number of epidermal cells that are capable of producing melanin, but genetics determines the amount of melanin that they produce. When the skin is

## OBJECTIVES

- Name and describe the layers of skin.
- Summarize how acne develops.

## VOCABULARY

- **dermis** the thick layer of living skin below the epidermis
- **epidermis** the thin outer layer of skin
- **melanin** the pigment that determines skin and hair color
- **sebaceous gland** a gland in the dermis that produces sebum
- **sweat gland** a gland in the dermis that produces sweat

### Human Skin

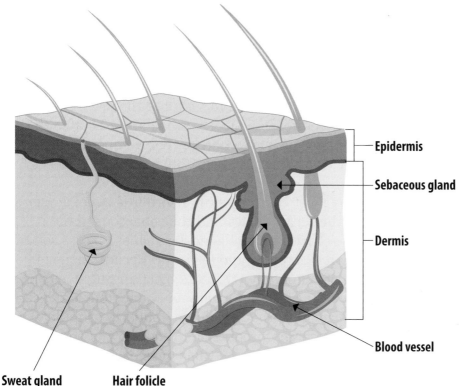

Epidermis

Sebaceous gland

Dermis

Blood vessel

Sweat gland          Hair folicle

Skin has pigment-producing cells called *melanocytes*, which produce melanin. When melanocytes become damaged by the sun, freckles appear. Freckles are abnormal areas of melanin pigment in the skin. People with fair complexions are more likely to get freckles.

Sports drinks can replace the electrolytes lost through sweating.

exposed to ultraviolet (UV) light, the melanin already present in the skin darkens, and the cells are stimulated to make more. This process is what causes skin to tan.

The **dermis** is a thicker layer of skin found below the epidermis. The dermis contains blood vessels, nerves, hair follicles, sweat glands, and sebaceous glands. **Sebaceous glands** are glands in the dermis that produce sebum and open into a hair follicle. The sebum—a white, oily substance consisting mainly of fat—is released into the hair follicle to protect the hair and skin from drying out. **Sweat glands** are glands in the dermis that produce sweat. Sweat contains small amounts of ammonia, urea, and electrolytes. Some sweat glands release sweat directly to the skin's surface through pores. Others release the sweat directly into the hair follicles. The body tries to resorb the electrolytes before the sweat is released. In very hot weather, a person sweats a lot—from 1.5 to 4 L per day. This increased sweating can throw the body's electrolytes off balance. Some sport drinks include salts to replenish the electrolyte salts lost through sweating.

Sweat is more than a waste-removal device; it plays an important role in keeping the body from overheating. In hot weather, the only way the body can get rid of extra heat is through evaporation, a process that uses heat. Sweat provides the body with its own liquid for evaporation and cooling purposes.

As you get older, your skin becomes thinner and less elastic. The sebaceous glands and sweat glands do not produce as much sebum and sweat, so the skin becomes drier and less able to regulate the body's heat. For this reason, elderly people are less able than younger people to handle high temperatures. They are at an increased risk during heat waves.

The most common skin disorder is acne, a disorder of the hair follicles and the sebaceous glands. Four factors affect acne: hormones, excess sebum, dead skin cells, and bacteria. Because acne is most commonly triggered by changing hormone levels, acne is very common during adolescence. Hormones produced during puberty stimulate extra sebum production. About 85% of young people have acne on their face, back, or shoulders. Acne begins when extra sebum mixes with dead

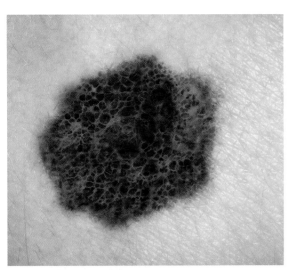

Moles are common and are usually harmless. Some moles can become cancerous. If you have a mole that changes in size, color, or shape, see a dermatologist.

**Acne Formation**

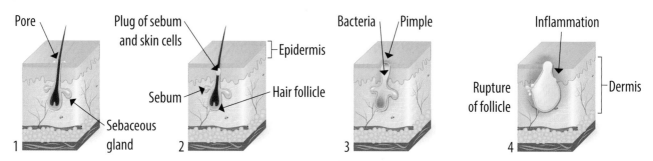

1 — Pore, Plug of sebum and skin cells, Sebum, Sebaceous gland, Epidermis, Hair follicle

2 — Bacteria, Pimple

3 — Inflammation, Rupture of follicle, Dermis

Acne is most common in teens, but it can affect people at any age. Fortunately, it is treatable by a dermatologist. (1) The sebaceous gland produces sebum, an oil that waterproofs the hair. (2) A blockage may prevent the oil from flowing freely out of the hair follicle. (3) Oil builds up in the shaft. Bacteria breed in the oil. (4) The hair shaft swells with pus, forming a pimple, or whitehead.

skin cells and clogs the skin's pores. Bacteria can grow in the mixture as well. Pus, a mixture of white blood cells and cell debris, collects in the follicle. The buildup of this mixture leads to inflammation—a pimple.

Environmental factors also can affect acne. Some drugs aggravate acne, and oily cosmetics can make skin follicle cells stick together, making acne worse. Pollution, high humidity, and stress can aggravate acne too. Friction from a backpack or a bike helmet can also contribute to acne. Rubbing, squeezing, or picking at blemishes can also aggravate the condition. Little truth has been found in the common beliefs that greasy foods and chocolate cause acne or that acne is caused by dirt. Serious cases of acne require treatment by a dermatologist—a physician who specializes in skin disorders.

Eczema and dandruff are other types of inflammation of the skin. Other skin disorders include psoriasis, which is caused by normal skin cells that grow very quickly, and skin cancer, which is the uncontrolled growth of abnormal skin cells. Most often skin cancer seems to be caused by exposure to the sun, but it can occur on areas of skin that are not exposed to sunlight. People with fair skin, whose skin produces less melanin, are at the highest risk for skin cancer.

## LESSON REVIEW

**1.** Describe the features of the epidermis.
**2.** Describe the features of the dermis.
**3.** Explain how acne develops. What is the most common cause of acne?

**FYI**

**Brrr!**
On cold days, the blood vessels of the nose and ears dilate, bringing more blood to those areas to prevent tissue damage from the cold. That is why ears and noses get red in cold weather.

The skeletal system contains 206 bones that support and protect your body. The muscular system is made of muscles that contract and relax in order to create movements, which range from running to talking to digestion to heartbeats. Together the skeletal and muscular systems give you shape and the freedom to move around. The nervous system sends signals to the leg muscles, heart muscles, lung muscles, and all the other muscles to move so that you can respond to your environment. In this chapter you will learn about these systems in more detail and appreciate how they work together to sustain your body.

All of the body systems would be a lumpy pile on the floor without bones and muscles to give shape and structure. The skeletal system protects the inner organs and tissues and allows the body to move around. God designed bones, which are extremely strong for their mass, to support the entire body. Bones are also a mineral storehouse for calcium and phosphorus, the body's two most abundant minerals. Blood cells are also produced by the bones.

Bones can be classified according to their shapes: long, short, flat, and irregular. Long bones are found in arms and legs. Short bones are found in wrists and ankles. The skull and ribs include flat bones. Irregular bones make up the backbone and certain bones of the face.

All mature bones, no matter what their shape, are about 70% inorganic matter and 30% organic matter. The minerals calcium and phosphorus are the inorganic matter, and most of the organic matter is a protein called *collagen*. If you think of bone structure like the reinforced concrete that is used to build skyscrapers, the collagen acts like the steel reinforcing rods, giving flexible strength, and the minerals act like the concrete, giving rigid strength.

You may think of bones as dry and dead. But bones in a living body are connected to nerves and blood vessels that bring them nourishment. Deep in the center of each bone is a place that is alive and active—the marrow. **Marrow** is the porous inner core of bones where blood cells are produced. Young children's bones all contain red

## Bones

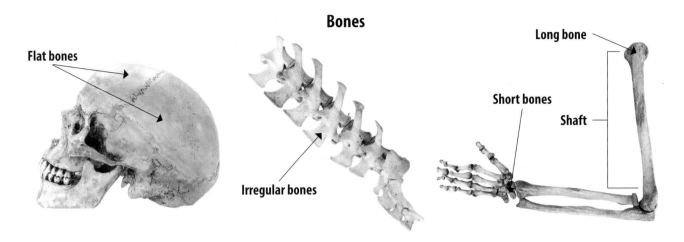

Flat bones

Irregular bones

Long bone

Short bones

Shaft

bone marrow, which produces red blood cells. Eventually much of the marrow of the arm and leg bones is replaced with yellow marrow, which is a supply of fat. Adults retain some red bone marrow. The body needs to continually make new red blood cells because red blood cells only live about 110 days. About 2.5 million of the body's 25 trillion red blood cells are destroyed every second, usually because they are defective or worn out. That is a lot of blood cells to replace each second!

Surrounding the bone marrow lies a layer called *spongy bone*. The spongy bone is not soft, but it has many spaces that are filled with bone marrow. The compact bone, the next layer, is the thickest layer. It is made of collagen and minerals. The outer surface of bone, which is called *the periosteum*, is a thin, dense layer of collagen through which the blood vessels and nerves run.

When babies are developing in the womb, most of their bones begin as cartilage. As a child grows, calcium is deposited onto the cartilage base, and bone forms. Bones can grow in two ways: thicker and longer. Bones become thicker as more and more minerals are deposited on the surface of existing bones. Your bones can continue to grow thicker throughout your life. The bones of children have growth plates made of cartilage. The cartilage growth plates gradually become bone as calcium and other minerals are deposited on them. Cartilage growth stops when you are in your twenties. Because the cartilage stops growing, your bones cannot get any longer.

Your skeleton has two basic sections. The axial skeleton is made of the bones that keep the body upright, including the cranium, the vertebrae, the rib cage, the sternum, and the pelvis. The appendicular skeleton includes the bones of the

### Inside a Bone

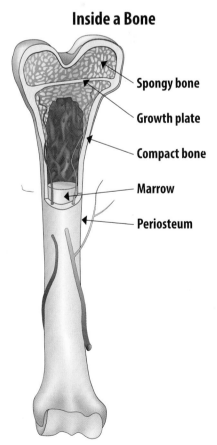

Spongy bone

Growth plate

Compact bone

Marrow

Periosteum

appendages (arms, hands, legs, and feet) and the bones where these attach to the axial skeleton (such as the shoulder bones and the pelvis).

Each of the 206 bones in your body has a purpose. Some of the larger bones protect sensitive parts of the body. The **cranium**, or skull, has several separate bones, and it protects the brain and eyes. The vertebrae are a series of bones stacked on top of one another. The vertebrae protect the spinal cord, which runs from the brain down through holes in the middle of the vertebrae. The rib cage, which protects the heart and lungs, includes 24 ribs. These ribs are positioned in pairs, most of which are attached to the sternum, or breastbone. The ribs attached to the sternum are called *true ribs*. The others are called *false ribs* and are attached to the sternum by a longer piece of cartilage. The bottom four ribs are the floating ribs. The **pelvis** consists of the two hip bones (each called *coxa*) and the sacrum, which is the lower part of the backbone. The pelvis supports the body's weight and protects the internal organs.

## Two Basic Sections of the Skeleton

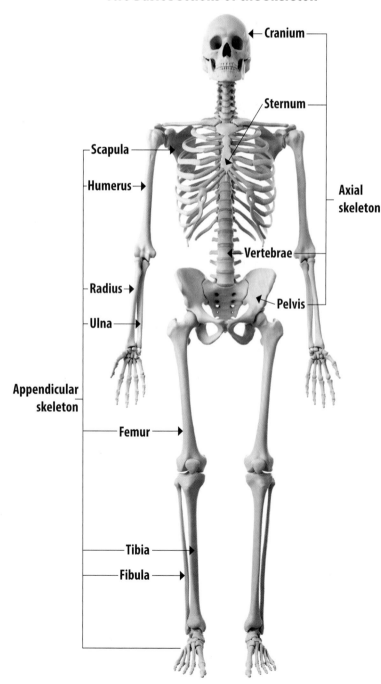

The longest bone in the body is the femur, or thigh bone. The lower leg has two bones side by side, the tibia and fibula. Similar to the legs, the arms have three bones as well. The longer upper arm bone is called *the humerus*. The radius and the ulna are the bones of the lower arm. More than half of the bones in your body are in your hands and feet! Both the hands and the feet are made up of many smaller bones, which gives them a large range of movement. Think about all the things your hands and feet can do. Brushing your hair or kicking a ball would be very difficult if there were fewer bones in your hands and feet.

The places where two or more bones connect are called *joints*. Joints are held together by connective tissue called **ligaments**. The connective tissues that attach muscles to bones are **tendons**. Both tendons and ligaments are composed mostly of proteins.

Joints are classified into three groups depending on the type of movement that they allow. The skull bones are connected by immovable joints, which are made of dense tissue. Partially movable joints are held together either by cartilage, which is slightly flexible, or by ligaments. For example, the two bones of the forearm (the ulna and the radius, which run from the elbow to the wrist) are held together by ligaments. They can move only slightly in relation to each other.

The most common—and the most complex—joints are the movable joints, or synovial joints. Because bones rubbing together in a movable joint would cause friction and pain, God designed the bones in these joints with protective cartilage cushions at the ends. Movable joints are enclosed in joint

**Moveable Joints**

Pivot

Ball and socket

Hinge

Gliding

**BIBLE CONNECTION**

"My frame was not hidden from You when I was made in the secret place, when I was woven together in the depths of the earth. Your eyes saw my unformed body" (Psalm 139:15–16a). Indeed the Lord knows your frame. The incredible strength of your bones, considering their relative light mass, is an engineering marvel.

The story of Ezekiel in the valley of the dry bones gives you a picture of the skeleton as the framework of the body on which the other body systems are built. That story is also a wonderful reminder that it is only as the Lord breathes life into people that their bodies actually live. The fantastic biological machine that is the body has no life of its own without the breath of life from God.

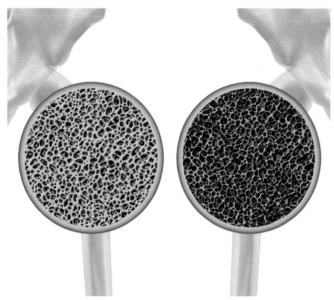

Normal bone density is on the left. Osteoporosis is on the right.

capsules made of fibrous tissue (and, in some cases, ligaments). Inside each joint capsule is a very slippery lubricating fluid called *synovial fluid*.

There are four basic types of movable joints. Each type provides for the particular movement at that joint. The shoulder and hip joints are examples of ball and socket joints, which can move in almost any direction. In the leg, the end of the femur, which is shaped like a ball, fits into a semicircular socket in the pelvis. Knee and elbow joints are hinge joints, which allow movement in only one direction. The end of the long bone is shaped like a cylinder and fits into a curved cavity at the end of the opposing bone. Pivot joints, such as the one in the neck, permit a rotating movement. Gliding joints are found between the flat surfaces of bones, such as the vertebrae of the backbone or the bones of the wrists or ankles.

Because of their high mineral content, bones are brittle and can break. Fortunately, God provided a repair mechanism. When a bone is broken, the blood vessels that feed it are injured, and a blood clot forms at the point of the break. Cells from surrounding tissues begin to invade the clot. Some cells produce a fibrous network between the broken ends of bone. Other cells produce cartilage, which becomes part of this network. Bone cells gradually add mineral deposits to calcify this network, forming new bone. This process takes four to six weeks and requires that the forming network is not disturbed, which is why most broken bones require a cast.

As people age, hormone levels change, which affects bones. Osteoporosis is a common disorder in which bones become more porous and brittle. That is why older people often break bones when they fall, even though a younger person could take the same fall without being hurt. An important way to prevent osteoporosis is to develop maximum bone mass early in life. About 45% of a person's bone mass is developed during adolescence, so the preteen years are an especially important time to provide the body with nutrients and exercise that strengthen bones.

Healthy bone growth and maintenance requires a good supply of certain nutrients. Vitamin C is essential for collagen production. Vitamin D is necessary for the transportation of calcium through the intestinal wall. Calcium is essential for bone growth and repair. The body sometimes borrows calcium from bones for other needs, such as muscle contractions. That means that even after your bones are fully formed you need calcium in your diet. Dairy products are particularly high in calcium, and many are fortified with vitamin D.

Exercise also plays an important role in bone health. Inactivity leads to loss of bone mass, which weakens bones. That is one of the reasons why patients in hospitals are encouraged to get up and walk around, even after surgery. God designed your body for activity and exercise, so it is not surprising that the benefits of activity and exercise even extend to the bones. Some activities, of course, lead to higher risks of falls, so using protective gear can help you take good care of your bones.

## TRY THIS

**Collagen**
Bake a fresh bone at 175°C for one hour. This will break down the collagen, which is the organic material, leaving only the minerals. Take the bone to an open area. Put on safety goggles and strike the bone with a hammer. Record your observations. Why is collagen important?

## LESSON REVIEW

**1.** List three of the functions of the skeletal system.
**2.** Name the four layers of bones.
**3.** Explain how bones grow.
**4.** Name four types of movable joints and give an example of each.
**5.** Name six major bones of the human skeleton.

## OBJECTIVES

- Identify and give examples of the three types of muscles.
- Describe how a muscle is structured.
- Explain how exercise affects muscles.

## VOCABULARY

- **striated muscle** a type of muscle marked by light and dark stripes or bands

## FYI

**Muscle Proteins**

Muscle cells contain two important kinds of proteins: actin and myosin. The actin and myosin filaments lie parallel to each other. When nerves send muscle fibers a message to contract, calcium ions and energy are released within the fiber. This action causes the actin and myosin filaments to slide in between each other, and the myosin filaments use the released energy and calcium to form bridges to hold onto the actin filaments. All of this happens in a fraction of a second!

Muscles are like miniature motors that run on nutrient fuel. Muscles take in nutrients and release heat and energy. The heat keeps the body warm, and the muscles convert the energy into movements. Many types of muscles make it possible for people to walk, talk, swallow, and even breathe.

Muscles are found in all the body systems. A network of nerves connects muscles to the brain. Nerve messages tell muscles when to relax and when to contract. The endocrine system contains glands that affect muscle activity and growth. The digestive system contains muscles for swallowing and digesting.

Your body has three distinct types of muscle tissue—skeletal muscle, smooth muscle, and cardiac muscle. About 40% of body weight is skeletal muscle. These muscles are attached to the bones throughout the body by tendons, contracting and relaxing like giant elastic bands to allow skeletal movement. Skeletal muscles are **striated muscles**, marked by stripes or bands. Your body has over 500 skeletal muscles, which come in a wide range of sizes. A thigh muscle can grow to over 60 cm long, whereas one tiny muscle in the inner ear is under 1 cm long. Skeletal muscles control breathing, posture, facial expressions, and locomotion (the ability to move around). Skeletal muscles are voluntary muscles; that is, you can control their use.

Smooth muscle, which is not striated, reacts slowly and tires slowly. Smooth muscle is found in glands, skin, eyes, blood vessels, and along the walls of hollow organs in the digestive tract, reproductive system, and urinary system. Some smooth muscle, such as that of the digestive tract, contracts spontaneously,

## Muscle Types

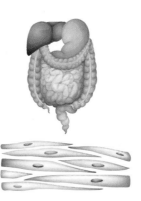
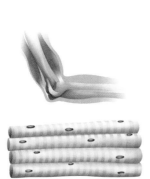
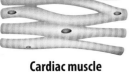

**Cardiac muscle**

**Skeletal muscle**

**Smooth muscle**

whereas other smooth muscle, such as that of the bladder, contracts only when nerves send messages. Smooth muscle is an involuntary muscle because hormones and nerves automatically cause it to contract and relax—you do not have to think about it.

Cardiac muscle is found only in the heart. Its regular and rhythmic contractions keep blood flowing through the body. Although cardiac muscle cells are striated like skeletal muscles, their contractions are involuntary. Because cardiac muscle needs to work constantly, cardiac muscle cells have many mitochondria, which produce energy for the cells. They also have an excellent blood vessel supply. Cardiac muscle cells metabolize very efficiently. The heart muscles rest only between beats and never get tired as skeletal muscles do.

## Skeletal Muscle

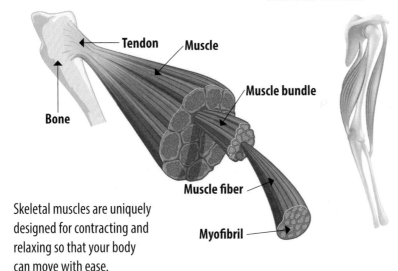

**Tendon** **Muscle**

**Muscle bundle**

**Bone**

**Muscle fiber**

**Myofibril**

Skeletal muscles are uniquely designed for contracting and relaxing so that your body can move with ease.

Skeletal muscles work in pairs. When your arm bends, the biceps muscle contracts and the triceps muscle relaxes. When the arm is extended, the triceps muscle contracts and the biceps muscle relaxes.

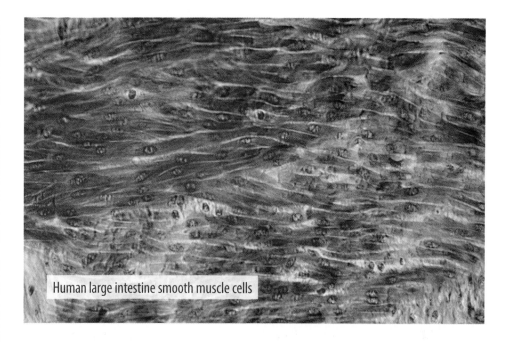

Human large intestine smooth muscle cells

Muscles are unique tissues because they can contract, or shorten. Muscle cells are long and cylinder shaped, which is why they are also called *muscle fibers*. Muscle cells are composed of slender threads called *myofibrils*. Each cell has many nuclei along its length rather than a single nucleus. Muscle cells also have many mitochondria. The mitochondria act as the power plants of the cell. They metabolize nutrients when a muscle needs energy to contract.

Muscle cells or fibers are organized into muscle bundles. Many bundles make up a skeletal muscle. The muscle fibers within a bundle take turns contracting and relaxing as long as the muscle contraction is sustained. Muscle fatigue is related to the energy supply available in the cells. Most skeletal muscles have a mixture of "fast-twitch" and "slow-twitch" fibers in their bundles. Fast-twitch fibers react more quickly to nerve messages. They tend to have fewer mitochondria and a smaller blood supply than slow-twitch fibers, so they fatigue relatively quickly. Because slow-twitch fibers are rich in mitochondria and in blood supply, they can resist fatigue, but fast-twitch fibers respond faster than slow-twitch fibers. Upper body skeletal muscles have more fast-twitch muscle fibers. The large muscles that hold the body upright have more slow-twitch fibers and tire less quickly.

Whether you are an athlete or not, your body is created for movement and exercise. Moderate, regular exercise helps keep your body healthy. Exercise benefits all three types of muscle. Although the number of skeletal muscle cells in the

**FYI**

**Muscle Problems**
Muscles shrink or waste away when they are not used. This decrease in size or wasting away of body tissue is called *atrophy*. In extreme situations, such as when a person is bedridden for a long time, muscle cells are replaced by connective tissue, and the muscle cells are permanently lost.

Cramps are painful muscle contractions triggered by an irritation within the muscle. They can occur in uterine muscle during a woman's monthly menstrual cycle or in skeletal muscle as a muscle seizes up in a condition sometimes called *a charley horse*.

Other muscular problems are more unusual. The disease muscular dystrophy, which is usually inherited, destroys skeletal muscle tissue. Fibromyalgia is chronic widespread pain in muscles or at the point where muscles join tendons.

**TRY THIS**

**Smile**
Place your hands lightly over your face as you make funny faces using some of the skeletal muscles that lie under the skin. Feel how many different ways the muscles of your forehead, eyebrows, cheeks, lips, and chin can move. Can you isolate the muscles? Does moving certain muscles seem to involve neighboring muscles? Smile, and then frown. Which seems to require more muscles?

body never changes, exercise can change the size of muscle cells. Exercise also causes fast-twitch muscle fibers to develop more mitochondria, which provide more energy to help them resist fatigue. Exercised muscle grows stronger, not only from the larger size of the muscle fiber, but also because exercising improves muscle metabolism, circulation, and respiration. Exercise also causes a loss of the fat that sometimes hinders muscle movement. It is important to remember that there are many kinds of exercise. Try something new!

A well-balanced diet is also important for healthy muscles. Growing muscle cells need amino acids to be able to process protein and carbohydrates as fuel for the mitochondria. Muscle cells also need calcium for muscle contraction. If the body's supply of calcium is low, it will pull calcium from the bones to keep the muscles working. That, of course, puts the bones at risk of weakening. So it is important to take in calcium through foods.

## LESSON REVIEW

**1.** Name three different types of muscle tissue and where they are found in the body.
**2.** Describe the structure of a muscle.
**3.** How does exercise affect muscles?
**4.** Describe the structure of a muscle.

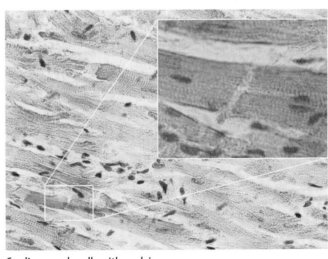
Cardiac muscle cells with nuclei

## OBJECTIVES

• Describe the three main parts of a neuron and explain their functions.
• Differentiate between the central and peripheral nervous systems.
• Identify the three major parts of the brain and discuss their functions.
• Name some diseases of the nervous system.

## VOCABULARY

• **axon** a long extension of a neuron that transmits messages
• **central nervous system** the brain and spinal cord
• **dendrite** the short-branched extension of a neuron that receives messages from other cells
• **peripheral nervous system** the nerves and sensory receptors
• **synapse** a gap between the axon of one neuron and the dendrites of another neuron

Imagine on a chilly autumn day you walk into your house and rub your cold hands together as you smell cookies that have just come out of the oven. You pick up a cookie and enjoy its warmth; then you take a bite and savor its flavor. Your little brother tries to grab your cookie, so you dance around, knocking your elbow on the "funny bone" and then holding it in pain. In the process you drop your cookie, which the dog gobbles up as you watch in irritation. All of these sensations—cold and warmth, smells and flavors, pain, and even emotional reactions—are monitored continually by your nervous system.

The nervous system is actually two systems in one. The **central nervous system** includes the brain and spinal cord. The **peripheral nervous system** consists of nerves and sensory receptors. The peripheral nervous system continually sends information (stimuli) to the brain about body temperature, blood pressure, pain, touch, smell, sight, sound, and more. The brain and spinal cord make decisions (responses) depending on what is going on inside and outside the body. They may decide to ignore the stimuli or to send signals to various body systems to act. Some of the stimuli and responses are voluntary. Voluntary responses are those that your thoughts can help control, such as deciding to move your arm to pick up the cookie. Other stimuli and responses are involuntary. Involuntary responses are those not affected by your thoughts—for example, you do not make decisions about which capillaries to send more blood to or how fast your heart should beat.

The nervous system includes two types of cells: neuroglia (which means "nerve glue") and neurons (cells that send messages throughout the body in the form of electrical energy). Neuroglia support and protect neurons. Neurons send and receive messages from one neuron to the next or from a neuron to an organ muscle.

A neuron has three main parts: the cell body, the axon, and the dendrites. The cell body contains the nucleus and many mitochondria. The cell body is the control center for the neuron. The **axon** is a long extension of a neuron that transmits messages. The messages are usually transmitted away from the neuron. Axons end in short branches tipped with chemicals. The endings are club-shaped endings and are called *terminals*. Axons can be very long—up to 1 m in length. **Dendrites** are the short, branched extensions of a neuron that receive messages from other cells.

The gap between the axon of one neuron and the dendrites of another is called the **synapse**. Neighboring neurons do not actually touch each other. Messages are sent along the axon of a neuron by an electrical impulse. At the end of the axon, the impulse triggers the release of chemicals known as *neurotransmitters*, which travel across the synapse to a neighboring dendrite. These chemicals trigger an electrical impulse in the dendrite. The impulse runs through that neuron and out through the axon. This process is how messages are sent along the nervous system—nerves function as one-way streets as they deliver messages.

Just as electrical cords on appliances have a layer of insulation around the wire, the axons of some neurons are covered with a myelin sheath, a protective layer that insulates the axon. The myelin sheath is made of neuroglia. The high fat content gives the myelin sheath a white color. Myelin sheaths are necessary for the nervous system to work properly. Diseases such as multiple sclerosis and diabetes gradually damage or destroy the myelin sheath. The electric impulses move much more slowly along the nerves without a myelin sheath, leading to poor control of skeletal and smooth muscle.

**Nervous System**

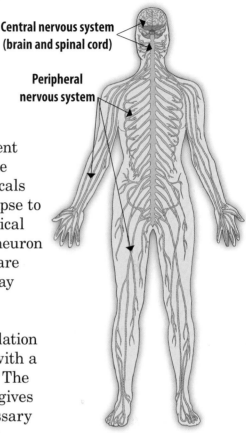

Central nervous system (brain and spinal cord)

Peripheral nervous system

## Anatomy of a Typical Neuron

Nucleus

Cell body

Axon

Dendrites

Myelin sheath

Terminals

### TRY THIS

**Test Your Reflexes**
Work with a partner to test how quick the reflexes in your hand are. Have your partner hold an index card. Hold your thumb and forefinger open around the card. Have your partner release the card without warning. Try to catch the card by pinching your thumb and finger together. Could you catch the card? Reverse roles.

There are several types of neurons. Sensory neurons send information from the sense organs (eyes, ears, nose, mouth, and skin) to the brain. Motor neurons carry electrical impulses from the brain or spinal cord back to muscles or glands. Association neurons (many of which are found in the spinal cord) function in reflexes, which transfer electrical impulses from a sensory neuron directly to a motor neuron.

The central nervous system runs from the brain down through the backbone. This system functions like a master computer that continually monitors conditions to keep the body operating well. The spinal cord is a critical communications link between the brain and the peripheral nervous system. The brain itself consists of three main parts: the brain stem, the cerebrum, and the cerebellum. Each completes its own special tasks while staying in communication with all other sections of the brain. The brain stem—which is at the base of the brain and connects with the spinal cord—is the portion of the brain that controls reflexes and involuntary mechanisms such as respiration, heartbeat, and swallowing. It also controls activities such as the sleep/wake cycle and some processing of sight and sound. Pathways that route messages to and from the brain run through the brain stem.

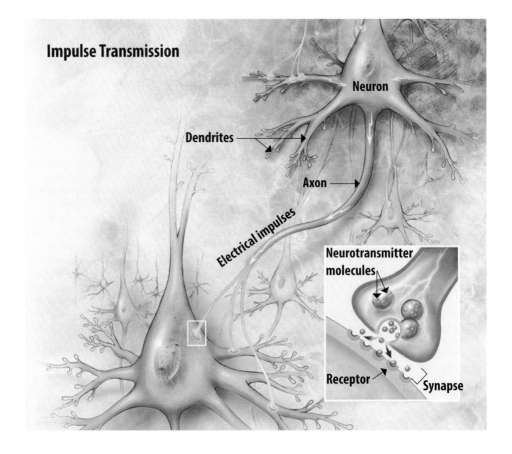

**Impulse Transmission**

Neuron

Dendrites

Axon

Electrical impulses

Neurotransmitter molecules

Receptor

Synapse

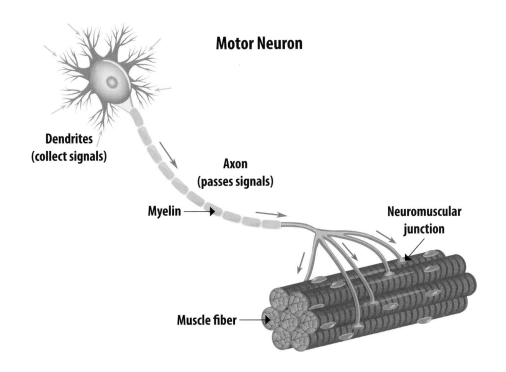

**Motor Neuron**

Dendrites
(collect signals)

Axon
(passes signals)

Myelin →

Neuromuscular
junction

Muscle fiber →

**FYI**

**Protect Your Nervous System**

Here are some things you can do to reduce the likelihood of brain or spinal cord injuries.

• Wear a lap and shoulder seat belt when you ride in a car.

• Wear a helmet when you ride a bike.

• Wear safety equipment properly when you play sports.

• Before you dive, check to make sure that the water is deep and that no large objects such as rocks are hidden beneath the water.

The thalamus and hypothalamus are glands that lie just above the brain stem. The thalamus influences mood and movement and is a major center for moving sensory messages along. The hypothalamus controls the function of the endocrine system, which regulates hormones. It also keeps the body in balance in terms of blood pressure, body temperature, hunger, and movement of food through the digestive system.

The part of the brain that keeps track of the position of the body is called *the cerebellum*, which means "little brain." The cerebellum lies at the back of the brain, just behind the brain stem. It is involved in balance, maintaining muscle tone, and coordination of muscle movement.

The cerebrum is the part of the brain that controls all thought and other voluntary acts and detects sounds, smells, sight, taste, pain, heat, and cold. The largest part of the brain, the cerebrum fills the top half of the skull. The wrinkled outer surface of the cerebrum is called *the cerebral cortex*. Different areas of the cerebrum have specific functions. The front is important for motivation, aggression, sense of smell, mood, and voluntary muscle movement. The center top section receives sensory information and evaluates it. Another area at the side of the cerebrum receives and evaluates smell and sound messages and plays an important role in memory, abstract thought, and judgment. A section at the back receives and evaluates visual messages. Two areas deep in the middle are important for speech. Still another area is involved in planning and organizing a series of muscle actions. Four areas in the cerebrum are crucial to association. For example, when you see and hear a bird, a message is sent to your cerebrum. The association areas compare the new message with past experience to determine what to do with the information.

A deep fold in the surface layer divides the cerebrum into left and right hemispheres. The left hemisphere receives sensory information from the right side of the body and controls the muscles on the right side of the body. The right hemisphere does the same for the left side of the body. Some learned activities (such as language, speech, and mathematics) involve the left hemisphere more than the right. Musical ability, spatial perception, and facial recognition involve the right hemisphere more than the left. However, the two hemispheres constantly share information through a variety of connections. God designed the cerebrum to give humans the abilities to speak, learn, judge, think, remember, create, and understand His love and purpose for people.

How do the two divisions of the nervous system work together? The central nervous system is like the computer headquarters for the body, whereas the peripheral nervous system provides the information for the central nervous system to process. The peripheral nervous system transfers messages from the body and the outside environment to the central nervous system and from there to the rest of the body. The nerves are linked together in a network that reaches all the muscles and organs of the body so they can send messages to and receive messages from the brain. Twelve pairs of peripheral nervous system nerves attach directly to the brain itself, and 31 pairs of peripheral nervous system nerves attach to the spinal

**Brain Anatomy**

Cerebrum

Thalamus
Hypothalamus

Cerebellum

Brain stem

column. For these sensory nerve endings to activate a response, there must be a stimulus. When something inside or outside the body stimulates an electrical impulse in a nerve ending, the impulse is conducted along nerves until it reaches the central nervous system, which translates it into meaningful information.

The peripheral nervous system is divided into a sensory division and a motor division. The sensory division is like a one-way street that relays messages to the brain. The motor division is like a one-way street that relays messages from the brain to the body. The motor division is further divided into pathways for two types of messages. Messages that travel along the somatic nervous system control mainly voluntary muscles, such as skeletal muscles. Messages that travel along the autonomic nervous system control involuntary activities such as heart rate, breathing, and digestion. You do not consciously send autonomic nerve messages.

Injuries to the spinal cord can cause paralysis.

Just as with the other body systems, injuries can affect the nervous system. If an axon is severed from the cell body, the frayed edges and the axon begin to degenerate. The myelin sheath eventually will rebuild itself, so if the loose axon is lined up opposite the new growing myelin sheath, the axon could reconnect and begin to function again. But damaged axons do not always reconnect. Injuries to the fragile spinal cord can have serious consequences including paralysis or even death.

## BIOGRAPHY

### Wilder Penfield

Canadian Wilder Penfield (1891–1976) was a devoted surgeon who mapped the brain. As a student, Penfield wanted to study medicine to make the world a better place, and he later became Montreal's first neurosurgeon, or brain surgeon. He established the Montreal Neurological Institute in 1934 so that surgeons and researchers could work and share their knowledge.

In the 1950s, Penfield worked with patients who had severe epilepsy. He knew that before patients have epileptic seizures they experience an aura, which is a warning that they are about to have a seizure. Penfield thought if he could use a mild electric current on the brain to bring about this aura, he would know which area of the brain was responsible for the seizures. Once he located it, he could destroy that tissue. He opened the skulls of fully conscious (though anaesthetized) patients to pinpoint the source of the seizures.

These experiments led to another exciting discovery: stimulation anywhere on the cerebral cortex caused a response of some kind. By stimulating the temporal lobes, which are the lower parts of the brain on each side, Penfield could produce a memory such as a certain song or the view from a childhood window. These memories were very distinct but were not necessarily about important things. If Penfield stimulated the same area again, the same memory occurred each time. He had found what may be a physical basis for memory!

## Multiple Sclerosis – Demyelination

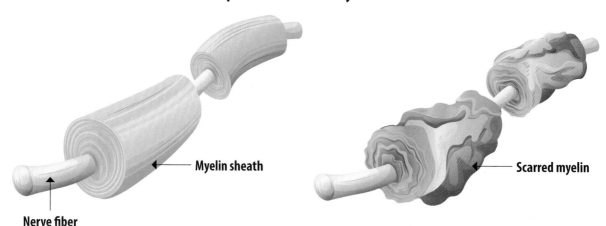

Myelin sheath

Scarred myelin

Nerve fiber

![icon]

### TRY THIS

**Getting on My Nerves**
Close your eyes. Have a partner hold the two tips of a hairpin either open or closed on the back of your hand, forearm, neck, back, and index finger. Can you tell whether the hairpin is open or closed? Which areas are more sensitive to touch? Repeat the procedure by having your partner place tips of the pin farther apart. At what point can you tell the distance between them? This activity gives you a rough indication of the distance between the nerve endings in different parts of the body.

There are also many disorders of the central nervous system. In multiple sclerosis (MS), the myelin sheath that protects the nerves hardens, limiting the travel of electrical impulses. In amyotrophic lateral sclerosis (ALS), which is also known as Lou Gehrig's disease, the motor neurons are gradually destroyed, and the muscles that depend on those nerves become useless. In epilepsy, neurons release their electrical charges suddenly, causing seizures. In meningitis, the layers of connective tissue that surround the brain and spinal cord swell because of a bacterial or viral infection. In Alzheimer's disease, the brain cells themselves degenerate and die, which causes a steady decline in memory and mental function.

Some disorders of the nervous system are unavoidable because they are caused by a virus or are genetic, but eating a balanced diet can help keep the nervous system healthy. The body needs vitamin $B_6$ to make neurotransmitters. Vitamin $B_{12}$ is needed to make the myelin sheath. Getting enough rest and exercise and making wise decisions about dangerous activities are other ways to care for your nervous system. Although accidents can happen in any setting, some activities entail a higher risk of spinal-cord injuries. It is important to take safety precautions to avoid injury and to be realistic about risks.

## LESSON REVIEW

**1.** Name the three main parts of the neuron and the function of each.
**2.** Explain the difference between the central nervous system and the peripheral nervous system.
**3.** What are the three parts of the brain and their functions?
**4.** Name three diseases of the nervous system.

If someone asked you to name the senses, you could probably easily list touch, sight, hearing, smell, and taste. But some scientists recognize 20 or more senses, including sensations such as pain, balance, itch, tickle, temperature, and pressure. Your senses help you interact with the world around you. The sensory receptors, sense organs, and the nerves that connect them to the brain work together in the peripheral nervous system to help you sense your environment.

How do the senses work? Consider the sense of touch. Some peripheral nerves end in deep tissues such as muscles, tendons, and ligaments, but many of them end at the skin. Some nerves are sensitive to hot, some to cold, some to pain, and some to touch. For these sensory nerve endings to work, there must be a stimulus—for example, the feel of an animal's fur as you pet it. The stimulus activates an electrical impulse that is conducted along the nerves. Eventually, the impulse reaches the central nervous system, which translates it into meaningful information. When the information reaches the brain, a message is projected back to the place the stimulus came from—in this case, your hand on the animal's fur. In an instant, the sensation is interpreted as being soft, furry, and warm.

## OBJECTIVES

- Identify the parts of the eye and describe how it enables sight.
- Label the parts of the ear and summarize how the parts work together to make hearing possible.
- Explain how the body detects smell.
- Name the five basic tastes.

## VOCABULARY

- **cone** a light-sensitive cell in the eye used for color vision and fine detail
- **rod** a light-sensitive cell in the eye used in noncolor vision and low-light situations
- **tympanic membrane** the eardrum

## Sense Organs in the Skin

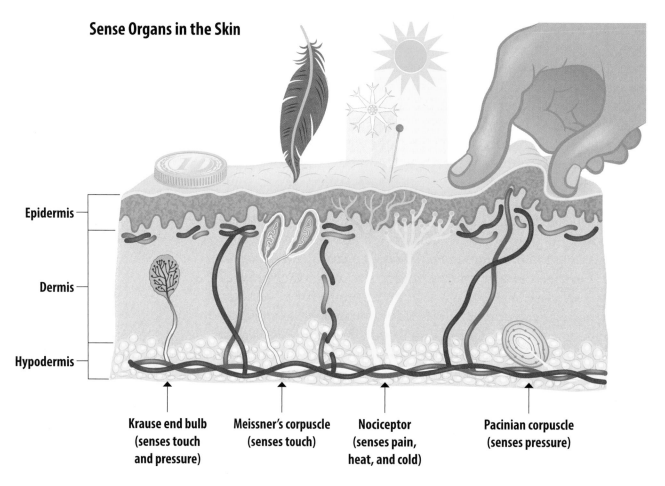

Epidermis

Dermis

Hypodermis

**Krause end bulb**
(senses touch
and pressure)

**Meissner's corpuscle**
(senses touch)

**Nociceptor**
(senses pain,
heat, and cold)

**Pacinian corpuscle**
(senses pressure)

Sight is another amazing sense. Even the structures around the eyes proclaim God's incredible attention to detail. Eyebrows prevent sweat from running down into eyes. Eyelashes trap dust and prevent it from irritating eyes. Glands near the corner of eyes lubricate the inside of the eyelids, which blink about 20 times per minute! Blinking spreads tears over the surface of eyes. Tears wash away foreign objects and contain an enzyme that kills certain bacteria. These structures around the eye work together to keep the eye healthy.

God designed the parts of your eye to work together as well. The cornea is a transparent layer of cells that light travels through into the eye. The iris—the colored part of the eye—lies just behind the cornea. The pupil is the opening in the center of the iris where light enters. The iris controls how much light can enter the pupil. Visual images are focused on the retina, which is the back lining of the eye. The retina actually covers the whole inside of the eye except for a small area at the front where the iris is and where the lens is attached. Although it is only about the size and thickness of a postage stamp, the retina allows people to sense the universe. There are millions of neurons in the retina that send messages to the brain. These messages come from two kinds of light-sensitive cells. **Rod** cells are used in noncolor vision and low-light situations. **Cone** cells are used for color vision and fine detail. The human eye has about 130 million rod cells and about 7 million cone cells.

As light rays travel through the cornea, pupil, and lens, they converge at a point called *the focal point*. The rays continue into the eye until they create an image on the retina. Because the rays meet at the focal point and cross, the image on the retina is upside down and reversed left to right. The light that strikes the retina activates the rod and cone cells, and they send electrical messages through the nerves. The optic nerve is the nerve that carries electrical impulses from the eye to the brain. Although the eye receives images upside down and backward, the brain sorts out the information so that the world is seen right side up.

Ears are amazing structures God gave people for hearing and for balance. Ears are a lot more than the visible skin and cartilage on the sides of your head. The ear has three parts: the outer ear, the middle ear, and the inner ear. The outer ear includes the visible part of the ear as well as the passageway that extends into the head. Sound waves travel through the ear canal to the **tympanic membrane**, or eardrum, which vibrates. The vibrating membrane makes tiny bones in the middle ear vibrate.

## FYI

**Red-Eye Effect**
The inner lining of the eye contains a dark pigment that absorbs the entering light. Looking into the eye through the pupil is like looking into a dark room, so the pupil ordinarily looks black. If a bright light is shone into the eye, the blood vessels on the surface appear red. That is why some people in photographs appear to have red eyes.

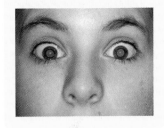

**Eye Anatomy**

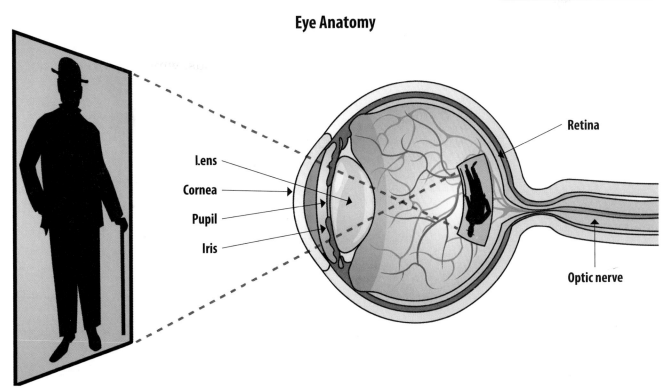

Lens

Cornea

Pupil

Iris

Retina

Optic nerve

These bones are called *the hammer, the anvil,* and *the stirrup.* The stirrup vibrates against the cochlea, a snail-shaped organ in the inner ear. The vibrations produce waves inside the cochlea, which is filled with fluid. Neurons convert the waves into electrical signals, which they send to the brain. The brain interprets these signals as sound. The sense of balance is controlled by the semicircular canals, which are filled with fluid. When the head moves, the tiny hairlike structures in the canals are moved as well. These movements produce nerve impulses that travel to the cerebellum. The cerebellum sends messages to the muscles to adjust for any loss of balance. The eustachian tube is the air passageway connecting the middle ear to the throat.

Although the nose is primarily a breathing organ, the receptors for smell are located high in the nasal cavity. These receptors react to chemicals that are inhaled and dissolved in the moist lining of the nasal cavity. Only a few molecules of a chemical are needed for the neurons to detect an odor. Tiny holes that extend through the bone in the sinuses allow the axons of these neurons to send messages through the olfactory nerve directly to the brain. The term *olfactory* is from the Latin word *olfactorius,* which means "to smell." Most other neurons are permanent cells, but the neurons that sense smell are replaced about every two months. As people age, these cells are not replaced as thoroughly, so some older people lose their sense of smell.

## Ear Anatomy

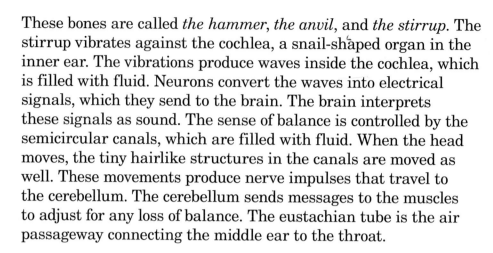
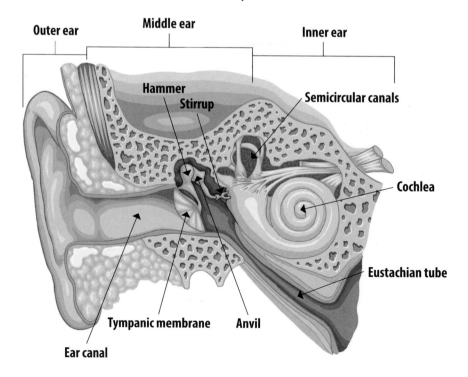

The tongue is where the receptors for taste are located. The tongue's surface has tiny projections called *papillae*, which contain taste buds. Taste buds are groups of 50–150 taste cells around a taste pore—a small opening on the surface of the papillae. When chemicals that have dissolved in the saliva enter a taste pore, the taste cells send signals to nearby nerves, which then send the taste message to the brain. The five basic tastes are bitter, sweet, salty, sour, and umami. Each of the thousands of taste buds can taste all five of the basic tastes. People used to believe that different parts of the tongue detected the different tastes, but this is not true with one exception. The taste buds at the very back of the tongue are the best at detecting bitter flavors.

## LESSON REVIEW

**1.** Explain how the eye sees objects.
**2.** How does the ear hear sounds?
**3.** How does the body sense smell?
**4.** How does the body sense taste?
**5.** Name the five basic tastes.

## FYI

**Taste Buds**
The body replaces taste buds after about 10 days. Newborns have taste buds on their tongues, on the roof of their mouths, and even on their lips and down their throats. As people age, they lose many of these taste buds. Their remaining taste buds become less sensitive. That is why older people often prefer foods with more powerful flavors or spices.

## TRY THIS

**Tasty**
Can you taste food if you cannot smell it? Pinch your nose with your fingers and close your eyes. Taste raw potato, apple, and onion given to you in random order on a spoon.

**Salty and Sweet**
What part does saliva play in taste? Dry your tongue with a paper towel. Dab some salt on your finger and touch your tongue. Do you taste it? Move your tongue around in your mouth to remoisten it. Did that make a difference in the taste? Repeat the process with sugar.

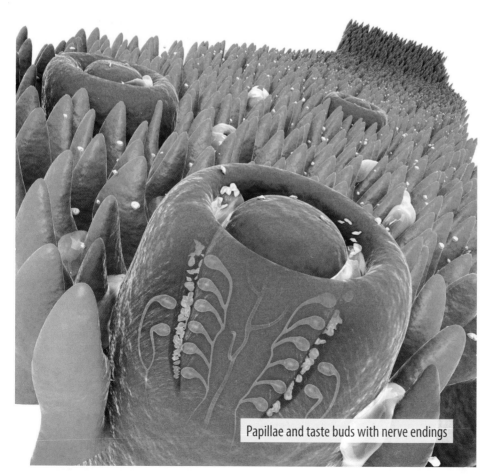

Papillae and taste buds with nerve endings

Your endocrine and nervous systems work together to regulate and coordinate all your other body systems. The endocrine system sends out chemicals called *hormones*. These chemicals carry information from one part of the body to another. Hormones travel through the bloodstream to various body tissues. The nervous system sends messages over a network like electrical wires. The endocrine system sends its messages out like different pieces of luggage on a conveyor belt. The tissue that recognizes each hormone's message picks it up, opens it, and follows the instructions.

Hormones control the body's metabolic rate and the growth and development of bones, muscles, and other tissues and organs. They also control sexual development and the reproductive system. Hormones cannot make cells do new things, but they can speed up or slow down certain cell activities. They can also make the cell membrane more or less permeable to certain molecules.

Hormone levels in the blood are not constant; they go up and down as necessary. Hormone levels are controlled either chemically by other hormones or chemicals, or electrically as neurons send messages. The cells that respond to specific hormones are called *target cells*. Target cells respond to specific hormones because they have hormone receptor molecules. These hormone receptor molecules fit with hormone molecules and affect cell function. The hormone receptor molecules act

as puzzle pieces that match particular hormones. When hormones have a message, they travel through the circulatory system looking for cells that have these specific matching pieces.

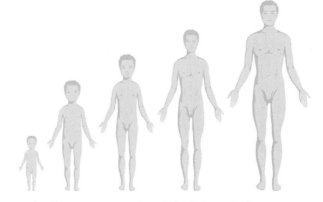

Hormones are secreted by organs called **glands**. The endocrine glands include the hypothalamus, pituitary gland, thyroid gland, parathyroid glands, adrenal glands, pancreas, thymus, pineal gland, ovaries, and testes. The hypothalamus is the link between the endocrine and nervous systems. It sends various signals to maintain homeostasis, the internal balance in the body. It also gives the signal for your body to begin puberty. The **pituitary gland** responds to signals from the hypothalamus to control the other endocrine glands. It is located in the center of the brain and secretes the most hormones of any gland, including growth hormone. It is known as *the master gland* because it controls the other endocrine glands.

Human development requires that all the different body systems work together to enable a baby to grow into a healthy adult.

The thyroid gland produces a hormone to regulate metabolism in most body cells. This hormone is essential for proper

**How Hormones Work**

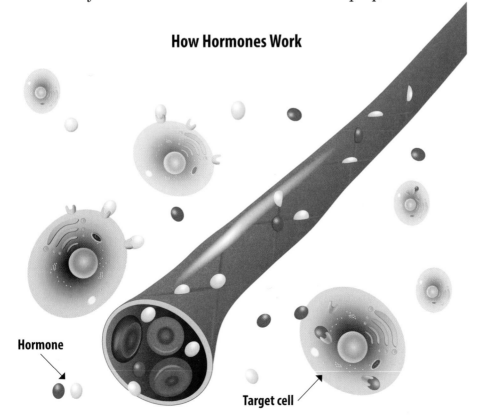

Hormone

Target cell

## FYI

**Androgen Hormones**
Some athletes consume androgens, synthetic male hormones called *steroids*, in order to help them increase their muscle mass. This practice can cause many serious side effects, which include the shriveling up (atrophy) of the testes, kidney damage, liver damage, heart attacks, and strokes. Most sports organizations ban the use of steroids. Several Olympic athletes have been stripped of their medals in recent years for taking androgens.

growth and maturity. The thyroid gland has two lobes of tissue that hug the trachea just below the throat. The *parathyroid glands* produce a hormone to regulate the amount of calcium in the body, particularly in the bones, the kidneys, and the small intestine. These four small glands are located behind the thyroid.

**Adrenal glands** are glands that produce the hormone called *epinephrine*, also known as *adrenaline*. One adrenal gland sits on top of each kidney. The adrenal glands release adrenaline in response to emotional excitement, stress, exercise, and injury. This hormone allows the body to respond quickly for "fight or flight." Adrenaline can help people do things that they would not otherwise be able to do—run faster or ignore pain if they are in danger or even lift very heavy objects that have fallen and need

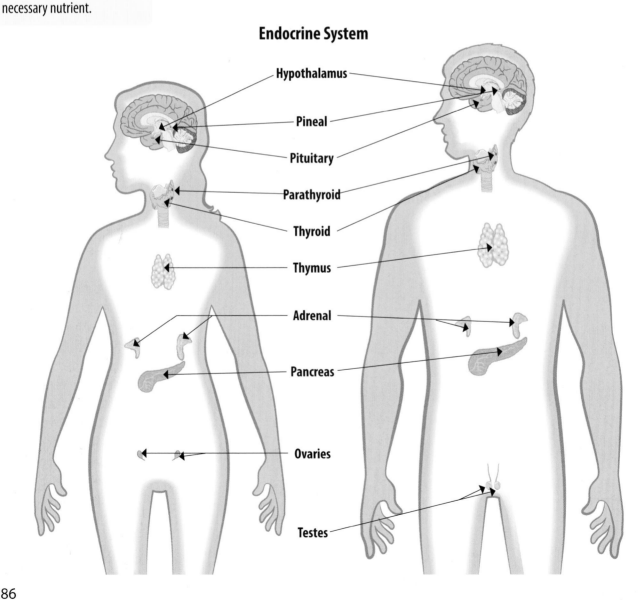

**Endocrine System**

- Hypothalamus
- Pineal
- Pituitary
- Parathyroid
- Thyroid
- Thymus
- Adrenal
- Pancreas
- Ovaries
- Testes

to be moved. The effects of adrenaline can be very dramatic, but they do not last very long.

Remember that the pancreas produces enzymes that help the small intestine during digestion. The pancreas is also the gland that produces insulin, which regulates the transport of sugar molecules into body cells. Insulin targets skeletal muscle, the liver, and fat tissue and increases the movement of glucose, or blood sugar, and amino acids into the cells. Insulin is released after meals to allow nutrients to be used. Insulin levels then gradually drop until, several hours later, little insulin is left in the bloodstream. At that point, the cells begin to use stored fat and protein for energy instead of sugar molecules. Insulin is important because the nervous system and the brain need glucose in order to function. Low blood-glucose levels can lead to a condition called *insulin shock*. This condition can lead to coma, seizures, and death—all because the nervous system does not have the glucose it needs.

The thymus gland, which is in the chest in front of the heart and behind the sternum, produces a hormone that aids in the production of T cells, which are important in the development of the immune system.

The pineal gland, which lies deep in the brain, produces a hormone that influences the development of the reproductive system and plays a role in such daily rhythms as sleep.

## FYI

**Pituitary Hormones**

| Target Tissue | Effect of Hormone from Pituitary |
| --- | --- |
| Kidney | Decreases the amount of water in urine |
| Most body tissues | Increases growth (via the growth hormone) |
| Uterus | Increases contractions during childbirth |
| Melanocytes (skin cells that produce melanin) | Darkens the skin |
| Fat tissues | Breaks down fat |
| Brain | Relieves sensation of pain |
| Mammary glands | Triggers milk production |
| Other endocrine glands | Regulates the levels of hormones produced |

Adrenaline can help people do things that they would not otherwise be able to do.

**Diabetes**

There are two types of diabetes. Type 1 typically starts in childhood; type 2 generally begins in adulthood. Both types involve difficulty with insulin. In type 1, the pancreas cells do not make any insulin or do not make enough of it. In type 2, the pancreas makes insulin but the body tissues do not respond to it. In both cases glucose (blood sugar) accumulates in the blood because it cannot get into the cells. The glucose level in someone with diabetes becomes so high that sugar spills into the urine. The sugar pulls more water into the urine, so the person's urine volumes are high. The person becomes

very thirsty, and the imbalance of ions can lead to insulin shock, which left untreated can result in a diabetic coma, brain damage, or even death. Today many people with diabetes check their blood sugar levels throughout the day and take insulin daily either by injection or by an insulin pump. They must also be careful about what they eat and should exercise to regulate their insulin levels. Left untreated, diabetes can cause damage to the kidneys, heart, nerves, and eyes.

The ovaries in females and the testes in males produce hormones that target most cells of the body. The ovaries are glands that produce the hormones estrogen and progesterone. These hormones affect the development of the female reproductive organs and breasts. Eggs also develop in the ovaries. The testes are glands that produce the hormone testosterone. Sperm cells are also produced in the testes. Testosterone affects the development of sperm production and the male reproductive organs. It also affects muscle mass development, hair growth, and voice changes.

## LESSON REVIEW

**1.** What are hormones and where are they made?
**2.** Which gland controls the other endocrine glands and has the greatest effect on the body?
**3.** List the endocrine glands and the functions of each.

Testosterone affects hair growth in males.

From the beginning of creation, God's creatures have been able to reproduce. Although most body systems are quite similar in both males and females, the reproductive system shows God's unique and yet complementary handiwork with males and females.

The first step of reproduction is making reproductive cells. Meiosis is the cell division process in which new cells with half the usual number of chromosomes are produced. An ordinary human cell has 46 chromosomes, the strands of genetic material made of DNA. The chromosomes exist in pairs in a cell's nucleus. During meiosis, pairs of chromosomes line up, and one chromosome from each pair ends up in the sperm or egg cell. As a result, each sperm or egg cell has 23 chromosomes.

Most of the reproductive organs in females lie within the body. As you learned in the previous lesson, the ovaries are the glands in a female body that produce the hormones estrogen and progesterone. Eggs also develop in the ovaries, which are two small, oval glands about the size of almonds. The **uterus**, the organ in which a fetus develops, is about the size of a pear. The uterus has strong muscular walls and stretches as the fetus grows. The neck of the uterus, the cervix, is where the uterus narrows and joins the vagina. The **vagina** is the tube that leads

## OBJECTIVES

- Name and explain the functions of the main structures in the female and male reproductive systems.
- Describe the events that take place during the menstrual cycle.
- Summarize the stages of development from infancy to childhood, adolescence, and adulthood.

## VOCABULARY

- **fallopian tube** the tube that connects each ovary with the uterus
- **ovulation** the release of an egg from the ovary
- **semen** a mixture of sperm cells and secretions
- **uterus** the muscular organ in which a fetus develops
- **vagina** the tube that leads from the uterus to the outside of a female's body

from the uterus to the outside of a female's body. The vagina, which is also known as *the birth canal*, is about 10 cm long.

When a girl is born, her two ovaries contain about 2 million immature egg cells. About 500 of these egg cells will eventually become mature eggs. They will then be released from an ovary, one at a time, approximately one month apart. The release of an egg from an ovary is called **ovulation**.

In the years following puberty, a woman's body experiences a monthly cycle. The menstrual cycle is the monthly cycle in which a woman's uterus produces a thick lining, which it sheds if she does not become pregnant. Menstrual cycles are generally about 28 days long. The word *menses* is from the Latin word for "month." The menses refers to the period of bleeding that occurs as the lining of the uterus is shed and drains out through the vagina. Day 1 of the menstrual cycle is the first day of bleeding. The menses usually lasts four or five days. During this time, the pituitary gland releases two hormones that cause some of the eggs to begin to mature. Between 10 and 20 eggs begin the process of maturing each month, although generally only one

## Female Reproductive System

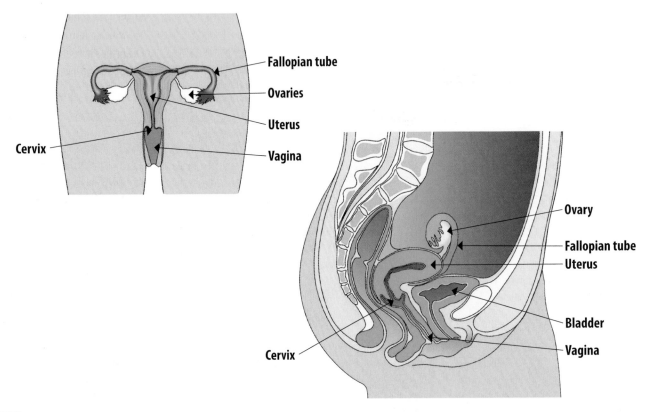

egg is released from the ovary. The pituitary gland also triggers the production of two female hormones in the ovary—estrogen and progesterone. These hormones send signals to the uterus to begin building a lining of delicate connective tissue that is rich in capillaries.

Levels of estrogen in the body peak just about the time of ovulation. Ovulation occurs halfway through the menstrual cycle on about day 14. As the mature egg is released from the ovary, it leaves behind a mass of cells that produces large amounts of progesterone. The egg itself travels down a **fallopian tube**, the tube that connects each ovary with the uterus. The egg may become fertilized if a sperm cell encounters it. A fertilized egg travels to the thick uterine lining and attaches itself there. The attached egg then sends out hormone messages that pregnancy has begun. If the egg is not fertilized by day 24 or 25, estrogen and progesterone levels fall, and the uterus lining begins to degenerate. The low levels of hormone also cause contractions in the muscle walls of the uterus, which help release the lining. These contractions contribute to the menstrual cramps that some women experience during menses.

**The Menstrual Cycle**

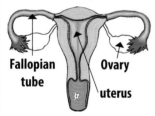

**Fallopian tube**    **Ovary**
**uterus**

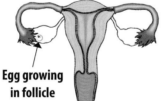

**Egg growing in follicle**

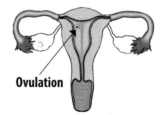

**Ovulation**

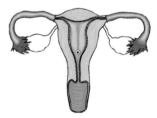

The menstrual cycle begins about 14 days after an egg leaves the ovary. An unfertilized egg passes out of the body. The hormones stop sustaining the lining of the uterus and menstruation begins.

A hormone causes the egg to ripen inside a follicle (tiny sac) within an ovary. Levels of the hormone estrogen rise, and menstruation stops. The lining of the uterus begins to thicken again for a fertilized egg.

About halfway through the menstrual cycle, ovulation occurs. The mature egg is released from the follicle and travels to the uterus through the fallopian tube, where it may be fertilized by a sperm—the male sex cell.

The hormone progesterone further prepares the uterus so that a fertilized egg may attach to it. Menstruation stops until the baby is born. If the egg is not fertilized, estrogen and progesterone levels fall, and the lining of the uterus begins to break down.

The lining of the uterus, which has become thick, is shed.

After menstruation the lining of the uterus is thin.

The lining of the uterus grows during ovulation.

The lining of the uterus is mature, ready to receive a fertilized egg.

## Male Reproductive System

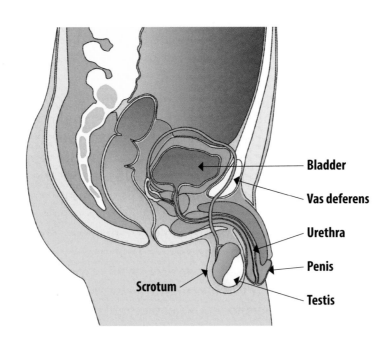

Many of the organs of the male reproductive system lie outside the body. As you learned in the previous lesson, the testes (singular: testis) are the glands in a male body that produce the hormone *testosterone*. Sperm cells are also produced in the testes. Sperm cells are male sex cells. The testes lie in a sac of skin known as the *scrotum*, which lies below the penis, the male reproductive organ through which sperm cells travel. The sperm cells develop in narrow tubes, which if stretched out would measure about 6 m. Each sperm cell develops a head that contains the nucleus of the cell, a middle piece that holds large numbers of mitochondria, and a tail that propels the sperm cell forward by spinning around.

## Stages in Human Embryonic Development

   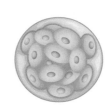 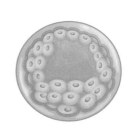 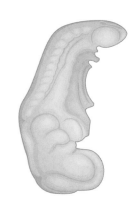

Each of the testes is attached to a duct called *the vas deferens*, the tube that carries sperm cells from each testis to the urethra. **Semen** is a mixture of sperm cells and secretions produced by a series of glands. The urethra is the tube within the penis through which semen and urine travel. The urethra is specially designed so that urine and semen never travel through it at the same time.

An egg that was released during ovulation and that has moved into the fallopian tube can be fertilized by a sperm in the semen. Even though thousands of sperm may reach the egg, only one sperm cell will penetrate the egg cell membrane. Fertilization is complete when the genetic material from the sperm and the egg merge. At this stage the new cell is called *a zygote*. The union of the sperm cell and egg cell is called *conception*.

After conception, the zygote begins to divide. Three days after fertilization, there are about 16 cells. About seven days after fertilization, the tiny ball of cells has reached the female's uterus. At this point, the developing child is called *an embryo*.

The placenta is an organ that is firmly attached to the wall of the uterus. It begins to develop after the embryo reaches the uterus. When fully developed, the placenta is a complex network of blood vessels linked to the unborn baby by the umbilical cord. The placenta exchanges oxygen, carbon dioxide, and nutrients between the mother and the fetus' blood.

The umbilical cord transfers these substances from the placenta to the fetus. The blood of both the mother and the child run through the placenta, separated only by thin membranes. Nutrients flow from the mother's blood to the child, and waste

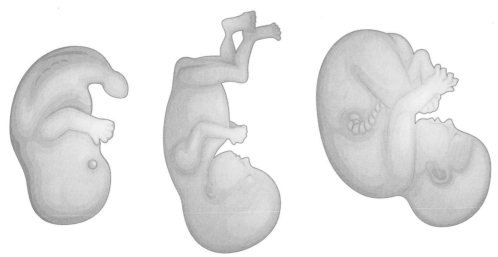

## BIBLE CONNECTION

God makes each person similar yet unique. God does the same with female reproductive systems. Some girls begin puberty as late as 16 or 17, others as early as 9. Although the menstrual cycle of many women is about 28 days long, some have a cycle as short as 18 days or as long as 40 days. Some women have hardly any cramps or discomfort during their menstrual periods; others have severe cramps. At a certain point in a woman's life (usually sometime between the ages of 45 and 55), eggs are no longer released from her ovaries, and the hormone balance of her body changes. At that point, progesterone levels are too low to trigger the buildup and release of the uterus lining, so the woman no longer menstruates or is able to bear children. That point of a woman's life is called *menopause*. The Bible says Sarah was 90 years old and well past menopause when God told Abraham that she would bear a son. That is why Sarah and Abraham laughed when they heard about Sarah's promised pregnancy. But nothing is impossible for God!

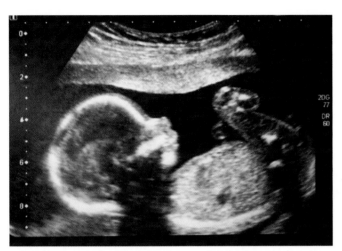
Sonogram image of a developing fetus in the womb

from the child flows back to the mother's body through the bloodstream. However, the blood of the mother and child never mixes.

After about two weeks, three layers of tissues have developed that will eventually form the body structures of the child. At this point, an amniotic sac is beginning to form. This sac, filled with protective fluid, surrounds the developing child.

During the next six weeks, the embryo develops quite rapidly. By 10 weeks after conception, all the organ systems have developed. The growing baby, now called *a fetus*, is about 2.54 cm long and weighs about 9.45 grams. From this point until birth, organ systems continue to develop, but the fetus' main job is simply to grow larger. The baby is usually born about 38 weeks after conception (40 weeks after the mother's last menstrual period).

Hormonal changes toward the end of pregnancy cause the muscle wall of the uterus to contract. This process is called *labor*. Most fetuses develop with their heads down against the cervix of the uterus. As the contractions progress, the cervix thins out and opens up. Once the birth canal is open, the mother uses her abdominal muscles to help push the baby down the birth canal and out as the uterine contractions continue. As the amniotic sac ruptures, the protective fluid is released and flows out of the mother's body. The final stage of the birth process happens after the baby is delivered. The umbilical cord is cut, which does not hurt, since the cord does not have any nerve endings, and contractions continue until the placenta detaches and emerges. Mammary glands located in the mother's breasts produce milk after the baby is born.

At birth, most infants are ready to breathe air and digest their own food. Still, the growth and development that occurs in the first two years is amazing—the helpless child learns to hold his or her head up, sit up, talk, walk, and feed himself or herself. This development is due in large part to the brain development that occurs. From birth to age 5, the child's brain triples in size. Even more important are the complex nerve pathways that develop as the child experiences stimuli and recognizes patterns.

After age 5, the brain continues to grow and develop in important ways, although more slowly than it did in infancy.

Personality and emotional characteristics develop during childhood. As the body grows, the child continues to develop language, motor skills, and social skills. The child can grow spiritually as well.

Puberty, which begins in adolescence, is triggered by the release of hormones. The hormones cause changes that lead to sexual development. The flood of hormones also brings emotional, physical, and chemical changes. Young people usually have a growth spurt during puberty and then continue to grow more slowly until they reach their adult size. Girls usually reach puberty when they are 9–13 years old. Boys usually reach puberty when they are 10–14 years old. There is, however, a wide range of normal ages for the beginning of puberty.

Adulthood is generally considered the stage of life after a person is 20 or 21 years old. Muscle and skeletal mass continue to increase until a person is about 30. But physical maturity is only one part of being grown up. Although some people think that by becoming adults they have finished growing, the process of Christian growth is never finished. Even though their bodies may have stopped growing, God wants all people to connect to Him through His Son, Jesus, and to grow spiritually and emotionally to become more and more like Jesus.

## LESSON REVIEW

**1.** What are the five main structures of the female reproductive system and their functions?

**2.** What are the five main structures of the male reproductive system and their functions?

**3.** Explain why a woman menstruates each month.

**4.** Explain where sperm cells develop and the route they travel on their way out of a man's body.

**5.** What are the roles of the placenta and umbilical cord during pregnancy?

**6.** What are the primary features of growth in each of the following stages of life: infancy, childhood, adolescence, and adulthood?

© Life Science

## 5.4.3  *Biological Rhythms*

People traveling to places far from home often suffer from jet lag. If you travel east across five time zones—for example, from Hawaii to Ontario—it takes your body a few days to adjust. The clock says that it is time for bed, but your body tells you that it is late afternoon, so you cannot sleep. In the morning the clock says that it is time to get up, but your body protests that it is the middle of the night. You even get hungry at the wrong times. After a few days, your body adjusts to the local clock, but when you go back home, you have to adjust all over again.

Your body follows a rhythm that tells you when to be awake, when to be asleep, when to eat, and when to be energetic. A rhythm is a pattern that occurs over and over. Not everyone follows exactly the same rhythm or cycle. For example, some people are more active in the morning and some like to stay up late, but these people have their own consistent rhythms.

Not only people but most organisms have God-given internal rhythms that regulate their behavior. Organisms use these rhythms to respond to time or season—by blooming, hibernating, or migrating, for example. For many years biologists believed that organisms used mostly external clues to separate day from night or spring from winter. They debated whether cosmic rays, the earth's magnetic field, or the earth's rotation

caused organisms' regular rhythms. They tested their ideas by taking plants and animals deep down into salt mines, transporting them to Antarctica, flying them around the world in airplanes, and even sending them into orbit. But none of these extraordinary conditions changed the rhythms of the organisms!

Most scientists agree that these rhythms are regulated by a **biological clock**, which is an inner mechanism that regulates an organism's biological rhythms. The biological clocks inside plants, animals, and microorganisms were created by God so that these organisms can take advantage of good environmental conditions and avoid dangerous ones.

Some biological clocks control relatively short periods of time. Most organisms have a biological clock to regulate their **circadian rhythms**—regular rhythms of growth and activity that occur in approximately 24-hour cycles. Some species of plants flower only in the morning and close at dusk. Other plants spread their leaves during the day and fold them at night. Some flowers secrete nectar at specific times of the day or night, and the biological clocks of pollinating insects tell them when to visit these flowers.

## TRY THIS

**Circadian Rhythms**

Plants have circadian rhythms, too. Keep a bean plant, a shamrock plant, and a prayer plant in a natural lighting environment with a constant temperature. Water the plants regularly. Observe the position of the leaves of each plant at the same time each morning, afternoon, and evening for three days. Sketch the leaf positions on the data charts.

Your body temperature is a circadian rhythm that you can monitor. Record your temperature for three days in a row at 7:00 AM, 10:00 AM, 1:00 PM, 4:00 PM, 7:00 PM, and 10:00 PM or bedtime, Read the thermometer carefully to note differences in your temperature. What pattern do you observe? Graph the results.

Circadian rhythms also cause some people to have inner alarm clocks. They wake up just before their real alarm clock goes off, especially if they are concerned about oversleeping that day. Does this happen to you? Test your inner alarm clock by telling yourself just before you go to bed that you want to wake up at a certain time. Do this several days in a row to see if you come closer to your expected time with practice. (Be sure to have a backup plan, such as an alarm clock or a family member, in case your inner alarm clock fails!)

What other circadian rhythms do you have? How could you test them?

**Circadian Rhythms**

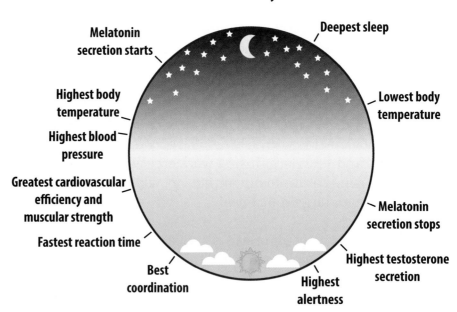

People have circadian rhythms too. Sleep patterns are circadian rhythms. The body's processes—including respiration, heart rate, and the secretion of hormones—fluctuate during a day. Generally, people sleep about 8 hours a night and are awake 16 hours. During the hours you are awake, your mental and physical functions are more active than when you are asleep. Even tissue cell growth increases when you are awake, so your body makes more white blood cells when you are awake than it does when you are sleeping.

Although humans do not experience annual rhythms like migrating birds or hibernating bears, they do experience monthly rhythms. The menstrual cycle of women is one example of a monthly biological rhythm. Research indicates that men experience changes in their weight, physical strength, and beard growth that follow a monthly pattern as well.

## LESSON REVIEW
**1.** How do biological rhythms help maintain creation?
**2.** What is the purpose of a biological clock?
**3.** What is the difference between an organism's biological clock and its circadian rhythms?
**4.** How do circadian rhythms help organisms?

The cells of your body are like little factories that make biochemicals and release the energy that your body needs to live, grow, and repair itself. Just as a factory needs many raw materials, cells need many chemicals to keep working. A substance that a living thing must take in to grow, maintain, and repair cells is called *a nutrient*. The cell gets its nutrients from a chemical fluid that surrounds each cell. Nutrients do not just magically appear in your body—they are supplied by food. If a critical nutrient is missing, either the cell's work stops or other methods, which may harm another part of your body, are used to complete the work. Different foods provide the nutrients your body needs.

Food can be divided into five groups: fruits, vegetables, grains, proteins, and dairy. Fats are not considered a food group, but they do supply essential nutrients and are part of a healthy diet. Many foods such as eggs, meats, and dairy products supply fats.

Your body uses six classes of nutrients: carbohydrates, proteins, lipids, water, vitamins, and minerals. Most foods contain a mixture of these six nutrient classes. Carbohydrates, proteins, and lipids contain the building blocks for cell growth and repair. They also provide the energy your body needs to function. The amount of energy available in a food is measured in joules in the metric system or in calories in the customary system.

Carbohydrates are organic compounds composed of simple sugars, or monosaccharides. Carbohydrates store and provide

> ## OBJECTIVES
> - Describe the roles of proteins, carbohydrates, and fats as providers of energy for growth and repair.
> - Distinguish between food groups and nutrient classes.

Carbohydrates are an essential part of the diet. However, it is good to monitor the amount of carbohydrates that you eat daily. It is best to eat more complex carbohydrates than simple carbohydrates.

the energy needed for your body to grow and function. Simple carbohydrates are commonly called *sugars*. Sugar occurs in nine common forms, including sucrose (the sugar that you buy at the store), fructose (found in honey and some ripe fruit), and lactose (milk sugar). Glucose, a sugar found in grapes, berries, and oranges, is also the form of sugar that travels in your blood.

Complex carbohydrates are made of sugar molecules linked together like long chains. They form the starch that is found in potatoes, pasta, and bread products. Some complex carbohydrates, such as cellulose, are not digestible. These indigestible carbohydrates are the main ingredients of fiber, which is necessary in your diet to keep your digestive system working. Vegetables, fruits, and whole grains contain fiber so they are complex carbohydrates.

Proteins are organic compounds made of chains of amino acids. Amino acids regulate the chemical reactions inside the body. They also transport and store materials and promote growth and healing. Like complex carbohydrates, proteins are long chains of molecules. The amino acids in the body are like the letters in the alphabet. Individual letters can be arranged in various combinations to form different words. In the same way, amino acids can form a variety of arrangements to make different proteins. The cells in your body need all 20 amino acids to make specific proteins for human growth and cell repair.

If it has the right ingredients, a cell can make 10 of the 20 amino acids by itself. The amino acids that the cell cannot make are called *essential amino acids*. They must be provided to your cells

Amino acids, the building blocks of proteins, are used in forming cellular structures, growing and repairing tissues, and aiding in chemical reactions.

## Fat Content of Oils and Shortening

| | Grams of Fatty Acids/100g | |
| --- | --- | --- |
| | Saturated | Monounsaturated |
| Butter | 51 | 21 |
| Coconut oil | 87 | 6 |
| Corn oil | 13 | 28 |
| Cottonseed oil | 26 | 18 |
| Peanut oil | 17 | 46 |
| Lard | 32 | 41 |
| Margarine, soft | 14 | 39 |
| Olive oil | 14 | 73 |
| Palm oil | 49 | 37 |
| Safflower oil | 3.7 | 4.8 |
| Soybean oil | 15 | 11 |
| Sunflower oil | 13 | 46 |

Foods with large amounts of saturated fats tend to raise cholesterol levels in humans, whereas foods with unsaturated fats tend to lower cholesterol levels.

through the food that you eat. The best sources of many of these essential amino acids are meat, fish, chicken, eggs, and dairy products.

Lipids are organic compounds that are fats and oils. Fats store energy and make up cell membranes, so they are essential for both growth and repair in the body. Fats are solid at room temperature, and oils are liquid at room temperature. Fats and oils provide more than twice the energy of protein or carbohydrates per unit mass.

Lipids are classified as either simple or compound. The most common simple lipids are triglycerides. Triglycerides contain a molecule of an alcohol called *glycerol* (made of carbon, hydrogen, and oxygen) that is combined with three molecules of substances called *fatty acids*. Each fatty acid is made of a long chain of carbon atoms with hydrogen atoms attached. If the fatty acids in a triglyceride contain all of the hydrogen atoms that they can hold, they are called *saturated fats*. Saturated fats are often found in foods that have animal origins—like butter. In contrast, triglyceride fats that do not have all of the possible hydrogen atoms that they can hold are called *unsaturated fats*. Unsaturated fats, such as olive oil, come from plants and are

Lipids provide more than twice as much energy per gram as carbohydrates. However, because some fats are harmful, it is necessary to keep your daily intake of fats to about 20%–30% of the food you eat.

## Vitamins—Uses and Food Sources

| Vitamin | Function | Some Food Sources |
|---------|----------|-------------------|
| A | Needed for vision, especially night vision | Milk, eggs, butter, yellow fruits and vegetables, dark green fruits and vegetables, liver |
| D | Helps calcium (a mineral) to be absorbed so bones and teeth grow properly | Salmon, sardines, herring, milk, egg yolks, sunflower seeds |
| E | Protects certain molecules from oxidation (combining with oxygen); prevents heart disease | Oils, eggs, wheat germ, sweet potatoes, nuts |
| K | Used in the liver to make blood-clotting chemicals | Green leafy vegetables, egg yolks, safflower oil, cauliflower |
| C | Promotes bone growth and healthy connective tissue | Citrus, cabbage, chili peppers, berries, melons, asparagus |
| $B_1$ (thiamine) | Needed in the cells to release the energy in carbohydrate molecules | Whole grains, brown rice, organ meats, egg yolks |
| $B_2$ (riboflavin) | Helps maintain the health of skin, eyes, and tissues that line the mouth and nose | Whole grains, legumes, nuts, organ meats |
| $B_3$ (niacin) | Essential for respiration inside cells | Yeast, wheat germ, meat |
| $B_6$ | Helps body use proteins, carbohydrates, and fats | Whole grains, yeast, egg yolks |
| Folic acid (a B vitamin) | Involved in cells' production of molecules such as DNA; needed to make new red blood cells | Liver, dark green leafy vegetables |
| $B_{12}$ | Needed by all cells to break down certain amino acids and fatty acids; critical in producing healthy red blood cells | Meats, milk, eggs |
| Biotin (a B vitamin) | Needed to help break down carbohydrates and fats to release the energy in them | Organ meats, yeast |
| Pantothenic acid ($B_5$) | Needed to release the energy in fats, proteins, and carbohydrates as well as in making certain lipids in the cell | Organ meats, green vegetables, egg yolks |

## Minerals—Uses and Food Sources

| Mineral | Function | Some Food Sources |
|---|---|---|
| Calcium | Builds bones and teeth, clots blood, active in nerve transmission | Dairy products |
| Chromium | Helps the body use glucose | Meat, cheese, grains |
| Copper | Needed for hemoglobin formation and bone development, releases energy from fats | Oysters, nuts, raisins, shellfish, liver, kidney, legumes |
| Fluoride | Strengthens teeth | Tap water with fluoride added |
| Iodine | Needed to make various hormones in the thyroid gland | Seafood, dairy foods, iodized salt |
| Iron | Critical in the structure of hemoglobin (the molecule that carries the oxygen on red blood cells) | Lean red meats, organ meats, whole grains, enriched cereals and breads, green leafy vegetables |
| Magnesium | Builds bones and teeth, helps transfer molecules through cell membranes, allows nerve messages to be transmitted | Cocoa, nuts, whole grains, legumes |
| Manganese | Needed to make cartilage, helps bone development, breaks down proteins and lipids | Nuts, grains, legumes |
| Molybdenum | Necessary part of enzymes | Organ meats, grains, legumes |
| Phosphorus | Builds bones and teeth, part of the DNA molecule, transfers energy | Milk, meat, fish, eggs |
| Potassium | Component of cells, important for nerve and muscle activity | Fruit, meat, vegetables |
| Selenium | Protects vital cell parts from damage | Seafood, kidney, liver |
| Sodium | Component of blood and fluid that surrounds cells, important for nerve and muscle activity | Table salt (NaCl) |
| Sulfur | Component of many amino acids and molecules in tendons, cartilage, skin, and bones | Milk, meat, eggs, legumes |
| Zinc | Needed to make DNA, helps healing, needed to break down proteins and glucose (blood sugar) in the cell | Meat, eggs, seafood |

Disease is a condition in which a body system, an organ, or a part does not work properly. Diseases have many causes. The causes can be divided into two categories: invasion by a pathogen and failure of a body system or tissue to function.

Diseases that are caused by pathogens are usually contagious and are called *infectious diseases*. Some examples of infectious diseases are influenza, chicken pox, measles, HIV, and the common cold. The microorganisms that cause these diseases can be spread by close contact with someone who has the disease or by breathing in microorganisms carried on moisture droplets coughed or sneezed by someone who has the disease. The body responds to these diseases using both innate and acquired immune responses. Bacterial diseases can be treated with antibiotics, which are drugs used to kill harmful bacteria and other microorganisms. Penicillin was the first antibiotic to be discovered. It can be used to treat strep throat and some types of pneumonia.

Antibiotics work in different ways. Some antibiotics damage the cell walls of the microorganism. Some keep the microorganism from synthesizing DNA. Others interfere with the cell membrane. All of these cause the microorganism to die instead of reproduce. Infectious diseases that are caused by viruses cannot be treated by antibiotics.

Chicken pox is an infectious disease caused by the varicella-zoster virus. Once you have had chicken pox, you will probably not get it again. However, the virus remains in your body for years after the illness is gone. If the virus becomes active again, it can cause a painful infection called *shingles*.

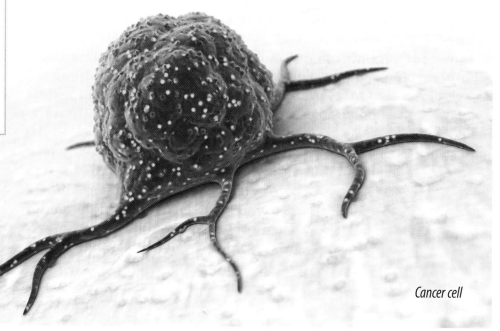

*Cancer cell*

Diseases that are caused by the failure of a body system or tissue and are not contagious are called *noninfectious diseases*. Noninfectious diseases include diabetes and cancer. Although diabetes may have a genetic link, it cannot be caught from someone with the disease. Cancer can affect various organs and body systems. Abnormal cells begin to multiply rapidly, creating a growth called *a tumor*. These cells lack the usual controls that tell a tissue to stop growing. Although T cells, natural killer cells, and macrophages recognize that tumor cells are not normal, they usually attack the virus or other invader that started the tumor instead of eradicating the tumor cells.

Cancerous tumor cells can move from the original site, travel through the lymphatic or circulatory system to a new location, and begin growing somewhere else. Cancer may cause changes or pain when tumors grow in tissues and organs, interfering with the way that they work. Benign, or noncancerous, tumors are not likely to spread through the body, but sometimes they create pressure on an organ and cause problems. For example, benign brain tumors can be dangerous because they can keep certain parts of the brain from functioning.

Some diseases occur when the immune system is not working properly. Such diseases can be grouped into hypersensitivity disorders, autoimmune diseases, and immunodeficiency. In hypersensitivity disorders, the immune system is overly sensitive and reacts to things that are harmless to most people. These reactions are called *allergies*. Antigens are substances that cause the immune system to respond. An antigen that triggers an allergic response is called **allergen**. Some allergies, such as reactions to plant pollens, cause rashes, swelling, and excess

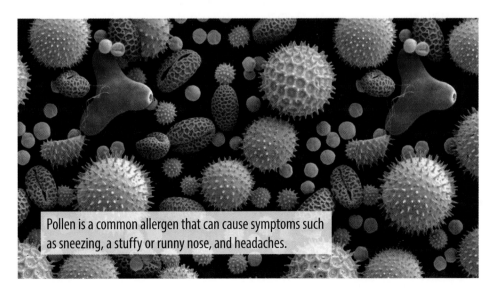

Pollen is a common allergen that can cause symptoms such as sneezing, a stuffy or runny nose, and headaches.

## BIBLE CONNECTION

The Levitical laws were instituted for the well-being of the Hebrew people. Today people understand the health basis for many of the laws regulating diet and disease. For example, the eating of pork was forbidden. Now it is known that swine can be hosts for the trichinosis nematode. By eating uncooked food wastes, hogs can pick up this worm, which then burrows through the hog's intestines and enters the muscular tissue. People who eat undercooked meat from an infected animal can be infected with this intestinal parasite. God's regulation about not eating pork relates to this possibility. The Levitical law also introduces the concept of quarantine, which isolates sick individuals and halts the spread of a disease. It also introduces the need to wash oneself or one's clothes after touching something dead or contaminated. Now it is understood that those items have bacteria that could easily spread disease.

## BIOGRAPHY

**Joseph Lister**

British surgeon Joseph Lister (1827–1912) pioneered antiseptic use in 1865. At that time, many surgical patients died from infection. The high rate of infection occurred because surgical instruments were washed only in soap and water, doctors operated in street clothes and with their bare hands, spectators were allowed to watch operations, and sawdust from the floors of sawmills was commonly used for surgical dressings.

Lister studied the clotting of blood and the inflammation that often affected injuries and surgical wounds. He tried to keep surgical instruments and rooms clean, but about 50% of the patients still died from infections. To keep the air clean, Lister sprayed it with carbolic acid, a chemical used to treat foul-smelling sewers. In 1865, he read Pasteur's germ theory that proposed the rotting of organic material was caused by microorganisms. This theory persuaded Lister to apply carbolic acid to instruments, wounds, and bandages. By 1869, the death rate from surgical infections had dropped to 15%.

### Cancerous Tumor Growth

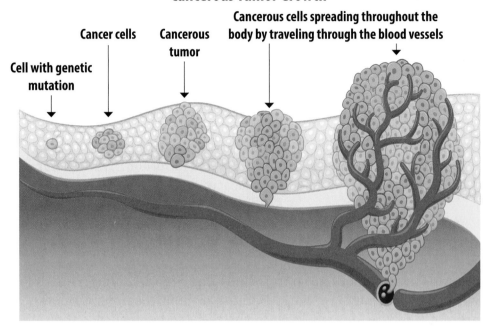

**Cell with genetic mutation** | **Cancer cells** | **Cancerous tumor** | **Cancerous cells spreading throughout the body by traveling through the blood vessels**

mucus production. Other allergies, such as reactions to bee stings or peanuts, can cause heart failure and a sharp drop in blood pressure.

Autoimmune diseases occur when the body cannot tell the difference between an invader and the body's own cells. As a result, the body starts to attack its own healthy cells. Autoimmune diseases include rheumatoid arthritis, lupus, rheumatic fever, and multiple sclerosis.

Immunodeficiency exists when some part of the immune system does not function properly. Acquired immune deficiency syndrome, or AIDS, is a fatal illness triggered by a virus known as HIV (human immunodeficiency virus). HIV attacks the immune system by binding to helper T cells and disabling them. Without helper T cells, the T cells and B cells cannot be activated to fight infection. The body becomes susceptible to many types of diseases because the immune system cannot function properly.

## LESSON REVIEW

1. Explain the difference between how people get infectious diseases and how they get noninfectious diseases.
2. How does cancer develop? Is it an infectious disease or a noninfectious disease?
3. How can you use your knowledge of diseases to take better care of your body and of other people?

Staying healthy requires that you take good care of the body God gave you. That, of course, means making good choices about the foods you eat, the rest you get, and the stresses you subject your body to. Choices about drugs also affect your health. A drug is any substance, other than food, that changes the way the body or mind works.

Many drugs are helpful when they are used properly, including antibiotics or drugs that relieve pain and fever. But other drugs are harmful. Some drugs can lead to an **addiction**, a physical or psychological dependence on a substance or behavior. Even prescription drugs can be abused—accidentally or on purpose. Sometimes this kind of drug abuse can happen when a person does not follow the doctor's direction. It can also happen when a person takes too much of a drug, uses the medication for too long, or uses it for the wrong reasons. Certain combinations of prescription drugs can also be dangerous. Other drugs have no legitimate use—there is no good reason to take them. These drugs are so dangerous that to use them even once is considered drug abuse; this is the case with most illegal street drugs. However, drugs are not the only thing a person can become addicted to. Behaviors such as gambling, shopping, sports, overeating, and pornography can become addictions too.

Every drug affects the human body in some way. Most drugs specifically affect the nervous system. As a person continues to use a drug, the body gradually builds up tolerance to the drug. Tolerance is a condition in which more and more of the drug is needed to get the same effect. Developing a tolerance raises the risk that someone may take enough of a drug to cause an overdose, which can cause serious physical or mental damage. An overdose can cause death.

A person who uses a drug begins to depend on that drug. **Dependence** is a condition in which the body relies either physically or psychologically on a substance or behavior to function normally. Physical dependence causes the person's body to become so used to the drug that it can only function normally if the drug is present. When the person stops taking the drug,

## OBJECTIVES

- Explain what an addiction is and give examples.
- Relate dependence, tolerance, and withdrawal to addiction.
- Recognize the five major groups of drugs.
- Defend why it is important not to get involved with addictive substances or behaviors.

## VOCABULARY

- **addiction** a physical or psychological dependence on a substance or behavior
- **dependence** the state of being either physically or psychologically dependent on a substance or behavior
- **withdrawal** the physical and psychological symptoms experienced when an addictive substance or behavior is discontinued

Drunk driving is an offense that will cost you thousands of dollars, time in jail, and loss of your driving license.

he or she may experience **withdrawal**, which is the physical and psychological symptoms experienced when an addictive substance is discontinued. These symptoms may include nausea, headache, vomiting, and shaking. People who abstain from their addictive behaviors will also experience withdrawal symptoms, but they are different from those of addictive substances. Not all drugs lead to physical dependence, but any drug, legal or illegal, can lead to psychological dependence. Psychological dependence causes people to think so much about the drug that they believe they cannot function without it. Sometimes, their lives become so centered on the drug that it causes serious problems at school, work, or with their families or friends.

People use drugs for a variety of reasons. Sometimes they start taking drugs because of peer pressure or curiosity. They may continue taking drugs because of social pressure, boredom, stress, insecurity, or rebellion—or because they have become dependent on the drugs.

Alcohol is a drug. It slows down the nervous system and can damage the liver if overconsumed. Alcohol is found in beer, wine, and hard liquor. Although the concentration of alcohol is greater in wine and liquor than in beer, the serving sizes of the drinks are different. A typical serving of wine, liquor, or beer contains about 14 grams of alcohol.

Behaviors such as overeating can be an addiction.

**FYI**

**Mystery Drugs**
Since street drugs are illegal, they are not regulated. This makes them even more dangerous. Some drugs are diluted with other substances, others are contaminated with additional or more potent drugs. One drug may be sold in place of another, sometimes with the seller's knowledge and sometimes without it. Almost two-thirds of the street drugs tested by the Addiction Research Foundation in Ontario were not what the seller said that they were.

In addition, people who abuse alcohol are often irresponsible or abusive, causing trouble for their families and friends. Drunk drivers cause many traffic accidents and kill thousands of people each year. In many places, the legal limit for being considered drunk can be reached with as little as one or two drinks.

Nicotine is a highly addictive, very poisonous chemical found in tobacco. It was once used in the United States to kill insects. Nicotine stimulates the nervous system and the heart. Ingesting a high concentration of nicotine can quickly become fatal. People who chew tobacco are as likely to become addicted to nicotine as those who smoke. Chewing tobacco also causes gum disease and cancer of the mouth, esophagus, and pancreas.

Tobacco smoke causes health problems as well. It has been linked to cancer of the lungs, larynx, esophagus, throat, bladder, stomach, colon, pancreas, and kidneys. Smoking tobacco also leads to high blood pressure, heart disease, and emphysema. Babies born to women who smoke are usually smaller because the nicotine in the mother's blood reduces the oxygen supply to the baby.

People who do not use alcohol and nicotine are less likely to try illegal drugs. For this reason, alcohol and nicotine are sometimes called *gateway drugs*, which are legal drugs that can lead to the use of illegal drugs. Cannabis is also considered a gateway drug.

Street drugs, which are illegal, are classified according to their effect on the body or the chemical family they belong to. The five major groups of illegal drugs are cannabis, stimulants, depressants, hallucinogens, and narcotics. The way that a drug affects a person depends on the person's age and size and the amount taken. The way the drug is taken (injected, smoked, inhaled, or swallowed) and the user's tolerance also influence how a person is affected.

Addiction can cause serious problems at school, work, or with families and friends.

Cannabis drugs are produced from various parts of the cannabis plant. These drugs include marijuana, hashish, and hash oil. Cannabis drugs increase the heart rate like stimulants do but generally have a depressant effect on the rest of the body. The effects of using these drugs vary with the user and dose, but most users lose memory and concentration and have an increased appetite. Some users withdraw from those around them and suffer from fear or depression, others experience anxiety or paranoia, and still others have hallucinations. Smoking cannabis drugs also damage the lungs. Cannabis drugs affect a person's coordination, so driving while intoxicated by these drugs is very dangerous. Cannabis drug use also affects a person's immune system, hormones, and heart function.

Drugs that speed up the actions of the nervous system are stimulants. Some stimulant drugs are nicotine, cocaine, and amphetamines. Users of these drugs develop tolerance very rapidly. Long-term use can lead to mental illness symptoms such as paranoia and delusions. Physical effects include kidney damage, seizures, lung problems, stroke, heart disease, and sudden death.

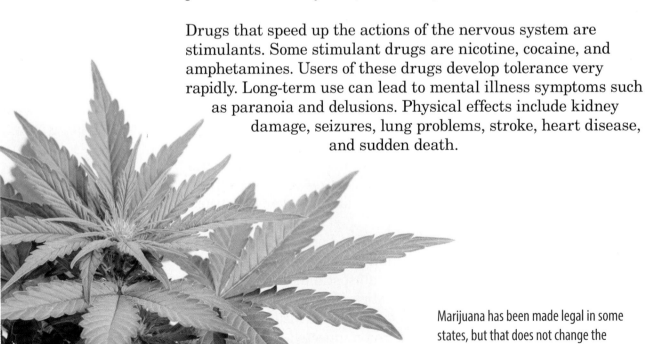

Marijuana has been made legal in some states, but that does not change the devastating effects it has on the brain and other parts of the body.

The opposite of stimulants are depressants, drugs that slow down the actions of the nervous system. These drugs are also called *sedatives* because they have a calming effect on people. Their effects make driving under the influence of a depressant drug extremely dangerous. The body does not build up a tolerance to the drugs' toxic effects. So, as users increase their dose to get the same effect, they run the risk of taking a fatal overdose. Depressants include alcohol, sleeping pills, and barbiturates.

Drugs that distort a person's senses, emotions, and mental processes are called *hallucinogens*. These drugs include LSD and PCP. When a person takes these drugs, the brain mixes up messages about what it sees, hears, tastes, or touches. Users of hallucinogens can also experience speech and memory problems or hallucinations for months or years after taking the drug.

Narcotics are highly addictive painkilling drugs. Some narcotics are opium, morphine, codeine, and heroin. While some narcotics are prescribed by doctors for patients who are in pain, even those can be abused. Taking these drugs leads to physical and psychological dependence. People who use these drugs build up a tolerance to them very quickly.

The consequences of a person's choices are very real. Some choices lead to wholeness, peace and health; others lead to brokenness and pain. In Deuteronomy 30:19, God set before His people two very clear choices: "I have set before you life and death, blessings and curses. Now choose life, so that you and your children may

## FYI

**Getting Help**

If you or someone you know needs help with drug abuse, help is available. Alcohol abuse is fairly common, and groups are available for alcoholics (such as Alcoholics Anonymous) and for their families (such as Al-Anon and Alateen). Similar groups are offered for narcotics addicts (Narcotics Anonymous) and for cocaine users (Cocaine Anonymous). Check your phone book, or ask your pastor or teacher for groups in your area that can help.

Some mushrooms have hallucinogenic properties and are very dangerous to ingest. They can cause permanent brain damage and even death.

## FYI

**Drinking and Driving**
Traffic accidents are the greatest single cause of death for young people. Alcohol is involved in almost half of these crashes. Never drink and drive, and never ride with someone who has been drinking.

Afghanistan is one of the world's largest sources of raw opium from cultivated poppies. Opium is a narcotic.

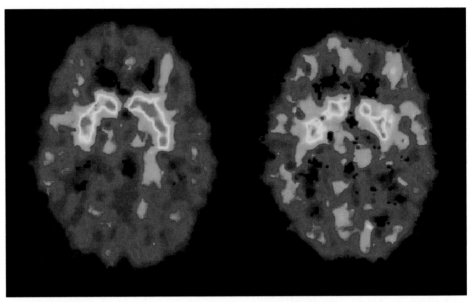

Notice the differences in the brain activity scans of a control subject (left) and a drug abuser (right).

live." God's desire for people is clear—He wants everyone to live according to His ways because He knows they bring abundant life. The lifestyle choices people make about nutrition, exercise, sexuality, and drugs point one way or the other–toward an abundant life or away from it. Now choose life!

## LESSON REVIEW

1. What is an addiction? Give two examples of substances and two examples of behaviors that people can become addicted to.
2. Explain how dependence, tolerance, and withdrawal relate to addiction.
3. What problems are caused by alcohol abuse?
4. Name the five major groups of drugs.
5. Why is it important to not get involved with addictive substances or behaviors?

Making choices according to God's way brings life.

314

**Genetics and Heredity**

## Chapter 1: *Genetics*
## Chapter 2: *Application of Genetics*

*Vocabulary*

allele
centromere
chromatid
chromatin
clone
codominant allele
gamete
genetically modified
  organism (GMO)
genetic code

genetic engineering
genetics
genome
genotype
heredity
heterozygous
homozygous
hybrid
inbreeding
karyotype

mutagen
mutation
phenotype
RNA
selective breeding
sex-linked trait
somatic cell
vector

*Key Ideas*

- Evidence, models, and explanation
- Change, constancy, and measurement
- Abilities necessary to do scientific inquiry
- Understandings about science and technology
- Structure and function in living systems
- Reproduction and heredity

- Regulation and behavior
- Diversity and adaptations of organisms
- Personal health
- Risks and benefits
- Science and technology in society
- Science as a human endeavor
- Nature of science
- History of science

### SCRIPTURE

From one man He made all the nations, that they should inhabit the whole earth; and He marked out their appointed times in history and the boundaries of their lands. God did this so that they would seek Him and perhaps reach out for Him and find Him, though He is not far from any one of us.

Acts 17:26–27

## 6.1.1   *Traits*

### OBJECTIVES

- Identify various genetic traits.
- Summarize the basic principles of genetics and heredity.

### VOCABULARY

- **genetics** the study of how traits are passed from parents to offspring, from one generation to the next
- **heredity** the passing of traits from parent to offspring

Did you ever wonder how there came to be so many different kinds of dogs? Have there always been so many kinds? For example, did poodles live in Bible times? Were collies always found on Earth?

The wide variety of dogs that people have as pets are members of one species: *Canis familiaris.* It is amazing to think that there is one species with so many possible variations. Dogs range from shaggy to hairless, from huge Irish wolfhounds to tiny Chihuahuas—and everything in between!

The variations of this one species are the result of conscious decisions by people to breed dogs with certain characteristics. At one time, all members of *Canis familiaris* were much alike. They were all "mutts." Then people learned how to breed dogs to preserve certain characteristics. This process is called *selective breeding.* Some dogs were bred to be small and fast so they could hunt foxes and rabbits. Dogs with heavy coats and quick feet were bred for herding sheep. Large dogs were bred to act as guards. Eventually, people began to breed dogs for racing, for looks, and for providing companionship.

Today there are over 150 different breeds of dogs, each bred to serve a certain purpose. Even if the dogs are not used for that purpose, their characteristics fit that purpose. Greyhounds are race dogs; collies are good herders; German shepherds are guard dogs and police dogs; and poodles are mainly show dogs.

Such variety within one species is the result of an understanding of **genetics**, the study of how traits are passed from parents to

offspring, from one generation to the next. In the coming lessons you will learn how God designed genetics to create both variety and order in creation.

"I could tell that you were your dad's boy just by the way you walk." "You look more and more like your mother all the time." Have you ever heard comments like these? Perhaps it is your dimples or your curly hair that remind people of your parents. Each of the things that makes you unique—your body shape, eye color, voice, and so on—is a trait. A trait is a distinguishing quality that all members of a species have in common. Your physical traits reflect the way your body is put together. Each trait may have variations. Variations are different forms of the same trait. Some variations for eye color are blue, brown, black, or hazel. Common variations for hair color include black, brown, blonde, and red.

Combinations of your parents' physical traits determine what you look like. **Heredity**, the passing of traits from parents to offspring, and environment, the social and cultural conditions in which you live, work together to influence who you are and who you will become. For many years people have discussed how heredity and environment affect a person's development. This discussion is often called *the nature* (heredity) *versus nurture* (environment) *debate*. In order to find out more about this subject, scientists have studied identical twins who were

## TRY THIS

**Rolling Your Tongue**
Tongue rolling is an example of a trait that is influenced by both genes and environmental factors. Try to roll your tongue (roll up the sides to form a valley in the middle). Test whether or not your parents, siblings, and other relatives can roll their tongues. Survey your classmates to see how many students can roll their tongues and how many cannot.

**My Physical Traits**

Here is a list of traits and their variations. Record your variation of each trait.

| Trait | Variation |
|---|---|
| tongue rolling | roller or nonroller |
| hair color | black, brown, blond, or red |
| hair texture | straight, tight curls, or wavy |
| eye color | black, brown, blue, or hazel |
| earlobe | free or attached |
| dimples | yes or no |
| bent little finger | yes or no |
| cleft chin | yes or no |
| index finger shorter than ring finger | yes or no |
| freckles | yes or no |
| handedness | left or right |

## BIBLE CONNECTION

"When God created mankind, He made them in the likeness of God. He created them male and female and blessed them. And He named them 'Mankind' when they were created. When Adam had lived 130 years, he had a son in his own likeness, in his own image; and he named him Seth" (Genesis 5:1–3).

From the beginning, God created humans to be like their parents. God also wants you to be more and more like Jesus, which includes being unselfish and compassionate. Developing Christlike characteristics is a lifelong process. The apostle Paul wrote to the Corinthians about the changes to character during the process of growing in faith: "And we all … are being transformed into His image with ever-increasing glory, which comes from the Lord, who is the Spirit" (2 Corinthians 3:18).

separated at birth and raised in different families. Because identical twins grow from the same egg and sperm, their heredity is identical and the environment is the only factor that is different. In many cases, identical twins who grew up in different homes (without knowing their twin) developed the same academic and professional interests. Scientists still do not know all there is to know about the relationship of heredity and environment, but their work is ongoing.

## LESSON REVIEW

**1.** What is a genetic trait and what is a variation?
**2.** Identify various genetic traits.
**3.** What is heredity?

God designed species to reproduce. The two basic types of reproduction are asexual and sexual. In asexual reproduction one parent is able to produce offspring. Sexual reproduction requires two parents.

Various green plants, fungi, protists, and a few animals can reproduce asexually. Since only one parent is needed to reproduce in asexual reproduction, the offspring are the same as the parent. Offspring produced by asexual reproduction are always identical to the parent because the parent is their only source of genetic material.

The four basic types of asexual reproduction are vegetative propagation, spore formation, budding, and fission. Vegetative propagation can occur in various ways. A common way is called *rooting*, in which a cutting is taken from the parent plant and rooted to create a new plant. Rooting is often used commercially to create plants with the parent plant's desirable features, such as resistance to drought. Fungi and some plants, such as ferns and moss, reproduce asexually through spores. Spores, like seeds, can grow into new organisms if conditions are suitable. Budding is a process in which a small bud forms on the parent, grows, and eventually breaks off to begin life as a new organism. Yeast and some invertebrates, such as hydra, use budding to reproduce. During fission, a strand of genetic material is copied, and a new cell membrane grows in between the two strands, dividing the cell into two daughter cells. Fission is the most common form of asexual reproduction. It is used by unicellular organisms such as bacteria.

Sexual reproduction requires male and female reproductive cells (sperm and eggs) called **gametes**. At the time of fertilization, the male and female gametes merge their genetic material. Each parent contributes genetic information that helps determine what the offspring will be like. Because sexual reproduction involves genetic information from two parents, sexual reproduction results in a much greater variety of traits in offspring than is possible in asexual reproduction.

During the process of reproduction, traits are passed from parent to offspring in

## OBJECTIVES

- Name and describe four types of asexual reproduction.
- Explain the role of gametes and genes in sexual reproduction.

## VOCABULARY

- **gamete** a male or female reproductive cell

A male gamete (sperm) fertilizing a female gamete (egg cell)

A V-shaped widow's peak at the hairline is an inherited trait.

Abraham Lincoln is believed to have had Marfan syndrome.

Fingerprint types showing whorl, arch, and loop patterns

small units called *genes*. A gene is a small segment of DNA that carries hereditary information. These small segments of DNA determine an organism's traits. Genes contain all of the information needed to build, maintain, and reproduce an organism. They also determine the features that make that organism unique.

Genes in a cell are linked together into long strands of genetic material called *chromosomes*. The instructions that genes carry are written in genetic code that the organism's cells can read. Just as you can read books over and over, the cells can read the coded instructions in the DNA over and over.

Some traits are determined by a single gene (one gene, one trait). For example, each of the following traits is determined by a different single gene: whether you have a straight hairline or a widow's peak, whether your little finger is bent, whether your earlobes are free or attached, and whether you have a cleft chin.

In other cases a single gene influences more than one trait (one gene—multiple traits). For example, Marfan syndrome, which is caused by a single gene, is a disorder that affects the heart, eyes, ligaments, and muscles. People with Marfan syndrome may have heart abnormalities and are often tall and thin; they have long bones and flexible joints.

Some traits are determined by the interaction of many genes (multiple genes—one trait). Skin color is an example of a trait determined by more than one gene. The pattern of your fingerprints—whether you have whorls, arches, or loops—is a trait determined by four to seven genes working together.

## LESSON REVIEW
**1.** Name and describe four types of asexual reproduction.
**2.** Which kind of offspring is more like its parent(s)—offspring produced by asexual or sexual reproduction? Why?
**3.** How do gametes and genes relate to sexual reproduction?
**4.** Name three of your characteristics that were determined by genes and three that were not.

Some variations of certain traits are more common than others. For example, brown or black hair is much more common than red hair. Why is this so? Why are there fewer people with red hair than there are with black hair or brown hair?

Much of the scientific understanding of this subject began in the 1800s when Gregor Mendel, an Austrian monk, studied the traits of pea plants. After years of careful study, he hypothesized that factors or units of inheritance were responsible for the traits he saw. These units are now called *genes*.

Peas reproduce sexually, but each plant has both male and female reproductive organs. Each pea flower contains both a pistil, which is attached to the ovary (the female organ), and many stamens (the male organs), which have pollen at the end. When the pollen travels down the pistil to the ovary, a seed forms from which a new plant later grows. The pollen from a plant may fertilize an ovary on the same plant in a process called *self-fertilization*. The pollen may also fertilize the ovary of a different plant. Cross-fertilization is when the egg of one individual is fertilized by the pollen (sperm) of another individual. The offspring produced by cross-fertilization are called **hybrids**.

## OBJECTIVES

- Explain the difference between genotype and phenotype, and homozygous and heterozygous.
- Give examples of Mendel's two laws.
- Predict the probability of genotypes and phenotypes using Punnett squares.

## VOCABULARY

- **allele** a different form of a single gene
- **genotype** the combination of an organism's dominant and recessive alleles for a trait
- **heterozygous** having different alleles from each parent for a particular gene
- **homozygous** having the same alleles from both parents for a particular gene
- **hybrid** the offspring produced as a result of cross-fertilization
- **phenotype** the observable traits of an organism

Pea flowers on a pea plant

To find out why certain variations seem to disappear and then reappear in the next generation, Mendel studied seven traits. Each of those traits has only two variations in pea plants. One of these traits was stem height: pea plants grow either on a tall stem or a short stem—not on an in-between length of stem. Another trait was pea texture: peas from a particular pea plant are either round or wrinkled—not a little bit wrinkled. A third trait was the color of pea flowers: pea flowers are either white or purple—not an in-between, lavender color. Another trait was the color of the peas—they were either green or yellow.

Mendel studied each of the pea plant's seven traits separately. He self-fertilized tall pea plants, which produced tall offspring. Then he self-fertilized those tall offspring. All of the resulting third generation plants were tall. He also self-fertilized short pea plants, and each generation that was self-fertilized produced short plants. He concluded that the tall plants had genes only for tallness and that the short plants had genes only for shortness. These two different forms of the same gene (tall and short) are called **alleles**.

Mendel cross-fertilized a purebred tall plant with a purebred short plant and all of the offspring were tall. Mendel wondered what had happened to the short characteristic, so he allowed these tall offspring to self-fertilize and produce a new generation of offspring. Three-fourths of these offspring were tall, and one-fourth were short. The short characteristic reappeared in the second generation of plants. A characteristic that disappears (as the short characteristic did in the first generation of offspring) is considered recessive. A recessive trait will only be physically present when there are two alleles for the recessive trait.

When two organisms with differing forms of a trait (such as one tall and one short) are crossed, and all of the offspring look like one of the parents (tall), the characteristic that shows up is

| | Plant height | Seed shape | Seed color | Pod shape |
|---|---|---|---|---|
| **Dominant** | Tall | Round | Yellow | Smooth |
| **Recessive** | Short | Wrinkled | Green | Wrinkled |

Mendel's pea plant traits

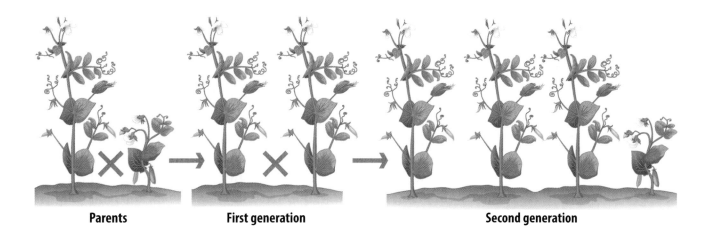

| Parents | First generation | Second generation |

called *the dominant trait*. The dominant trait always masks the recessive trait. If the dominant trait is present, it will always be expressed.

Scientists use a system of letters to indicate an organism's genetic makeup. The combination of an organism's dominant and recessive alleles for a trait is called a **genotype**. A capital letter indicates the allele for a dominant trait, and a lowercase letter is used to indicate the allele for a recessive trait. For example, in plant height, the letter $T$ can identify the dominant allele (tall), and the letter $t$ can identify the recessive allele (short). Because each organism has two alleles for each gene (one from its male parent and one from its female parent), two letters are used to indicate the genotype. For example, genotypes could be expressed as $TT$, $Tt$, or $tt$.

A genotype of an individual has two alleles for each trait, but a gamete has only one allele for each trait because of meiosis (for example, $T$ or $t$). If the genotype of an individual is $Tt$, half of the gametes it produces will contain the $T$ allele, and half of them will contain the $t$ allele.

According to Mendel's first law, the law of segregation, each trait is determined by factors, or genes. The parent has two alleles for

# MENDEL'S LAWS

## LAW OF SEGREGATION
EACH PARENT CONTRIBUTES ONE FACTOR TO A PAIR OF FACTORS.

## LAW OF INDEPENDENT ASSORTMENT
PAIRS OF FACTORS NEEDED FOR EACH TRAIT ARE PASSED ON TO OFFSPRING INDEPENDENTLY OF EACH OTHER.

Your genes were passed to you from your parents and they determine your genetic traits. But how do your cells know what to do to express those traits? Is there a code? Actually, there is and God designed it into your DNA.

In 1968, Robert W. Holley, Har Gobind Khorana, and Marshall W. Nirenberg received the Nobel Prize in Medicine for discovering how DNA sends coded messages to cells. Scientists had known that genetic information was carried by DNA, but they did not know how DNA communicated with the cells. The **genetic code** is the set of chemical instructions that cells use to translate the genetic information from DNA. Scientists use three-letter codes, or triplets, to show the instructions to the cells. The order of the triplets in the message specifies the order of the amino acids, which determines the type of protein that is made. The genetic information is translated into proteins in the cell.

Protein molecules are long chains of smaller molecules called *amino acids*. Just like the letters of the alphabet can be arranged into thousands of different words, the 20 different amino acids can be arranged into thousands of different proteins. The number of amino acids and the order in which they are arranged are

different for each protein. Each intricate protein is produced using the information in DNA.

Proteins are essential for life. Every cell in your body uses them. Hair, muscle tissue, and the cartilage in your nose and ears are all different types of proteins. Proteins make up hormones (which act as chemical messengers), various enzymes (which are involved in many chemical reactions), antibodies (which help your body fight infection), and receptor molecules in your brain (which allow you to think). The information held in one gene tells a cell how to make a certain protein, which triggers a specific action.

Genes are composed of DNA (deoxyribonucleic acid), which consists of nucleotides. Each nucleotide is made of a sugar, phosphate, and combinations of four nitrogen bases: adenine, thymine, cytosine, and guanine. Different combinations of these bases give people different characteristics.

By controlling what proteins are made in a cell, DNA controls all the characteristics of the cell. DNA works with another molecule to make proteins. Like DNA, the molecule RNA is a nucleic acid. Nucleic acids are the main information-carrying molecules of the cell. By directing what proteins are made in every cell, nucleic

## TRY THIS

**Decoding a Gene**
Bradykinin is a protein in blood that helps regulate blood pressure.

What is the sequence of amino acids that makes up bradykinin? (Hint: First determine what RNA molecule would be made from the DNA molecule.)

 **HISTORY**

**Hemophilia in the Royal Family**
Alexei Romanov was the son of Tsar Nicholas II of Russia. Alexei suffered from hemophilia. The disease was inherited through his great-grandmother, Queen Victoria of Great Britain. Hemophilia prevents blood from clotting, which means that even the smallest cut could cause severe problems. Alexei spent much of his childhood sick. Alexei's illness frightened his parents because he was the only son born to Nicholas and Alexandra. Only sons could inherit the throne in Russia at the time. Hemophilia was not understood and there was no treatment. The royal family thought a man named Grigory Rasputin knew how to heal their son. The country did not like Rasputin and did not approve of the tsar trusting a man many knew nothing about. Russians were divided in opinion over Rasputin's closeness to the tsar and hatred of the throne began to grow. A revolution took place before Alexei could ever become tsar. Alexei's case of hemophilia is studied because of its relation to the fall of the Russian imperial throne.

## FYI

**Mutation and Amino Acids**

All amino acids have three-letter codes. A mutation will not affect the amino acid if the conversion of information from the RNA remains the same. For example, if the U is replaced by a C in the three-letter codes below, the sequence for the amino acid tyrosine will still be produced.

UAU → Tyr

UAC → Tyr

acids determine what traits every living thing will inherit. **RNA** (ribonucleic acid) is a molecule in the cells of plants and animals that makes protein production possible. The RNA molecule is similar to the DNA molecule but it is much shorter and it is a single strand, not a double helix. Another difference is that RNA has the base uracil instead of thymine as in DNA.

The process of making a protein is called *protein synthesis*. There are several steps to this process, which starts in a cell's nucleus:

**First, the DNA molecule unzips.** The molecule separates into two strands the same way it does during DNA replication. One of the unzipped DNA strands becomes a template, or pattern, with the genetic information.

**Next, RNA nucleotides pair up with the DNA nucleotides.** During synthesis, the DNA's adenine pairs up with uracil rather than thymine. The pairing forms a special kind of RNA called *messenger RNA* (mRNA). The newly formed mRNA molecule copies the DNA code and then detaches from the DNA strand. It is now ready to carry the coded message to the cell's ribosomes, where proteins are made.

**Then, the mRNA leaves the nucleus with its message.** The mRNA travels out of the nucleus to one of the ribosomes. In the ribosome, another type of RNA called *transfer RNA* (tRNA) is ready to receive the coded message. The mRNA pairs up with the tRNA to transmit the message.

**Finally, the tRNA reads the instructions and gathers materials.** The tRNA takes the specific amino acids to the cell's ribosome and puts them together according to the directions to build the protein. Then that protein is used to make the characteristics that make you unique!

## FYI

**Amino Acids and Their Abbreviations**

| Amino Acid | Abbreviation |
| --- | --- |
| Alanine | Ala |
| Arginine | Arg |
| Asparagine | Asn |
| Aspartic Acid | Asp |
| Cysteine | Cys |
| Glutamine | Gln |
| Glutamic Acid | Glu |
| Glycine | Gly |
| Histidine | His |
| Isoleucine | Ile |
| Leucine | Leu |
| Lysine | Lys |
| Methionine | Met |
| Phenylalanine | Phe |
| Proline | Pro |
| Serine | Ser |
| Threonine | Thr |
| Tryptophan | Trp |
| Tyrosine | Tyr |
| Valine | Val |

God gave your genes the capacity to store the information needed for an incredible variety of traits. Because DNA is such an amazingly long molecule, thousands of different genes are possible. That is why, even though billions of people have lived on Earth, no two people are genetically identical except for identical twins. The DNA molecule shows how just a few substances are needed to make the wide variety of organisms on Earth.

Sometimes changes in DNA occur. Some changes can cause a **mutation**—a change in the sequence of one or more nucleotides in a DNA molecule. Several kinds of mutations are possible. For example, the disease sickle cell anemia occurs when one nucleotide is substituted for another. This tiny substitution in the DNA sequence means that one amino acid is switched for another in the hemoglobin in the blood. The change in amino acids causes sickle-shaped red blood cells. Another mutation occurs when chunks of DNA are either added or taken away from the original molecule. This sort of mutation causes Huntington's disease, a disease of the central nervous system. Occasionally, a section of the DNA ladder is cut out, turned around, and re-inserted backwards. Hemophilia, a disease in which a person's blood does not clot, is caused by this type of mutation.

Mutations can occur in any type of organism.

Mutations can change the information that is in the DNA. But sometimes mutations are not expressed in any way. Because a DNA molecule is so long, there are stretches of coded information between genes that are not needed for protein production.

### FYI

**Genetic Code Triplets — RNA Sequences and Their Coded Amino Acids**

| Code | Amino acid | | | | | | | | | | | | | | |
|------|------------|-----|-----|-----|-----|-----|-----|-----|-----|-----|-----|-----|-----|-----|-----|
| | | AUC | Ile | UCG | Ser | GCC | Ala | CAG | Gln | UGC | Cys | AGG | Arg |
| UUU | Phe | AUA | Ile | CCU | Pro | GCA | Ala | AAU | Asn | UGA | Stop | GGU | Gly |
| UUC | Phe | AUG | Met | CCC | Pro | GCG | Ala | AAC | Asn | UGG | Trp | GGC | Gly |
| UUA | Leu | GUU | Val | CCA | Pro | UAU | Tyr | AAA | Lys | CGU | Arg | GGA | Gly |
| UUG | Leu | GUC | Val | CCG | Pro | UAC | Tyr | AAG | Lys | CGC | Arg | GGG | Gly |
| CUU | Leu | GUA | Val | ACU | Thr | UAA | Stop | GAU | Asp | CGA | Arg | | |
| CUC | Leu | GUG | Val | ACC | Thr | UAG | Stop | GAC | Asp | CGG | Arg | | |
| CUA | Leu | UCU | Ser | ACA | Thr | CAU | His | GAA | Glu | AGU | Ser | | |
| CUG | Leu | UCC | Ser | ACG | Thr | CAC | His | GAG | Glu | AGC | Ser | | |
| AUU | Ile | UCA | Ser | GCU | Ala | CAA | Gln | UGU | Cys | AGA | Arg | | |

Albino strawberry

## CHALLENGE

If you have four letters in your genetic code alphabet, and you want to make combinations that are three letters long, how many different combinations can you make?

Mutations that occur in those areas of the DNA molecule do not affect the organism. A mutation only affects an organism if it falls within a gene.

Gene mutations can have a wide variety of effects. Some mutations do not have a noticeable effect on an organism's phenotype. Some mutations do not affect the organism's health. Albinism, for example, is a condition in which the organism lacks pigment. Pigment is what gives color to the organism. Plants, animals, and people can have albinism. People with albinism can be as healthy as other people, although their eyesight is often poor and they have to be especially careful to avoid sun exposure.

Other mutations are harmful, such as the mutation that causes cystic fibrosis. In cystic fibrosis, a mutation in a gene changes a protein that controls the movement of salt in and out of cells. As a result, people with the disease have thick, sticky mucus, which clogs passageways in their respiratory, digestive, and reproductive systems.

Mutations that are caused by a mistake in DNA replication for no known reason are called *spontaneous mutations*. Sometimes the effects of these mutations can take years to develop. For example, some cancers eventually develop from a mutation.

A mutation in this hedgehog's DNA caused a lack of pigmentation known as *albinism*. Animals that have this mutation have white or very light colored hair, light-colored skin, and eyes that often appear red or pink.

A **mutagen** is a physical or chemical agent that causes mutations to occur. Mutagens include cigarette smoke, industrial chemicals, pesticides, drugs, ultraviolet light, and radiation from atomic weapons' tests, nuclear power plants, or unprotected exposure from medical and dental X-rays. For example, some of the people exposed to radiation after the bombings of Hiroshima and Nagasaki in World War II gave birth to babies with various mutations. Sometimes drugs or medicine that a mother takes when she is pregnant can cause mutations in the baby. Although not all spontaneous mutations can be prevented, it is important to avoid known mutagens.

## LESSON REVIEW

**1.** Describe the process of protein synthesis.
**2.** State the roles of mRNA and tRNA.
**3.** In what process is the genetic code used to translate information from DNA and RNA into proteins?
**4.** How can a mutation occur?
**5.** Give three examples of human genetic mutations.

### FYI

**Specific Interests**
You cannot read every book in the library at once; you have to read them one at a time. In the same way, the cell reads only certain sections of the DNA molecule at any one time. And not all cells read all of the genes. For example, blood cells read instructions on how to make hemoglobin, the protein in red blood cells. But blood cells do not read the instructions that tell how to make keratin, the protein in your hair and fingernails. God created cells to read only the gene parts that they have to know about.

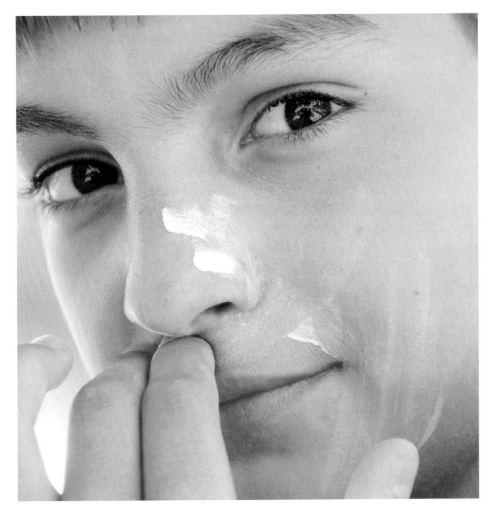

Sunscreen protects skin from the sun's harmful ultraviolet light, a mutagen that causes some skin cancers.

## OBJECTIVES

- Distinguish between mitosis and meiosis.
- Name and describe the phases of mitosis and meiosis.

## VOCABULARY

- **centromere** the point at which sister chromatids are attached to each other
- **chromatid** one of the two identical strands of a replicated chromosome
- **chromatin** the mixture of DNA and protein that makes up a chromosome
- **somatic cell** a cell that has the full number of chromosomes for that organism

You, along with every other living thing, have genetic information coded in your DNA. But how does that information get passed along to the next generation? How did your cells get the message to give you the traits that you have?

When cells divide, the DNA must be copied so the new cells also have the same genetic information. Humans have two types of cells. One type is **somatic cells**, which are cells with the full number of chromosomes for that organism. Somatic cells, also known as *body cells*, are not involved in reproduction. Examples of somatic cells include brain cells, skin cells, and liver cells. The second type of cells are gametes. Gametes are the reproductive cells (sperm and eggs) that divide to make new cells. Mitosis is the cell division process in which new somatic cells are formed. Meiosis is the cell division process in which new gamete cells are formed. In mitosis, each daughter cell receives an identical copy of the parent's chromosomes, which contain the DNA. In meiosis, the new cells have half the usual number of chromosomes.

When a cell is not actively dividing, its DNA looks like tiny threads inside the nucleus. The mixture of DNA and protein that makes up a chromosome is called **chromatin**. When a cell is dividing, the chromatin gathers into more compact bundles called *chromosomes*. Chromosomes are the structures found in the nuclei of cells that carry most of a cell's genes. By using very powerful microscopes, it is possible to see chromosomes dividing.

In organisms that reproduce sexually, chromosomes always come in pairs—one from the father and one from the mother. Members of a chromosome pair are called *homologous chromosomes*. Each member of this pair of chromosomes carries the same genes. For example, the gene determining a fruit fly's eye color is carried on chromosome 2. The fruit fly has two copies of chromosome 2—one from the father, the other from the mother. So a fruit fly's eye color depends on the combination of dominant and recessive alleles on the two copies of chromosome 2.

A cell's life cycle goes through three stages: interphase (when the cell grows and its nucleus produces a copy of each chromosome), mitosis (when the parent cell's chromosomes condense and the formation of two new daughter cells

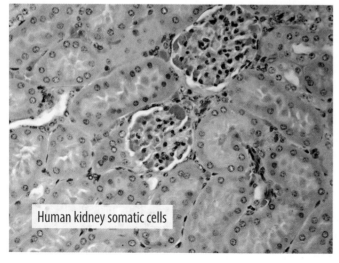

Human kidney somatic cells

develops), and cytokinesis. During interphase, the cell prepares to divide by replicating, or copying, its DNA. After replication, each chromosome is made up of two identical sister chromatids. A **chromatid** is one of the two identical strands of a replicated chromosome. These sister chromatids are attached to each other at a point called the **centromere**. The loose ends of the two chromatids resemble an X. The two chromatids of the pair are designed to attach to each other in a characteristic place. Some attach right in the middle, whereas others attach closer to one of the ends.

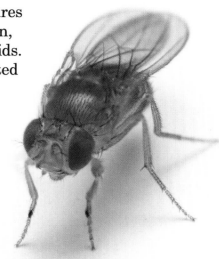

In mitosis, the chromosomes of a cell are replicated exactly and passed on to two daughter cells. Each daughter cell receives the same number of chromosomes that its parent cell has. Mitosis is a continuous process. Scientists have divided the events of mitosis into four phases: prophase, metaphase, anaphase, and telophase. A helpful way to envision the four phases is to think of the round cell as a globe with an equator and two poles.

The gene determining a fruit fly's eye color is carried on chromosome 2.

When interphase is completed, each chromosome consists of two identical sister chromatids. The first phase of mitosis is called *prophase.* During this phase, the X-shaped chromatids condense into rodlike structures (the chromosomes), and the nuclear membrane disappears. Spindle fibers start to grow from the poles of the cell and connect to the sister chromatids. These fibers look like the lines of longitude on a globe. They connect to the centromere, the point of the X where the sister

## Mitosis

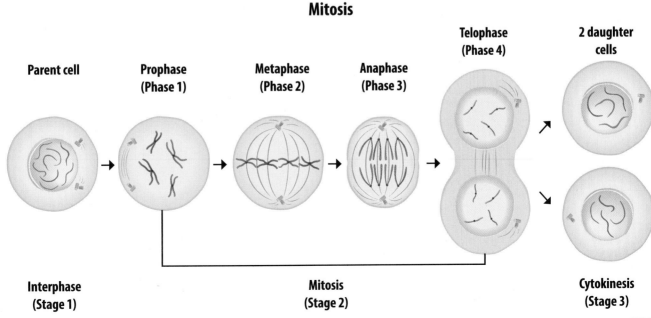

# Meiosis

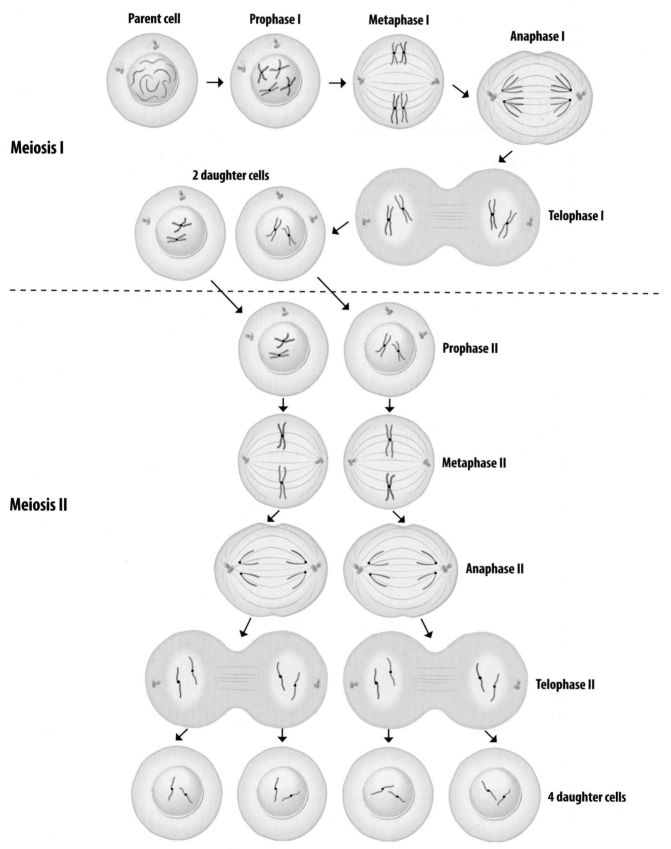

Parent cell

Prophase I

Metaphase I

Anaphase I

**Meiosis I**

2 daughter cells

Telophase I

Prophase II

Metaphase II

**Meiosis II**

Anaphase II

Telophase II

4 daughter cells

chromatids are joined. Each sister chromatid ends up attached to a fiber—one fiber from one pole and one fiber from the opposite pole.

The second phase of mitosis is called *metaphase*. All of the chromosomes line up at the equator (center) of the cell in random order. The chromatids are still attached to the spindle fibers at the centromeres. The third phase is called *anaphase*. The sister chromatids separate and move to opposite ends of the cell. It is as though the spindle fibers are pulling chromatids apart. The fourth phase is called *telophase*. Now that half of the chromatids are at each pole, the nuclei reorganizes there and the spindle fibers disappear. A new nuclear membrane begins forming around each group of daughter chromosomes. The cell begins to pinch in the middle.

Cytokinesis is the final stage of cell division. In animal cells, the cell membrane constricts at the equator of the cell until the cytoplasm is divided in half and two new daughter cells are formed. In plant cells, the division of the cytoplasm begins in the center of the cell, moving out to the cell membrane; cell formation follows this. Once they are surrounded by a nuclear membrane, the chromatids lose their rodlike form, and they unravel to become the threadlike chromatin strands again. The result is two new individual daughter cells with DNA identical to the parent cell.

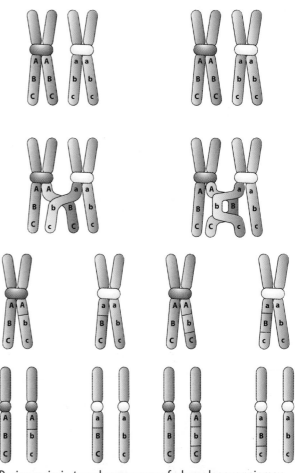

An original human somatic cell has 23 pairs of chromosomes (a total of 46 chromosomes). New cells created by mitosis end up with the same number of chromosomes. Each skin cell, brain cell, blood cell, and other somatic cell has the same number of chromosomes as the parent cell.

God designed a different process to produce gametes. In a process called *meiosis* the number of chromosomes found in each gamete is reduced to half the original number. Just as in mitosis, the chromosomes are doubled when they are replicated during interphase. But in meiosis, the cells that produce eggs and sperm divide twice in succession.

During meiosis, two chromosomes of a homologous pair may exchange segments in a process like this one. This produces genetic variation. This process is called *crossing-over*.

**Chromosome Number**
Different organisms
have different numbers
of chromosomes. Each
organism has a characteristic
number of chromosomes in
each of its cells.

| Species | Number of pairs of chromosomes |
|---------|-------------------------------|
| human | 23 |
| dog | 39 |
| carp | 52 |
| fruit fly | 4 |
| potato | 24 |
| onion | 8 |
| pea | 7 |
| red ant | 16 |
| broad bean | 6 |
| corn | 10 |
| frog | 13 |

The two cell divisions are called *Meiosis I* and *Meiosis II*. Each of the two cell divisions has four phases: prophase, metaphase, anaphase, and telophase.

In Meiosis I, the homologous chromosomes are bound together. First, homologous chromosomes (for example, chromosome 1 from the father and chromosome 1 from the mother) join. Because each chromosome has two chromatids joined at its centromere, four chromatids are actually attached together. This stage is called *prophase I*. At this point, crossing-over may occur. Small parts of homologous chromosomes break off and are exchanged with the homologous partner. If you think of these chromosomes as being different colors—for example, red and blue—after crossing-over, the result is a mostly red chromosome with a short blue section at one end. The other chromosome becomes a mostly blue chromosome with a short red section at the end. Crossing-over is another way that the genes of a mother or father are mixed so that the offspring has a combination of traits from each parent. It produces more genetic variation.

During metaphase I, the spindles are completed and extend from the poles. The chromosomes all line up on the equator. This stage is similar to what happens in mitosis except for one important difference. Instead of the sister chromatids attaching to opposite poles on the spindle fibers, entire chromosomes attach to one pole or the other with the sister chromatids still attached. For example, chromosome 1 from the father attaches to one pole, and chromosome 1 from the mother attaches to the opposite pole.

During anaphase I, the chromosomes separate and gather at separate poles. In telophase I, a membrane divides the two new nuclei. Now there are two cells, each with half the chromosomes.

But unlike mitosis, the process is not yet complete. Both new cells now go through Meiosis II. However, the chromosomes

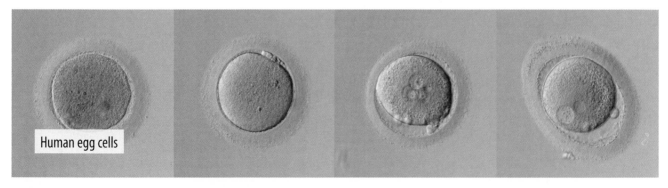

Human egg cells

do not replicate before Meiosis II, so this stage divides the number of chromosomes in half.

In prophase II, new spindles form at the poles of the cells. Then during metaphase II, the spindle fibers attach to the chromatids, just as they do in mitosis. A fiber from one pole attaches at the centromere of each sister chromatid. Another fiber, from the other pole, also attaches at the centromere. The chromosomes all line up along the equator.

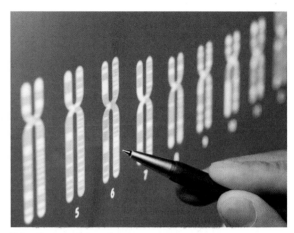
Scientist examining chromosomes

In anaphase II, the sister chromatids move to the opposite poles. During telophase II, the nuclear membrane forms around the chromosomes. The membrane begins to pinch in the middle and the cell divides into two cells. There are now four daughter cells, which are gametes, or sex cells.

In meiosis, then, one cell with the full number of chromosomes becomes four cells, each with half the number of chromosomes. The purpose of these gametes will be to find other gametes to combine with. When they do, they will form a new organism.

Why is it important that gametes end up with only half the number of chromosomes as somatic cells? Think about how a new organism is created. When an egg cell and sperm cell unite, the sperm contributes half the chromosomes that the offspring needs, and the egg contributes the other half. If gametes were made by mitosis, the number of chromosomes in offspring would keep doubling. God, in His wisdom, planned a different type of cell division for gametes so that each new organism has the same number of chromosomes that its parents have.

## LESSON REVIEW
1. What is the difference between somatic cells and gametes?
2. What are homologous chromosomes?
3. What are the three stages of a somatic cell's life cycle?
4. Name and describe the phases of mitosis and meiosis.
5. Why is it important that meiosis involves two divisions of chromosomes?

## OBJECTIVES

- Summarize how an individual's gender is determined.
- Analyze chromosomal differences on karyotypes.

## VOCABULARY

- **karyotype** a tool used by geneticists to analyze the chromosomal characteristics of an individual cell

You have 23 pairs of chromosomes; 22 of those pairs can be matched up with a look-alike partner. These are called *the autosomal chromosomes*, or *autosomes*. An autosome is any chromosome that is not a sex chromosome. Sex chromosomes, which determine the sex, or gender, of an individual, are in the twenty-third pair of chromosomes. The two types of sex chromosomes are X chromosomes and Y chromosomes. The medium-sized X chromosomes look like an X because the sister chromatids join in the middle. The very small Y chromosomes also have sister chromatids, but their centromere is close to the end of the chromosome, which makes it look more like a Y.

Because both of a female's sex chromosomes are Xs, when her body makes gametes during meiosis, each of her eggs carries one X chromosome. A male, however, has one X and one Y chromosome, so when his body makes gametes, half of the sperm carry an X chromosome and the other half carry a Y chromosome.

A somatic cell, such as a cell from inside your cheek, that is undergoing cell division can be stained with dyes and placed on a microscope slide. When the cell's chromosomes are examined under a microscope, some of the areas of the chromosomes have what appear to be bands. The two chromosomes of a pair can be recognized by their characteristic banding pattern. The position of the centromere and the length of the chromosome also vary. When banding, length, and centromere position are examined, the chromosomes can be identified.

## HISTORY

**I Want a Son!**
In the 1500s King Henry VIII of England wanted a son to be king after him. He divorced his first wife and executed his second wife because they gave birth to daughters instead of sons. Was King Henry correct in thinking that his wives were responsible for not having sons? Why?

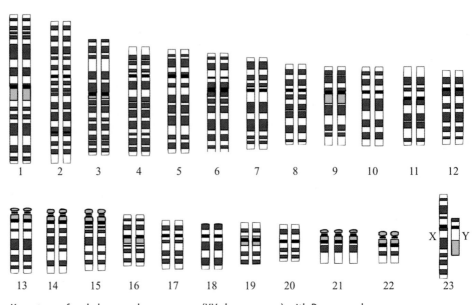

Karyotype of male human chromosomes (XY chromosome) with Down syndrome

**X and Y Chromosome Distribution**

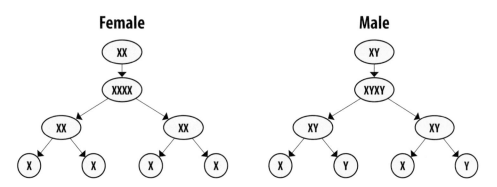

To study your chromosomes, a geneticist would collect and stain some of your somatic cells. A camera attached to a microscope would be used to take a picture of the chromosomes from a single cell. The individual chromosomes would then be cut out of the photograph with scissors, sorted, and lined up in pairs, starting with the longest pair (designated as chromosome 1). This arrangement of chromosome pairs by shape, size, and chromosome number is called a **karyotype**, a tool geneticists use to analyze the chromosomal characteristics of a cell. A karyotype can tell a lot about a person's genetic makeup or explain why a person has a disease caused by a chromosomal abnormality. Look at the karyotype on the previous page. Notice the twenty first set of chromosomes. What is unusual about it?

Trisomy 21, or Down syndrome, is a genetic disorder in which all the body cells have an extra twenty-first chromosome. Children born with Down syndrome may have various physical problems and some degree of mental impairment, but many people with Down syndrome lead productive lives.

## LESSON REVIEW

1. Explain how the gender of a child is determined by the father.
2. Contrast the twenty-third pair of chromosomes of a male and a female.
3. What is a karyotype?
4. Why might a karyotype be helpful?

## BIOGRAPHY

**Nettie Stevens**
Nettie Stevens (1861–1912) was a United States biologist whose experiments with beetles showed that sex is determined by a specific chromosome. Stevens found that the chromosomes X and Y determined the sex of individuals. Until her discovery, scientists did not know whether sex was determined by heredity or by other factors. Stevens also proved that chromosomes influence human traits. Her experiments were the first direct evidence that the units of heredity suggested by Gregor Mendel are related to chromosomes. Stevens' discovery paved the way for research in science and medicine that continues to this day.

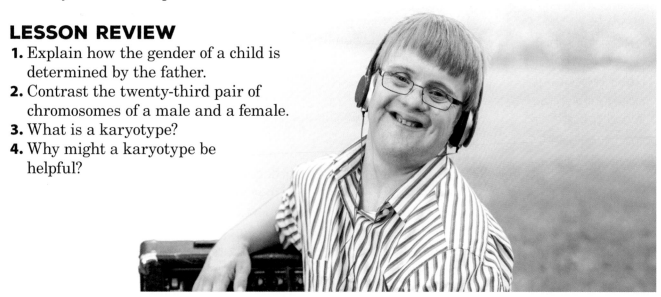

## OBJECTIVES

- Identify codominant alleles.
- Explain what sex-linked traits are and why they affect males more than females.

## VOCABULARY

- **codominant allele** an allele that is neither recessive nor dominant
- **genome** the total genetic material of an organism
- **sex-linked trait** a trait that is carried on the X chromosome

## FYI

**Blood Type**

People each inherit one blood type allele from the mother and one from the father. A pair of alleles in your DNA is called *the genotype* for that trait.

**Blood Type**

| Phenotype | Genotype |
|-----------|----------|
| A | iAiA or iAiO |
| B | iBiB or iBiO |
| AB | iAiB |
| O | iOiO |

Scientists use the term **genome** to describe the total genetic material of an organism. God wonderfully created the human genome with about 100,000 genes from about 3 billion nucleotides in the DNA. You have already learned that physical traits—such as a widow's peak hairline—are determined by genes and that the genes in each pair may be dominant or recessive. In humans, traits such as blood type are determined by genes.

Some alleles are codominant. **Codominant alleles** are alleles that are expressed jointly. That means that one allele is not dominant over the other. For example, three different alleles, *iA*, *iB*, and *iO*, determine blood type. Scientists sometimes put the *i* before the blood type to signal that they are all types of alleles. The alleles *iA* and *iB* are both dominant over *iO*. When the alleles *iA* and *iB* are found in the same cell, however, both alleles are expressed. The individual will have type AB blood. In regard to phenotype (blood type), there is no difference between a cell with the genotype *iAiA* and one with *iAiO*. Both of these genotypes produce the blood type A because the *iA* allele is dominant and the *iO* is recessive. The same is true for the *iB* allele. A person will only have type *iO* blood when 2 *iO* are present because *iO* is recessive. Commonly the genotypes are written as AA, AO, BB, BO, AB, and OO.

To determine if a blood type is positive or negative, scientists determine if the Rh factor is present. The Rh factor is an additional protein found in some people's blood. If the factor is present, that person has a positive blood type: A+, B+, O+, or AB+. If the protein is not present, the person has a negative blood type: A−, B−, O−, or AB−. The most common blood type is O+ and the least common blood type is AB−. Considering the dominance relationships, how many possible phenotypes (blood types) are there? Check your results with those in the box to the left.

Genes can also be the source of certain diseases. Some of these inherited diseases result from a defect in a single gene. These are called *single-gene genetic disorders*. Phenylketonuria, or PKU, is a single-gene disorder. The disorder occurs when a person inherits two recessive genes for PKU. A person with PKU cannot produce the important enzyme that breaks down the amino acid phenylalanine. Very high levels of phenylalanine accumulate in the tissues, causing brain damage. In the 1950s, scientists reasoned that if a person with PKU ate less of certain proteins that are high in phenylalanine, brain damage could be reduced or prevented. Their reasoning was correct, but this diet therapy works only if the disease is diagnosed within a few days of birth. Now newborns are routinely tested for this disorder, using just a few drops of blood.

Some inherited diseases like PKU affect both males and females. Others affect mostly males because the Y chromosome is much shorter than the X chromosome. It does not always have a corresponding gene to counteract the effects of a mutated gene on the X chromosome. The X chromosome has over 1,000 different genes. Most of these genes have nothing to do with whether a person is male or female. These genes affect traits such as teeth, skin, the nervous system, blood groups, hearing, vision, and the muscular system. Traits whose genes are carried on the X chromosome are called **sex-linked traits**. Sex-linked recessive genetic traits affect males more often than females. Because girls inherit two X chromosomes, they exhibit a recessive trait only if the alleles on both X chromosomes are recessive. Because

## CAREER

**Genetic Counseling**
Families who have children with genetic abnormalities benefit from the support and information that genetic counselors can provide. Genetic specialists work together in a team that usually includes a physician, a genetic counselor, a molecular geneticist (who understands the function of the DNA in the genes), and a cytotechnologist (who makes the karyotype). Together these experts can answer questions about the genetic condition and explain if it can be treated. They also are able to determine the probability of the parents having another child with the same condition. Genetic counselors can also connect the family with others who are facing the same challenges.

 **FYI**

### Family Pedigree Charts
Family pedigree charts are family trees that allow geneticists to see and understand how traits and their variations are passed down from parents to their offspring. Each pedigree chart shows the variations of one trait. Females are represented as circles; males are represented as squares. By shading in the circle or square of each family member who shows the recessive variation, geneticists can trace how a trait variation is inherited. For example, eye color is one inherited trait. A pedigree of this trait would identify the eye color of family members. Those with blue eyes, the recessive variation, would be shaded in.

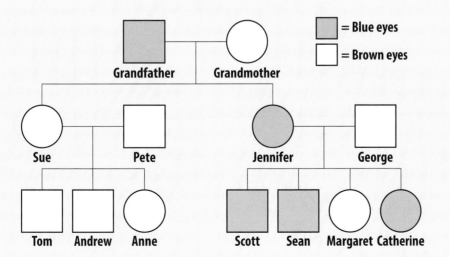

**Hemophilia: A Sex-Linked Trait**

<sup>h</sup> = hemophilia

|  | X | X<sup>h</sup> |
|---|---|---|
| X<sup>h</sup> | X<sup>h</sup>X | X<sup>h</sup>X<sup>h</sup> |
| Y | XY | X<sup>h</sup>Y |

This Punnett square reveals the probability that the offspring of a father with hemophilia (X<sup>h</sup>Y) and a mother who is a carrier (XX<sup>h</sup>) will have the disease.

X<sup>h</sup>X female carrier = 25%  X<sup>h</sup>X<sup>h</sup> female hemophilia = 25%
XY male normal = 25%  X<sup>h</sup>Y male hemophilia = 25%

boys inherit only one X chromosome, however, they always exhibit the recessive trait when the allele on their one X chromosome is recessive.

Sex-linked genetic disorders include hemophilia and color blindness. Even though color blindness affects both males and females, more males than females are color-blind. Most people who are color-blind see a range of color, usually blues and yellows. However, they have difficulty distinguishing between reds and greens. The gene for green-sensitive and red-sensitive pigments has three alleles. The dominant allele makes the normal pigment. The other two alleles—one making a defective pigment and one making no pigment at all—are recessive. Boys who inherit the allele that makes either the defective pigment or no pigment have some degree of color blindness. Girls who inherit two recessive alleles are also color-blind.

Scientists have determined that traits such as hair color, skin color, and height are determined by multiple genes. Some genetic disorders also result from the complex interaction of several genes. Examples include club feet (feet turned inward) and a cleft lip (a split upper lip).

Other disorders exist because of errors not in genes but in whole chromosomes. These errors may happen occasionally when sister chromatids fail to separate and form daughter cells during anaphase. As a result, some cells have an extra chromosome or

People with normal vision will see the numbers 12, 42, 6, and no number. Those with color blindness may see nothing at all because they cannot tell the difference between the colors. One male in twenty suffers from some form of color blindness, but only one in several hundred females is color-blind.

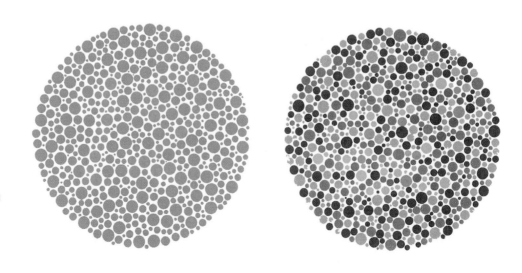

are missing a chromosome. In mitosis, the consequences may not be very significant. But in meiosis—that is, if an egg or sperm with the improper number of chromosomes is fertilized—the new organism will have an abnormal number of chromosomes. Several genetic disorders are related to having either too many chromosomes or too few. A female who has a missing X chromosome has Turner syndrome, which can result in multiple physical differences.

Some diseases are said to run in families. This means that people in those families are more likely to get them. However, many of these diseases are the result of both genetic and environmental factors. It is often difficult to determine what factor actually causes a disease. For example, cancer is one disease that appears to have both genetic and environmental causes. Many people who suffer from lung cancer are or were smokers. Yet some smokers never get lung cancer, and some nonsmokers develop lung cancer. Scientists do not fully understand why or how this happens. Still, they do know that choices can affect the chances of getting a disease. It is important to make choices that reduce the likelihood of these diseases, such as eating the right foods and not smoking. Making healthy choices during your life helps nurture and protect your intricate genetic design.

## LESSON REVIEW

**1.** What are codominant alleles?

**2.** Give two examples of sex-linked traits.

**3.** Why do sex-linked traits affect males more than they do females?

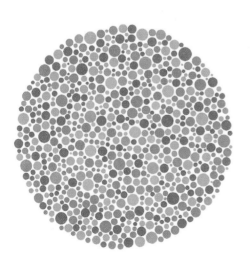

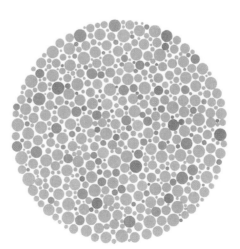

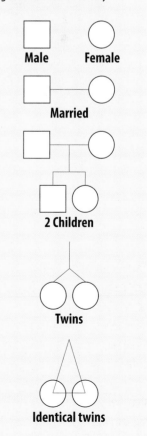

### TRY THIS

**Make a Pedigree Chart**
These symbols are used in making pedigree charts. Make a pedigree chart showing how traits can be passed down from generation to generation. Choose six traits and make charts to show how they run through three or four generations of a family.

Male    Female

**Married**

**2 Children**

**Twins**

**Identical twins**

### OBJECTIVES

- Distinguish between selective breeding and inbreeding.
- Paraphrase the process of genetic engineering.

### VOCABULARY

- **genetic engineering** the process used by scientists to transfer genes, or parts of DNA, from one organism to another
- **inbreeding** the mating of closely related organisms
- **selective breeding** the intentional crossing of plants or animals that have desirable traits to produce offspring that have those traits
- **vector** a carrier used to transfer genetic material from a donor organism to a target cell

For many years farmers have been using the principles of genetics to produce better crops. Cross-fertilization, for example, helps produce crops that are stronger, that bear bigger or better-tasting fruit, or that have a longer growing season. Geneticists are now using the principles of genetics to insert genes from one type of organism into other organisms to produce benefits. For example, they are inserting genetic material from humans into tobacco plants. The modified plants release a protein used to make cancer-fighting drugs. They have also developed a variety of rice that produces vitamin A and a variety of potatoes with leaves that kill such pests as beetles.

You might wonder if these developments are too good to be true. Some people think that they are. They think that mixing any genes will cause big problems. Others believe that such technology is a blessing that can help people solve many health or agricultural problems. Whatever the opinion, it is important to use the knowledge of genetics wisely.

Over the years, dog breeders have developed more than 150 different varieties of dogs, ranging from 1 kg Chihuahuas to 65 kg French Mastiffs. Some of these different breeds have been developed to enhance certain traits, such as the ability to herd sheep or cattle, hunt birds or foxes, guide blind people, or pull sleds or wagons. For example, traits such as ear shape, coat color, or fur texture are important for purebred show dogs. Breeders select a male and female dog that express the traits they want and mate the dogs to get puppies. The breeders further develop those traits through **selective breeding**, the intentional

Cassava is a staple food for hundreds of millions of people around the world. Unfortunately, this important crop is frequently destroyed by cassava virus diseases. The Virus Resistant Cassava for Africa (VIRCA) project is working to develop virus-resistant cassava varieties. The project's genetically engineered cassava would provide a more reliable source for food and income for farm families in East Africa.

crossing of plants or animals that have desirable traits to produce offspring that have those traits.

Farmers have selected animals to breed and seeds to plant for thousands of years. But over the last two centuries, selective breeding has been done more scientifically. Luther Burbank, a famous plant breeder, worked for more than 50 years to produce new varieties of fruits, vegetables, and flowers. By crossbreeding different plants, he developed varieties that had more desirable characteristics—bigger, sweeter fruit and vegetables that matured faster and could be harvested earlier. Producing all these new varieties was challenging work. To develop his raspberry-blackberry hybrid, Burbank grew and discarded about 65,000 berry bushes!

Cotton seeds contain a chemical that keeps bugs away but makes the seeds inedible. Researchers are producing genetically engineered seeds that do not contain the inedible part, but are still maintaining the plant's protection from insects. The seeds can be used to make tasty flour.

Today, selective plant breeding is used to increase crop yields and to give plants a natural resistance to certain diseases or pests. Over the years, plant breeders noticed that hybrid plants—those developed by crossing different varieties—were unusually strong and yielded a better crop. This strength is known as *hybrid vigor*. Today, most of the grain crops in North America—corn in particular—are grown from hybrid seed.

Even though crossbreeding can produce vigorous varieties, **inbreeding**—the mating of closely related organisms—can produce weaker varieties. Because the genes of closely related living things are naturally quite similar, doubling these genes in breeding leads to a lack of diversity. This lack of diversity causes a weakening of the offspring—the opposite effect of hybrid vigor. Inbreeding in humans can cause serious problems. If a small group of people intermarry for many generations, their offspring express a greater number of recessive traits. These traits would otherwise probably be masked by dominant genes. Some of these traits may be harmful. Inbreeding in animals can also cause problems. Dogs, for example, are often inbred to produce certain traits, but another result is that various harmful traits become common in certain breeds. Labrador retrievers, for example, often inherit a hip defect called *hip dysplasia*.

Selective breeding of plants and animals is limited, of course, by the species barriers that exist. For example,

A horse and a donkey can breed and produce a mule, but mules are rarely able to breed and produce offspring.

it is not possible to crossbreed a dog and a cat. In rare cases similar organisms have been able to crossbreed, but their offspring are sterile, meaning they cannot effectively produce offspring.

Organisms from different species cannot be bred in the usual way, but since the early 1970s geneticists have been able to splice parts of chromosomes from one organism into the DNA of another organism. This process in which genes, or parts of DNA, are transferred from one organism to another is called **genetic engineering**. Genetic engineering is possible because the structure of DNA in all living things is a double helix made up of the same four nucleotide bases. It is the sequence of the nucleotides in an organism's DNA that results in the variety of all of God's creatures. But how does genetic engineering work?

**First, scientists decide what gene they want to copy.** For example, human insulin is mass-produced through genetic engineering. People with diabetes do not produce enough insulin, so they need to take it to control the amount of sugar in their blood. A scientist who wanted to produce human insulin would first identify what human gene was coded for insulin production.

**Next, a carrier is selected.** A **vector** is a carrier that is used to transfer genetic material from a donor organism to a target cell. Two basic types of vectors are plasmids and bacteriophages. Plasmids are small rings, or loops, of DNA found in a bacterium. Bacteriophages are viruses that infect bacteria. The selected vector will be used to carry the DNA fragment. For insulin production, scientists use a plasmid as the vector.

**Then, the gene and the vector are cut into pieces.** Scientists use an enzyme to cut the gene into very specific pieces. The enzyme also acts like glue on the ends of the piece of DNA. In insulin production, the ring of plasmid is cut and a piece is removed.

**Now, the piece of DNA is fitted into the vector.** The fragment of DNA is inserted into the ring of plasmid. The newly created molecule

Soybeans are often genetically modified to make plants that are resistant to pathogens and produce larger crops in response to the increased demand for soy products.

is called *recombinant DNA* because the DNA has been combined with the plasmid.

**Finally, the recombinant DNA is inserted into a bacterium.** Scientists use a very small needle to place the recombinant DNA into the target cell. The bacterium now has the gene for producing human insulin. As the bacteria grows and divides, the human insulin gene is also reproduced in the cell. All of the offspring of the bacterium will produce human insulin protein molecules. Eventually, these molecules will be collected and purified for use in insulin injections for people with diabetes. Unlike insulin from cows or pigs, which was used in the past, insulin produced using recombinant DNA is genetically the same as the insulin produced by the human pancreas.

Using recombinant DNA techniques, scientists have been able to make bacteria and yeast cells into little protein factories. Changing the DNA of cells by using recombinant DNA and a plasmid vector is called *transformation*. Bacteria are capable of very fast reproduction rates—some can divide every 20 minutes to produce new cells. Within a few hours, millions of bacteria that make the protein can be produced.

## BIBLE CONNECTION

Since God created genes and knows the negative effects of inbreeding, He gave the Israelites laws to govern their sexual relationships. These laws protected the children of the Israelites. An example of one of these laws can be read in Leviticus 18:6.

**Transformation**

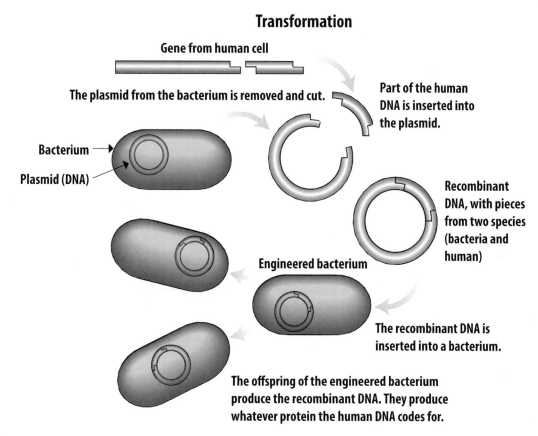

Gene from human cell

The plasmid from the bacterium is removed and cut.

Part of the human DNA is inserted into the plasmid.

Bacterium

Plasmid (DNA)

Recombinant DNA, with pieces from two species (bacteria and human)

Engineered bacterium

The recombinant DNA is inserted into a bacterium.

The offspring of the engineered bacterium produce the recombinant DNA. They produce whatever protein the human DNA codes for.

Science can help keep creation looking beautiful.

Scientists also use bacteriophages, which are viruses that attack bacteria, as vectors. Bacteriophages, or phages, are very simple—a DNA molecule covered by a protein coat. They make good vectors because viruses are naturally designed to introduce DNA into a cell. Scientists insert the new gene into the bacteriophage's DNA. When a bacteriophage invades a bacterium, the bacterium becomes a factory that makes proteins coded for the bacteriophage's DNA, including the inserted gene. In that way, the bacteriophage ensures that only the protein for which it codes is made. The process of changing a cell's genes is called *transfection*.

There are other methods of genetically engineering organisms such as plants. Particle gun technology is also used to genetically engineer corn, rice, wheat, soybeans, and other crops. New DNA is coated onto tiny particles of gold, tungsten, or platinum. The metal particles are put into a particle gun and fired into the plant cells. Once inside the cell, the DNA is integrated into the plant cell's own DNA and becomes active.

Plants and animals that have genes from more than one species are labeled *transgenic*. The first transgenic plant was made in 1983. Since then many plants have been genetically engineered to provide resistance to insects, viruses, and bacteria. Many varieties of transgenic crops have been approved by the United States Food and Drug Administration (FDA) and Health Canada. These include varieties of corn, canola, soybeans, potatoes, cotton, flax, wheat, and tomatoes.

The Rainbow papaya is the original transgenic papaya. It is resistant to the papaya ringspot virus and was developed using a particle gun.

Scientists have produced transgenic goats that make a human blood-clot dissolving enzyme. Scientists surgically removed eggs from a mother goat and fertilized the eggs with the sperm of a male goat. Then they inserted a human gene into the DNA of the mother goat's fertilized eggs. They implanted the eggs back into the female goat, where the eggs developed into baby goats. The DNA of the offspring contained the gene with instructions for making the human enzyme, so the milk that these goats produced contained the human enzyme. The enzyme was then extracted from the milk and used to dissolve blood clots, which helped prevent heart attacks and strokes in humans.

Several human proteins have been made by genetic engineering in addition to insulin. Growth hormone is another protein made by genetic engineering. Children whose pituitary glands do not produce growth hormone need it to prevent dwarfism. A protein clotting factor made by genetic engineering is also now available for people with hemophilia, a disease that keeps their blood from clotting.

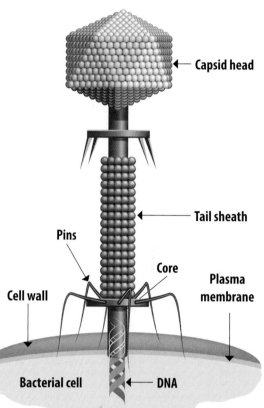

**Bacteriophage**

Capsid head

Tail sheath

Pins

Core

Plasma membrane

Cell wall

Bacterial cell

DNA

Genetic engineering may eventually provide a cure for people with cystic fibrosis—a disease caused by an abnormal gene that produces a thick mucus in the lungs. The mucus makes breathing difficult and lung infections common. In the laboratory, scientists have successfully stopped the production of excess mucus by inserting the normal gene into lung cells.

Various microorganisms developed through genetic engineering are already being used to help clean up contamination at waste sites. Genetically engineered bacteria are able to break down oil and could be used to clean up oil spills. Scientists are also working on bacteria that can digest plastic, which does not break down under ordinary conditions. A wise use of genetic engineering can help people be better caretakers of God's world.

## LESSON REVIEW
**1.** What is selective breeding?
**2.** What is inbreeding? What problems can inbreeding cause?
**3.** What is genetic engineering? How is this process possible?
**4.** Explain the process of genetic engineering.

A **clone** is a group of genetically identical cells that usually contains identical recombinant DNA molecules. For decades science fiction writers and filmmakers have produced stories about clones—people or even dinosaurs from the past that come back after having been cloned from bits of DNA.

In real life, Dolly the sheep made history in 1996 as the first mammal to be successfully cloned from an adult. Scientists at the Roslin Institute in Scotland cloned Dolly using a technique called *nuclear transfer*. Nuclear transfer is the removal of the nucleus (and its DNA) from a recipient cell and the replacement with the nucleus of a donor cell. When this technique was used to create Dolly, the recipient cell was an unfertilized egg taken from a sheep soon after the egg left the ovary. The donor cell was the one to be copied. In Dolly's case, the donor cell was a somatic cell. Somatic cells have a full set of DNA rather than the half set of DNA that egg cells and sperm cells have. The nucleus of the egg cell was removed from the recipient cell. Then the nucleus of

**Cloning**

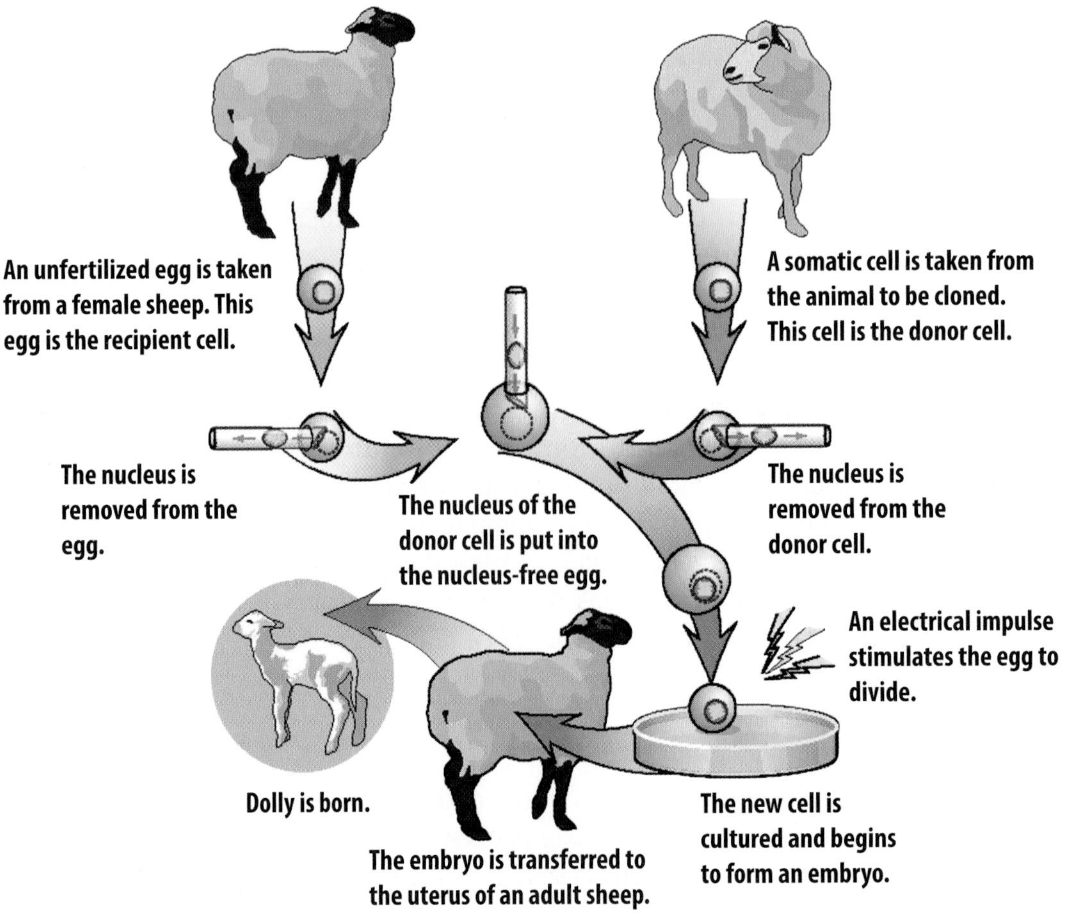

An unfertilized egg is taken from a female sheep. This egg is the recipient cell.

A somatic cell is taken from the animal to be cloned. This cell is the donor cell.

The nucleus is removed from the egg.

The nucleus of the donor cell is put into the nucleus-free egg.

The nucleus is removed from the donor cell.

An electrical impulse stimulates the egg to divide.

Dolly is born.

The embryo is transferred to the uterus of an adult sheep.

The new cell is cultured and begins to form an embryo.

350

the donor cell was injected into the nucleus-free recipient cell. An electrical pulse triggered the development of the egg (with its new nucleus). The egg was then implanted in the uterus of another female sheep, and the embryo grew into Dolly. Unlike other animals, Dolly did not have a typical mother and father. Her chromosomes were not half from her mother and half from her father because no egg and sperm united to form her. Dolly was a clone because all her DNA was identical to the female sheep from which the somatic cell was taken.

Various mammals, such as mice, sheep, cattle, cats, deer, dogs, horses, mules, oxen, rabbits, and rats, have been cloned from somatic cells.

So far only plants and animals have been cloned. As you might expect, some people are interested in cloning humans. But all life is sacred and created by God, and the cloning of humans raises some serious ethical problems. First of all, when scientists clone animals, many of them die before birth or at a young age. Many of the clones that survive have mutations. To clone humans would be to knowingly put cloned babies' lives at risk. Even if a baby did survive and seemed healthy, there is no way to know if cloning would lead to unforeseen health problems later in life. This is one reason why cloning humans could be considered irresponsible and risky.

Another question is this: Who would be considered the clone's parent? The clone embryo would be implanted in a human mother where it would develop. Would that person be the child's mother? Or would the person from whom the child was cloned be the parent?

Yet another issue with cloning has to do with the reasons that people would choose to clone someone. Cloning could be seen as a means of producing people with specific traits: intelligence, good health, good looks, athletic ability, or artistic or musical talent. But all people are valuable—not just people with certain traits or talents. Using cloning in this way implies that some people are more valuable than others. It would not be a wise use of the knowledge of genetics. Can you think of any other issues related to cloning humans?

## LESSON REVIEW
**1.** Explain how Dolly was cloned.
**2.** Should cloning be allowed? Defend your answer.
**3.** What ethical issues are raised about cloning humans?

## 6.2.3 Genetically Modified Organisms

### OBJECTIVES

- Explain what a genetically modified organism is.
- Express opinions about whether GMOs are ethically questionable.

### VOCABULARY

- **genetically modified organism** a living thing whose DNA has been altered by the addition or deletion of genes

People are very protective of their food and water supplies—and with good reason. Foods that are not handled properly can cause illness or even death. So when scientists change the genetic makeup of some foods, people are quick to question what they are doing and why.

Scientists can alter the genetics of a living thing by adding or taking away certain genes. A **genetically modified organism**, or GMO, is a living thing whose DNA has been altered by the addition or deletion of genes. Organisms that have been developed by inserting new genes into their DNA are genetically modified organisms. Some people argue that plant and animal breeders have been genetically modifying the food supply for thousands of years. Some are suspicious of the new methods and argue that humans are "playing God" by changing the genetic profile of various life-forms. Others argue that people are made in God's image and have the ability to "create." They say that genetic engineering is an application of God-given abilities. Of course, as with any other ability, people have a responsibility to be good stewards.

If handled the right way, genetically modified foods can be safe and beneficial. Some genetically modified foods are created by exposing plants to a mutagen. For example, DNA is damaged (mutated) if it is exposed to X-rays, gamma rays, or ultraviolet light. This technique has been used to produce ruby red grapefruits with enhanced coloring and a variety of rice that has highly desirable traits.

Other organisms are modified by inserting new genes through recombinant DNA techniques. The controversy about these organisms hinges on the fact that genes are added more quickly and from a wider variety of other organisms than is possible with conventional plant and animal breeding.

Some of the GMOs that are approved by the US Food and Drug Administration (FDA) are plants to which a gene from a bacterium has been inserted. For example, the Flavr Savr tomato was one of the first genetically modified foods approved by the FDA. The added gene delayed the softening of the tomato, making it spoil less quickly. However, consumers felt it lacked flavor. It was also expensive to produce, so the Flavr Savr tomato is no longer grown. Not all GMO projects are successful.

Other gene changes make plants resistant to harmful insects and viruses. The benefit to this change is that farmers do not have to use as much pesticide and herbicide on their crops. Using fewer pesticides and herbicides is better for the environment. Many North American farmers have grown such genetically modified corn, soybeans, and canola crops since the late 1990s.

Researchers have developed genetically modified apple varieties that do not turn brown when cut.

Some foods produced from genetically engineered organisms are more nutritious than those made from ordinary varieties. For example, vitamin A deficiency often leads to blindness in parts of the world where rice is the main food. Vitamin A is found in dark green leafy foods and in orange and deep yellow foods. The body can also make vitamin A from two molecules of a chemical called *beta carotene*, the dark yellow pigment found in carrots and pumpkins. In the summer of 2000, it was announced that golden rice, a GMO, had been developed. Golden rice carries the gene to make beta carotene. It can provide people in poorer parts of the world with an important vitamin that is missing from a diet that is primarily rice. Other varieties of rice have been developed with more than three times the normal level of iron. In the future, the higher iron rice could help eliminate iron deficiency in countries in Africa and Asia that have a rice-based diet.

Despite all of the apparent blessings of genetic engineering, it still raises many ethical questions. For example, some people want to patent certain human genes or to create patentable living GMOs. A patent is a license that gives a person or a small group of individuals the sole right to use, make, or sell a certain item. Do you think it would be right to have a patent on a human gene?

Many plants that people eat, such as corn, are now GMOs.

GMOs also raise scientific and political concerns. Most of these center on three areas—the impact on environment, the impact on human health, and the impact on poorer countries.

Some environmentalists are concerned that the new genes in plants might escape through pollen into nearby weeds or other plants. The weeds might be able to cross-pollinate with the genetically modified organisms. For example, corn pollen can cross-pollinate with other plants. If weeds picked up a modified gene from genetically modified corn, the weeds could become just as resistant to herbicides as the corn.

Many people are concerned that GMOs will affect human health. For example, genetically modified crops might produce resistance to antibiotics, which could increase disease. The new proteins in genetically modified foods might cause allergic reactions in people who eat them. Some people are also concerned that genetically modified foods could contribute to cancer or harm the immune system. These fears may or may not have a real basis. Scientists continue to research the long-term effects of GMOs, but the ripple-effect of genetically modified organisms is not yet known.

Genetically modified foods also may reduce food production in poorer countries. Farmers in many regions cannot afford to buy seeds every year, so they save seeds from their crops to plant the following year. But many genetically modified crops produce seeds that do not germinate. Farmers would have to buy seeds every year, which could reduce crop production in poorer countries.

## LESSON REVIEW
**1.** What is a genetically modified organism (GMO)?
**2.** Give three reasons why genetically modified organisms are controversial.
**3.** Would you willingly eat a food that has been genetically modified? Defend your answer.
**4.** Do you think GMOs are ethical? Why?

## CHALLENGE

**GMO Options**
What kind of regulations do you think should exist to control designing, growing, and exporting new genetically modified organisms? What questions do you have about the safety and ethical issues related to genetically modified crops? Write to a company that produces genetically modified organisms and ask them to discuss these issues and answer your questions.

Just as a map indicates all the cities along a highway, the genes along a strand of DNA can be mapped. Scientists have been determining the exact locations of specific genes on a strand of DNA. For example, the genes that determine color blindness and those that cause hemophilia are both X-linked genes. By studying how these genes cross over during meiosis, scientists can figure out just how far apart those genes are on the chromosome.

Each of your genes is made up of a series of nucleotide bases. They are the As, Ts, Gs, and Cs that make up your DNA. The 46 human chromosomes contain approximately 3 billion nucleotide bases. Identifying the 21,000 genes in the human body and finding out the sequence of all the nucleotides in the entire length of human DNA became the most ambitious project in the history of biological research. The project, called *the Human Genome Project* (HGP), began in 1990. It involved scientists at 16 institutions in France, Germany, Japan, China, Great Britain, and the United States. These scientists used automated machines that ran continuously to gather data. The final draft of the HGP was completed in 2003. The information on human genes gained by the HGP will have a major impact in the fields of medicine, biotechnology, and the life sciences.

What are scientists doing with this information? Scientists are continuing to analyze the data and anticipate that it will help them discover which genes are responsible for various diseases and conditions. Identifying genetic-disease genes could lead to genetic testing and possibly to cures for these genetic diseases!

## OBJECTIVES

• Evaluate the social and ethical issues of the Human Genome Project's findings.

The Human Genome Project

Beyond

On April 14, 2003, Dr. Collins announced the completion of the Human Genome Project at the National Institutes of Health, in Bethesda, Maryland, USA. Dr. James Watson, who along with Dr. Francis Crick discovered the double helix structure of DNA, worked with Collins and various other scientists on the HGP.

**Single Nucleotide Polymorphism (SNP)**

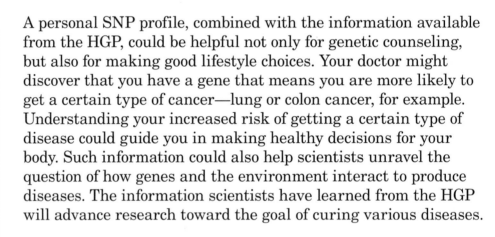

Partial gene sequence from A blood group

**The G is omitted.**

The same gene sequence from O blood group

Unless you are an identical twin, your DNA is different from that of everyone else. The point along your DNA where one of your nucleotide pairs is different from another person's is called *a single nucleotide polymorphism* or *SNP* (pronounced "snip"). People generally have a SNP about once every 300 nucleotides along their DNA. That adds up to about 10 million SNPs in the human genome. The SNPs differ from person to person and contribute to what makes each person unique. It is now possible to get a genetic profile of someone showing the person's own set of SNPs. Scientists are working to identify and map SNPs.

A personal SNP profile, combined with the information available from the HGP, could be helpful not only for genetic counseling, but also for making good lifestyle choices. Your doctor might discover that you have a gene that means you are more likely to get a certain type of cancer—lung or colon cancer, for example. Understanding your increased risk of getting a certain type of disease could guide you in making healthy decisions for your body. Such information could also help scientists unravel the question of how genes and the environment interact to produce diseases. The information scientists have learned from the HGP will advance research toward the goal of curing various diseases.

People can choose to use DNA information harmfully, however. The Human Genome Project has included research on the ethical, legal, and social issues that relate to mapping the genetic code. As always, humans must use their knowledge wisely. How do you think God would have people use this knowledge?

At creation, God surveyed His handiwork and proclaimed it very good. Humans are still the crown of God's creation, stamped with God's image. The issues raised by the new scientific capabilities for gene manipulation with the human genome should challenge all people to think critically about these issues and to make wise choices.

This is Zach, a patient with progeria, formerly known as *Hutchinson-Guilford progeria syndrome*. At the time of the photo, Zach was only four years old. Progeria is a rare and fatal genetic disorder caused by a mutation in the gene called *LMNA*. This disorder results in the premature aging of children. They often die by the age of 14. The information learned from the Human Genome Project helped scientists to develop a genetic test that enables doctors to diagnose and initiate treatment early in the disease process.

## LESSON REVIEW

**1.** What was the purpose of the Human Genome Project?
**2.** How could the Human Genome Project be used for good?
**3.** What Human Genome Project findings do you feel pose an ethical dilemma? Explain your answer.

356

**Chapter 1:** *The Ecological Landscape*
**Chapter 2:** *Ecosystem Dynamics*
**Chapter 3:** *Biomes*

### Vocabulary

abiotic factor
acid rain
adapt
benthos
biodiversity
bioluminescence
biomass
biome
biosphere
biotic factor
carrying capacity
commensalism
community
conifer
conservation
deciduous tree
desertification
detritivore

ecological
  succession
ecology
ecosystem
exotic species
habitat
intertidal zone
limiting factor
moment-of-truth
  defense
native species
nekton
neritic zone
oceanic zone
pampas
permafrost
photic zone
pollutant

population
population density
prairie
preferred habitat
primary consumer
primary succession
savanna
secondary consumer
secondary
  succession
sediment
steppe
tertiary consumer
trophic level
water hardness
wetland

### Key Ideas

- Systems, order, and organization
- Evolution and equilibrium
- Abilities necessary to do scientific inquiry
- Understandings about scientific inquiry
- Populations and ecosystems
- Matter, energy, and organization in living systems
- Behavior of organisms

- Regulation and behavior
- Form and function
- Diversity and adaptations or organisms
- Abilities of technological design
- Understandings about science and technology
- Populations, resources, and environments
- Natural hazards
- Risks and benefits

### SCRIPTURE

"Look at the birds of the air; they do not sow or reap or store away in barns, and yet your heavenly Father feeds them. Are you not much more valuable than they? ... See how the flowers of the field grow. They do not labor or spin. Yet I tell you that even Solomon in all his splendor was not dressed like one of these."

Matthew 6:26, 28

### OBJECTIVES

- Explain the nature of an organism's habitat.
- Describe the needs that a habitat should meet for an organism.

### VOCABULARY

- **habitat** the place where an organism lives and where all of its needs are met
- **preferred habitat** the place where an organism has an advantage over other organisms

To distinguish a living thing from a nonliving thing, there are common questions to ask. Does it move? Is it breathing? Can it grow? Does it eat? When scientists want to know if something is living or not, they look for certain characteristics common to all living things. For instance, all living things are made of cells, reproduce, grow and develop, respond to their environment, and use energy. If something has all of these characteristics, it is called *an organism.* God designed living things with characteristics that help them survive as part of a larger system.

Organisms need certain gases and energy to stay alive. Some organisms need oxygen; others, such as plants, need carbon dioxide. Most organisms get the energy they need from their food. After an organism eats, it digests the food and absorbs the nutrients. When the food is broken down, energy is produced. Energy is used to perform all the activities needed to stay alive. All organisms obtain food to carry out these activities.

An organism is usually smaller in its beginning stages of development than it is as an adult. As it grows and develops, it uses materials, such as food, water, and gases, and it uses energy. When it is mature, it has the ability to reproduce, or make more of its kind. The offspring may be similar or identical to its parents, depending on how the organism reproduces. God designed organisms to react to change. These normal

reactions define its routine behavior. Some organisms hibernate whereas others migrate. Some grow thick fur to prepare for colder weather. The routine behavior of an organism helps identify what kind of living thing it is. By studying the behaviors of living things, scientists can better understand God's creation.

Sometimes an ecosystem undergoes a sudden change that threatens an organism's survival. When such a change occurs, the organism must adapt quickly, regardless of whether its body is made of one cell or millions of cells. If a change is so drastic that the organism cannot adapt, it dies.

Zoos try to recreate the natural habitat of the organisms that live there as best as possible.

If you wanted to find a rabbit, where would you search? You might look at the edge of a wooded area near a field. In a field, a rabbit can find plants to eat. In the woods a rabbit can find safety from predators—perhaps in a hollow log, where it can also shelter its young. All organisms have specific needs, and this habitat meets the needs of rabbits.

A **habitat** is the place in which an organism lives—the place where all its needs are met. A good habitat provides an organism with an acceptable climate, enough space, and a good supply of food, water, and gases. An animal's habitat also provides shelter for hiding, nesting, and raising young. A plant's habitat provides sufficient sunlight and space to grow.

Habitat is a broad term because organisms have a wide variety of habitats. The habitat of a chimpanzee might be a tropical rain forest, the habitat of a slug might be under a rock, and the habitat of a dandelion might be a few centimeters of soil. Some organisms live in very specific habitats: the nine-lined goby fish often lives near sea urchins, and the spotted cleaner shrimp can be found between the tentacles of sea anemones. Other organisms, such as mice, can live in a wide range of habitats.

The habitats of various organisms are different because organisms require different physical and chemical conditions. For example, a fern needs less light than a palm tree. A tuna needs more salt in its water than a trout does. A lizard needs

### TRY THIS

**Bring on the Butterflies!**
The preferred habit of butterflies is often a garden in which certain flowers grow. Butterflies feed off the nectar and pollinate the flowers. Research which flowers attract butterflies and create a garden of those flowers at your school as a habitat for butterflies. If your school already has a butterfly garden, start one at a nursing home, community center, church, or another place where people might appreciate watching these beautiful creatures.

Barracudas travel in schools to avoid predators and to find food.

insects to eat and an otter needs fish to eat. Temperature, the ingredients in the soil, the amount of acid in the water, and the presence of other living things all help determine a good habitat for an organism.

An organism's **preferred habitat** is the place where the organism has an advantage over other organisms. For example, a rabbit's preferred habitat will include a place to hide from weasels or other predators. As an organism's needs change, its preferred habitat may change as well. For example, the preferred habitats of salamanders in their larval stage are ponds and streams. The preferred habitat of adult salamanders is beneath decaying logs. Young barracudas hide in the roots of mangrove trees, spend their early adulthood near coral reefs, and swim the open ocean as full-sized adults, returning to the reef only to hunt. Even throughout the course of a day, an animal may visit more than one habitat to meet certain needs. It may visit other habitats to hunt for food or look for water. One of these habitats, however, will be the animal's preferred habitat, where it spends most of its time.

## LESSON REVIEW
**1.** What are organisms and what are their characteristics?
**2.** What makes a good habitat for an organism?
**3.** Why did God vary the habitats of organisms?
**4.** What is a preferred habitat?

### TRY THIS

**Habitat Search**
Go outside and see how many species of organisms you can find. Using a field guide, write the name of each species of organism in a data chart. Record where you found each species, what its preferred habitat is like, and what basic needs are met in its preferred habitat.

The Victoria crowned pigeon (*Goura victoria*) is a turkey-sized bird that lives in New Guinea's rain forest. Eight smaller species of pigeons live in the same forest. Because pigeons eat similar foods, you might expect that some of the species would replace the others. But this ecosystem has a place for all nine species. The larger species perch on heavier tree branches and eat the larger fruit. Smaller species eat smaller fruit hanging from thin branches; their beaks are too small to open the larger fruit. The pigeons live in harmony.

Organisms of the same species that live in the same place at the same time are called a **population**. For example, the robins in your backyard are a population. The dandelions in a field are also a population. Individual organisms come and go, but a population remains year after year.

Populations have several properties that scientists study to determine the health of the group of organisms. Such studies also help scientists predict whether a certain population can thrive. These properties include population density, carrying capacity, age structure, birth rate, and death rate.

**Population density** is the number of individual organisms of one species in a given area. Examples of population density are

The *Goura victoria* is native to Indonesia and Papua New Guinea.

## OBJECTIVES

- Describe the factors that affect population density and carrying capacity.
- Explain how age structure affects populations.

## VOCABULARY

- **carrying capacity** the maximum number of organisms in a population that an environment can support over a long period of time
- **community** the populations of different species living in the same place
- **population** a group of organisms of one species that live in the same place at the same time
- **population density** the number of individual organisms of one species in a community

## TRY THIS

**Population Density**
Determine the area of your classroom in square meters. To determine the population density of the room, divide the number of people in the room by the area. What is the population density of the school? The city? The county?

the number of deer thriving in every five hectares of forest or the number of amoeba found in a drop of water or the number of humans living in one city. Population density is limited by many factors, including light, temperature, nesting spaces, availability of food, and a shortage or excess of nutrients. For example, light limits the population density of plants in a forest. Nesting spaces limit the density of bluebirds in an area. Available food limits the population density of deer in a forest. What other limiting factors can you think of?

**Carrying capacity** is the maximum number of organisms in a population that an environment can support over a long period of time. Factors like food and shelter help determine the carrying capacity for animals. The carrying capacity for plants depends on factors such as available sunlight, water, and space.

Each population has an age structure—the proportion of individuals of different ages. Ecologists look at the age structure of a population to predict what the population will be like in the future. For example, if most of the frogs in a pond are old, that is a bad sign. When the old frogs die off, there will not be many frogs left. But if most of the frogs are young, they will continue to reproduce, and the population will probably continue to thrive.

Age structures are determined by birth rates and death rates. The birth rate is the average number of organisms that are born in a population in a period of time. In general, if the birth rate is higher than the death rate, the population grows. If it is lower, the population shrinks. But population growth also relates to age structure. For example, if half the deer population in an area is less than two years old, the population will grow even if the birth rate and death rate are the same.

The death rate is the average number of organisms that die in a population in a given amount of time. Biologists often study the death rates of different species by

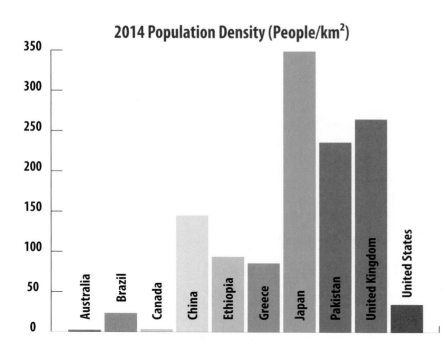

**2014 Population Density (People/km²)**

using mortality patterns. A mortality pattern is the number of individuals in a species that die at certain ages. The mortality pattern for many animals, including most vertebrates, shows a high death rate for the first year of life, followed by a stable death rate for those that live beyond that first year. For instance, many eggs are lost before hatching. In addition, several of the young may die before reaching maturity for various reasons.

The tropical rain forests have the greatest carrying capacity of any land region on Earth.

Every population in an area is important because many populations depend on each other. In many ways these different populations are like parts of one body—if one part is affected, other parts will be as well. Thousands of relationships like this are found in every community. A **community** is the populations of different species living in the same place. A community does not include abiotic factors such as water, soil, or gases. So the pigeons and fruit trees in New Guinea's rain forest are part of the same community. The birds, squirrels, trees, grasses, flowers, weeds, and insects in your backyard make up a community. Communities also include the thousands of organisms that are often overlooked, such as fungi and bacteria.

## LESSON REVIEW

**1.** What is a population?
**2.** Why do population densities vary?
**3.** Why do ecologists examine the age structure of a population?
**4.** What is a community?
**5.** What affects carrying capacity?

### FYI

**Shoo Fly**
Beginning with one female, a fly population would grow to be huge if all offspring survived to reproduce.

| Generation | Number of flies that survive |
|---|---|
| 1 | 120 |
| 2 | 7,200 |
| 3 | 432,000 |
| 4 | 25,920,000 |
| 5 | 1,555,200,000 |
| 6 | 93,312,000,000 |
| 7 | 5,598,720,000,000 |

## OBJECTIVES

- Differentiate between biotic and abiotic factors and identify several abiotic factors and cycles that determine the nature of ecosystems.
- Explain the need for understanding the complexity of the biosphere to better care for God's creation.
- List the five levels of environmental organization.

## VOCABULARY

- **abiotic factor** a nonliving component of an ecosystem
- **biosphere** the part of Earth and its atmosphere in which living organisms exist
- **biotic factor** a living component of an ecosystem
- **ecology** the science of how organisms interact with each other and their environment
- **ecosystem** a system in which organisms and their environment interact

Think about your home and all of the interactions that take place there. Everything runs smoothly when you and your family members communicate and support each other. The members of your household play different roles and contribute to the household in different ways—talking and listening, shopping, making food (and eating it), cleaning the house, taking out the garbage, paying the bills, caring for young children, and repairing things that are broken. Each contribution is important.

When people take care of the earth so birds and fish and all the other organisms can live, they are fulfilling God's command to rule over the earth as He intended. In doing so, they show gratitude to God for the wonderful home that He gave everyone on this planet.

Sometimes people do not preserve creation effectively because their needs require using land in certain ways. For example, the Cuito River flows in the war-torn African country of Angola and helps form the Okavango River in Botswana. All along these rivers are grasslands and tropical forests, filled with wildlife. Some Angolans are now returning to live beside the Cuito River and clearing land to live on. Such clearing causes a negative impact to the environment along the river. In contrast, Botswana is working to restore cleared woodlands, burned brush, and drained wetlands around the Okavango Delta. Farmers had used these damaging methods to eradicate the tsetse fly, a parasite that causes deadly sleeping sickness in humans and in cattle,

The Cuito River is the main and longest river that feeds into the Okavango River, which ends in the Okavango Delta (pictured here), an area protected as a UNESCO World Heritage Site.

## TRY THIS

**Biotic and Abiotic Factors**
Biotic and abiotic factors both play roles in the biosphere. In a large jar or aquarium, build a biosphere model in which biotic and abiotic factors interact. Place a few centimeters of pond water in the jar or aquarium. Add rocks, plants, and snails from the same pond. Seal your "biosphere," and place it in an area where it will receive 12 hours of light per day. Observe what happens over a period of several weeks. How do the biotic and abiotic factors work together? Record the dates and the changes that you observe.

which grazed on the grasslands near the water. Now tsetse-fly traps are being used instead, which control the spread of disease without damaging the environment.

These situations are good examples of the conflict that exists between the needs of a growing human population and concerns for wildlife. Many people worry about the animals and plants that are disappearing in countries where industrialization is changing the habitats native to each region. There is pressure on countries to preserve wildlife, but conflict arises when a choice has to be made between feeding people and saving wildlife.

**Ecology** is the science of how organisms interact with each other and with their environment. The root word for *ecology* is the Greek word that means "home"—*oikos*. Just as at home, people have different roles and make different contributions to the care of the earth. When people care for the earth, they are caring for the home that God gave them—the home that in Genesis 1:28 He commanded people to care for.

The study of ecology gives people the knowledge and tools that are needed to be better stewards of God's world. You can learn how pollution from factories affects the fish in nearby rivers, how draining wetlands to build a mall affects the birds that live there, how recycling reduces pollution and saves natural resources, and how being careful about what you buy can help the whole Earth.

© *Life Science*

A burned or dead tree is an ecosystem—a system in which organisms and their environment interact.

If a tree in a field is blown over, what do you think you will see if you passed by a year later? Do you think you would see just a dead tree or a place where many living things grow and thrive? Caring for the earth first comes with recognizing all the forms of life God placed on it. In this case, moss now grows on one side of the tree. Insects have made their homes inside and outside of the rotting log. A shelf fungus pokes out from the side of the tree. A chipmunk darts inside a hollow space. Even though you cannot see them, bacteria are breaking down the wood. And if you look under the rotting log, you will see various slugs and grubs.

That tree is an **ecosystem**—a system in which organisms and their environment interact. The earth contains a wide variety of ecosystems. An ecosystem can be as large as an ocean or as small as the rotting log. The ecosystem includes living and nonliving things—the interactions of these factors determine what an ecosystem is like.

Many interactions happen in an ecosystem. The water cycle is the movement of water throughout the biosphere. God designed solar energy to drive this cycle, which consists of cold and warm ocean currents, clouds, and precipitation, as well as groundwater and surface water. The water cycle, nitrogen cycle, oxygen and carbon dioxide cycle, and other cycles interact constantly. This

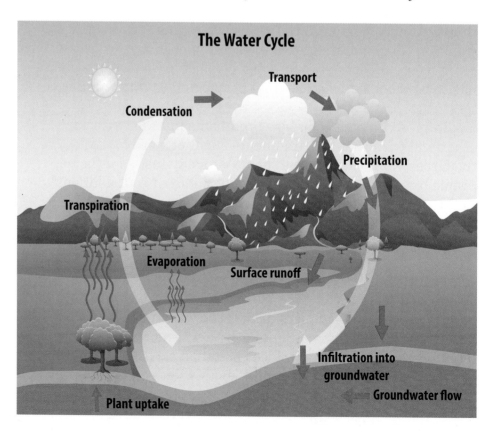

The Water Cycle

Condensation

Transport

Precipitation

Transpiration

Evaporation

Surface runoff

Infiltration into groundwater

Groundwater flow

Plant uptake

interaction maintains the important balance between the biotic and abiotic factors in an ecosystem. For example, plants and animals recycle carbon, oxygen, nitrogen, and other minerals. Meanwhile, bacteria recycle important nitrogen compounds for plants to use and grow.

Ecosystems are woven together. They are not closed, separate systems divided by sharp boundaries. Most overlap in what are called *transition zones*. For example, transition forests in Canada form a boundary between the coniferous forests to the north and the forests of broadleaf trees to the south. Wetlands,

 **FYI**

### Biosphere 2
Biosphere 2 is a 3.14 acre closed system in Oracle, Arizona, that was built to study how the earth works and how people interact with the biosphere. Biosphere 2 contains five biomes: an ocean, a rain forest, a desert, a mangrove wetland, and a prairie (savanna grassland). The system—made of glass, steel, and concrete—is 27.7 m tall at its highest point. Inside, over 1,000 sensors monitor the air, soil, and water.

Four women and four men entered Biosphere 2 in 1991 and lived there until 1993. A second crew of biospherians entered Biosphere 2 in 1994. Each crew grew and harvested crops by hand. They also ate certain animals living in Biosphere 2.

The biospherians were all highly trained in specialized fields such as chemistry, engineering, SCUBA diving, marine biology, computers, and agriculture. This knowledge was important for living in Biosphere 2. However, many technical problems caused the project to fail. In 2011 the University of Arizona took over Biosphere 2 and is using it as a laboratory to study climate change and water.

## TRY THIS

**Survey an Ecosystem**
Survey an ecosystem near your school. Observe and record biotic factors and the way they interact with one another. For example, a shrub may provide shelter for a rabbit, or an insect may feed on a piece of grass. Also record the abiotic factors in the ecosystem. How is each biotic factor and abiotic factor important? Do you think that the ecosystem you have chosen will continue to thrive?

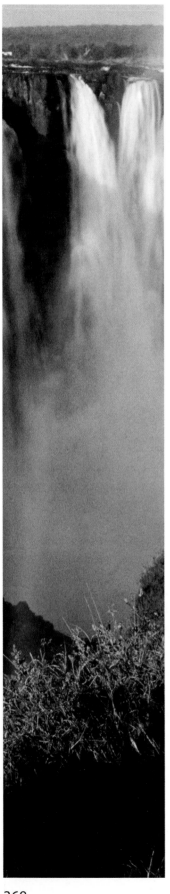

### Rachel Carson

Rachel Carson (1907–1964) was a famous biologist who wrote about the effects of pollution on the natural world. Even as a young child Carson was interested in the outdoors. She went to college to become a writer; however, she later decided to study science instead. Eventually she combined science and writing, publishing books about the sea: *Under the Sea-Wind*, *The Sea around Us*, and *The Edge of the Sea*.

In the 1940s and 1950s, the chemical DDT was widely used to kill mosquitoes and other pests. When a friend sent Carson a letter describing how several songbirds had fallen dead after DDT was sprayed, Carson decided to research the pesticide. She discovered that insects and other pests were becoming resistant to DDT and that DDT was killing wildlife and contaminating crops. The bird population was dropping, and DDT was responsible.

In 1962 Carson wrote a book called *Silent Spring*, which warned people that DDT kills birds. Many scientists supported Carson's research. *Silent Spring* became a best seller, and President John F. Kennedy set up the President's Science Advisory Committee to investigate pesticides. The Environmental Protection Agency (an organization that was Carson's idea) banned DDT in the United States in 1972. *Silent Spring* was translated into a dozen other languages, and many countries now regulate pesticide use as well.

In honor of her conservation efforts, the Rachel Carson National Wildlife Refuge was established in 1966. She was also the recipient of the National Audubon Society Medal. Today, the National Audubon Society presents the Rachel Carson Award honoring American women whose work has greatly advanced conservation locally and abroad.

Rachel Carson said, "Conservation is a cause that has no end. There is no point at which we will say, 'Our work is finished.'"

meadows, and young forests also often mesh together. In some cases, however, ecosystems do change drastically within a short distance. For example, at a certain altitude called *the tree line*, constant wind and other harsh conditions separate mountain forests from lichens and shrubs. In Hawaii, only one mountain separates a desert from one of the world's wettest rain forests! Because neighboring ecosystems are connected to each other, caring for even one part of one ecosystem helps not only that ecosystem but the ones nearby. The same is true for harming one part of an ecosystem. For example, imagine what would happen to a field ecosystem if a neighboring pond were drained.

You might wonder why you should care about what someone in Botswana does to the trees on the other side of the planet. It is because the entire Earth is one large, connected system called *the biosphere*. The **biosphere** is the part of Earth and its atmosphere in which living organisms exist. Except for the energy from the sun, Earth is a closed system. All of the gases, food, fertile soil, and water that living things need to survive is found in this system.

Earth's biosphere extends from the atmosphere to the bottom of the oceans.

**Five Levels of Environmental Organization**

**Organism**
A member of a species

**Population**
Two or more members of the same species living together

**Community**
Populations of different species living side by side

**Ecosystem**
A community of organisms living in an area plus the nonliving things such as air, water, and soil

**Biosphere**
The parts of the earth that can support life

What biotic and abiotic factor can you identify in this picture of Patagonia, Chile?

Two sets of factors work together in the biosphere. **Biotic factors** are things that are living or that were once living in an ecosystem. **Abiotic factors** are the nonliving things in an ecosystem. These factors include the water, soil, light, and temperature. Some interactions between biotic and abiotic factors help maintain the biosphere. For example, plants change carbon dioxide and water into oxygen and sugar. People grow and eat plants. People plant forests. Animals eat plants and each other. Bacteria convert dead matter into nitrogen and phosphorus, and they change nitrogen in the air into chemicals that plants need in order to grow.

However, some interactions between biotic and abiotic factors can be harmful. People reroute rivers, plow up land, and cut down trees. Pesticides like DDT harm birds and other wildlife.

People can help maintain balance in the biosphere by thinking twice about using paper they do not need and carpooling or walking when possible. They can even spend time researching the products they buy to make sure that none of the materials were obtained in ways that damage God's Earth. It is the responsibility of everyone on Earth to care for God's creation.

Ecologists are just beginning to understand how different ecosystems stay in balance. This knowledge helps people keep the biosphere healthy. When people study the complexity of the ecosystems, they increase their awe of God's creative powers. Greater understanding of the ecosystems and the biosphere help people to be better stewards of God's creation. No one but God could create such an intricate system and keep it running smoothly.

## LESSON REVIEW
**1.** What is an ecosystem?
**2.** Give three examples each of biotic and abiotic factors.
**3.** What are the five levels of environmental organization?
**4.** What cycles determine the nature of any ecosystem?
**5.** Give two examples of how your choices affect the biosphere.

If you have ever gone swimming in the ocean, you may know what it is like to get a mouthful of saltwater. Once that happens, you are careful not to let it happen again. But you do not have that problem swimming in freshwater. Freshwater is water without salt. Freshwater does not necessarily lack pollution, however.

Approximately 3% of the world's water is freshwater. Two-thirds of the earth's freshwater is stored in the ice and snow of the polar ice caps. Much of the remaining one-third of freshwater is groundwater. Groundwater is the slow-moving water trapped in the soil, which acts like a giant sponge for rainwater. The final bit of Earth's freshwater is found in freshwater lakes, rivers, ponds, streams, and in the atmosphere.

God designed water to sustain His world. Unlike most other substances, solid water, or ice, is less dense than liquid water, so ice floats on water. If it did not, freshwater life in temperate zones of the world could not exist because ice would build up on the bottom of ponds, streams, and rivers, where many creatures normally hibernate for the winter. Water is also transparent, allowing sunlight through, which is necessary for underwater photosynthesis.

Much of the life in a freshwater environment is plankton. Plankton are microscopic organisms that live near the surface of the water. A wide variety of protists, rotifers, crustaceans, and water fleas also live in freshwater. Many worm species, including

## OBJECTIVES

- Name and describe four freshwater ecosystems.
- Explain the interaction between the abiotic and biotic factors in freshwater ecosystems.

## VOCABULARY

- **wetland** an area of land where the water level is near or above the ground surface for most of the year

**The World's Water**

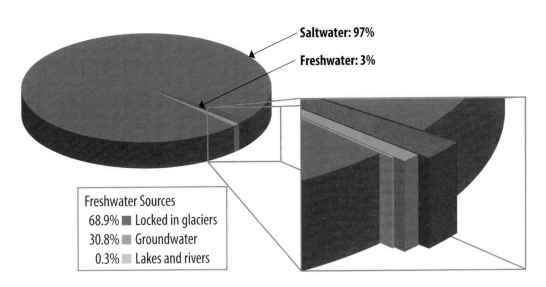

Saltwater: 97%

Freshwater: 3%

Freshwater Sources
68.9% ■ Locked in glaciers
30.8% ■ Groundwater
0.3% ■ Lakes and rivers

*Austropotamobius pallipes*, a European freshwater crayfish

microscopic worms, leeches, and planarians, live in freshwater.

Freshwater also provides habitats for many vertebrate species. Many different species of fish, frogs, salamanders, turtles, and snakes live in freshwater. Mammals such as beavers, otters, and muskrats make homes in streams and lakes. Raccoons, marsh rabbits, and many species of mice find food and protection near freshwater. Waterfowl, such as ducks and geese, along with dozens of other bird species, such as warblers and osprey, depend on these habitats for food and water.

Common freshwater ecosystems include streams, ponds, lakes, and wetlands. A stream is characterized by its flowing water. A stream usually flows fastest in the spring because of extra rainfall or melted snow. The more the water is tossed about in a stream, the more oxygen it has. The flow and the turbulence of the water also determines how much food washes down a stream.

God designed the organisms that live along a stream to have features that keep the water from sweeping them away. The plants have strong roots and flexible stems. The aquatic creatures have streamlined bodies that allow water to easily flow around them. Hooks and claws help organisms cling to objects in fast-moving water. Gills allow fish and aquatic insect larvae to breathe underwater. Some organisms that live in or near streams are moss, snails and clams, freshwater crustaceans (crayfish, scuds, and isopods), and insects. Many insect species spend up to two years living in freshwater before developing into adults. Insect larvae and nymphs are the most important food source for stream fish. Craneflies and mayflies are two common stream fly species, and water striders, water boatmen, and back swimmers also live in streams.

A pond is a shallow body of water with a muddy or silty bottom. Because the water in ponds is still, it has less oxygen than the water in streams or rivers. Many living things thrive in and around ponds. Ponds are often covered with floating algae and ringed with cattails and other vegetation. Ponds provide many animals, such as frogs and

The more the water is tossed about in a stream, the more oxygen it has.

turtles, with moist homes and plenty of food. Ponds also house countless microscopic organisms. Oxygen levels in ponds vary greatly during the day because the many pond animals and bacteria produce a lot of carbon dioxide in the water.

Deep lakes are also important freshwater ecosystems. In the summer, these lakes form a warm layer of water that does not mix with the deep cold layer. These layers mix in the spring and the fall (when water temperatures drop at the surface), refreshing the deeper parts of the lake. Bass, bluegill, leeches, hydras, crayfish, trout, and phytoplankton live at depths where light penetrates the water. Catfish, carp, fungi, bacteria, worms, and insect larvae live in the depths where light does not reach. Grasses, such as sedges and rushes, line the shores, along with ferns, water milfoil, water lilies, and buttercups.

A **wetland** is an area of land where the water level is near or above the ground surface for most of the year. A variety of plants and animals live in wetlands. Wetlands also play a very

A pond in South Australia

A swamp is a wetland where trees and vines grow. Swamps are often found in low-lying areas and near slow-flowing rivers. Swamps are usually only flooded for part of the year, depending on rainfall. Willows, elms, oaks, and cypresses are some swamp trees. Swamp vines include poison ivy. Other swamp plants include water lilies, swamp iris, and Spanish moss. Swamps are also rich in algae. Swamp animals include cottonmouth snakes, bobcats, foxes, blue heron, pileated woodpeckers, and raccoons.

A swamp in the Everglades of Florida, USA

important role in flood control by soaking up excess water. The water seeps underground, replenishing groundwater supplies.

A marsh is a treeless wetland where grassy vegetation grows. Freshwater marshes are often found along the shallow shores of lakes, rivers, streams, and ponds. Common marsh plants include grasses, reeds, rushes, and wild rice. Marsh animals include turtles, frogs, muskrats, and birds such as red-winged blackbirds.

Life in freshwater ecosystems depends on a balance of gases and nutrients. Algae provide both oxygen and food, but most aquatic life also depends on the surrounding land. For example, riparian vegetation—waterside plants—adds nutrients to streams. These plants drop leaves, fruits, and twigs into streams. Bacteria then decompose the plant parts into nutrients. The shade provided by riparian vegetation also lowers the water temperature. Because cold water holds more oxygen than warm water does, riparian vegetation helps maintain the necessary balance of gases for the water-dwellers. In addition, these plants and their roots provide homes for many insect larvae, worms, and other stream organisms. Many stream fish then eat these insects and worms.

Wetlands are very important ecosystems. In addition to the abundance of wildlife, wetlands also filter out pollutants and buffer inland areas from storms, waves, and flood damage. Unfortunately, wetlands are often drained so the land can be used for houses or businesses. Wetland destruction can upset the important balance that God intended in the biosphere.

## LESSON REVIEW

1. Name four freshwater ecosystems.
2. Compare and contrast a pond and a stream.
3. Aside from pollution, why would a stream running through a city not be as healthy as one running through a more unpopulated ecosystem?
4. Differentiate between a marsh and a swamp.
5. Why is the draining of wetlands a problem?
6. Give an example of an interaction between biotic and abiotic factors in a freshwater ecosystem.

Humans need clean water to live. In the past, people depended on local wells for their water. Today most urban areas have water-treatment plants to help assure that the water from the faucet is clean and safe to drink. In rural areas, soil filters contaminants from well water. People in many places take clean water for granted. Sometimes the water in a region becomes temporarily unsafe to drink because of pollutants in the water. A **pollutant** is a substance or condition in the air, soil, or water that is harmful to living things. When pollutants are found in drinking water, people buy bottled water or boil water from the faucet until the problem is solved. In some regions very little freshwater is available. People in these regions use different technologies to get freshwater from saltwater.

**Water hardness** is the term used to describe the amount of the minerals magnesium and calcium present in water. Water hardness is not considered pollution. Fish and mollusks use these metals in small amounts, so their water needs to have some hardness. The hardness of a body of water depends on the minerals that make up the soil that the water flows over and through to get to the pond or stream. The water hardness of freshwater can vary from 10 ppm (parts per million) to 800 ppm. The water in mountain and forest streams is often soft. Streams that flow through prairies and limestone landscapes have much harder water because of the minerals that the water picks up.

## OBJECTIVES

- Explain how water hardness, sediments, pollution, and acid rain affect the chemistry of water.

## VOCABULARY

- **acid rain** a type of rainfall made acidic from pollution in the atmosphere
- **pollutant** a substance or condition in the air, soil, or water that is harmful to living things
- **sediment** the solid material that settles to the bottom of a liquid
- **water hardness** the amount of magnesium and calcium in water

The wearing away of rocks and soil by water or wind is called *erosion*. When water flows across the ground, it picks up sediments. **Sediments** are solid materials that settle to the bottom of a liquid. Soil from construction or farming is a sediment. You might not think that soil, a natural substance, could cause pollution, but it does. It can pollute water and turn it brown. Sediments can be the worst form of stream pollution because they block light from the water, making it difficult for plants to photosynthesize. Sediments also add extra nutrients to the water, which causes too much algae to grow. When the extra algae die, bacteria decompose them, which uses up the oxygen fish and other organisms need to live. Sediments also fill in the habitats of insects, worms, and other stream invertebrates that are food for fish. Sediment particles can clog fish gills. Then the fish cannot breathe normally.

| Water Hardness Scale | | |
|---|---|---|
| Grains/Gallon | mg/L and ppm | Classification |
| Less than 1 | Less than 17.1 | Soft |
| 1 – 3.5 | 17.1 – 60 | Slightly hard |
| 3.5 – 7 | 60 – 120 | Moderately hard |
| 7 – 10 | 120 – 180 | Hard |
| Over 10 | Over 180 | Very hard |

Temperature changes can also cause problems in a freshwater ecosystem. Remember that cold water holds more oxygen than warm water. When people cut down trees and clear vegetation from the stream banks, the stream receives less shade. When this happens, the water warms up and loses oxygen, causing many organisms to die from lack of oxygen. Water temperatures rise when factories and electricity-generating power plants pump hot water into lakes and rivers.

**Acid rain** also causes water pollution. Acid rain is the result of sulfur dioxide and nitrogen oxide reacting in the atmosphere with water and falling to Earth as rain, fog, snow, or sleet, or condensing as dew. The pH scale is a scale that measures the acidity of a solution—such as water. The pH scale ranges from 0 to 14. A solution with a pH of less than 7 is acidic. A pH of more than 7 is basic. A neutral pH is 7, the desired balance of acid and base for drinking water. Acid rain lowers the pH of streams, lakes, and rivers, which kills aquatic life.

The water on the left contains sediments.

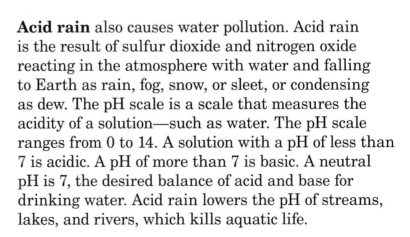

Ammonia, a basic chemical, comes from the waste products of fish and from decaying plants and animals. Too much ammonia raises the pH level so the bacteria and fungi can thrive. These bacteria and fungi cause fin and tail rot, infections, sores, cloudy eyes, body slime, and mouth fungus in fish. Ammonia also prevents a fish's blood from carrying oxygen. Even small amounts of ammonia can harm aquatic creatures.

The nitrates from sewage and fertilizers can also pollute water. These pollutants can cause plants and algae to grow excessively. Excess plants and algae affect the dissolved oxygen levels in water and create stressful conditions for fish. Nitrate levels are also harmful to people. Nitrates prevent red blood cells from carrying oxygen.

## LESSON REVIEW

1. Which would you expect to have harder water—a stream or a lake? Why?
2. Why is soil, a natural substance, sometimes considered a pollutant? How do you think this pollutant finds its way into bodies of freshwater?
3. Which would you expect to have more dissolved oxygen—a cold stream or a warm lake? Why?
4. How does acid rain change the pH of water?

Acid rain causes chemical weathering.

# 7.2.1 Food Chains and Food Webs

## OBJECTIVES

- Determine the various relationships between producers and consumers in a food web.
- Identify the trophic levels in an ecosystem.

## VOCABULARY

- **biomass** the total mass of all the organisms in a given area
- **detritivore** an organism that feeds on dead matter
- **primary consumer** an herbivore
- **secondary consumer** an animal that eats herbivores
- **tertiary consumer** an animal that eats primary carnivores
- **trophic level** a group of organisms that share the same position in a food chain

*There is a time for everything, and a season for every activity under the heavens:*

*a time to be born and a time to die,*
*a time to plant and a time to uproot. . . .*

*He has made everything beautiful in its time. He has also set eternity in the human heart; yet no one can fathom what God has done from beginning to end.*

*—Ecclesiastes 3:1–2, 11*

God designed the world to run in cycles. The earth's orderly rotation ensures 24-hour days. The earth's revolution around the sun provides predictable seasons. The Arctic tern's instinct about environmental clues tells it when to migrate. In every ecosystem, flowers bloom with masterful precision. Some trees regularly drop their leaves, birds follow flight patterns hundreds of kilometers long, salmon swim thousands of kilometers to spawn in the same creek where they hatched, and desert frogs are inactive for months as they wait for rain. Every part of God's creation depends on the intricate timing He established.

## Food Web

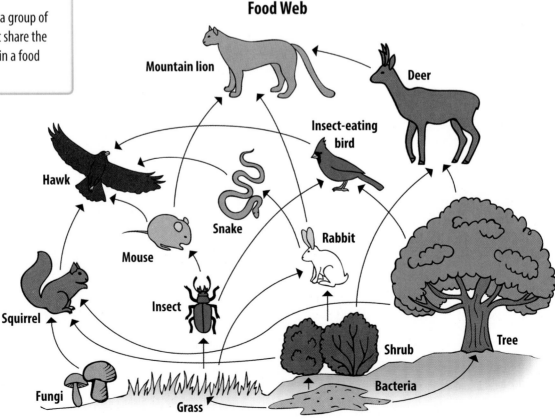

Have you ever heard the song "I Know an Old Lady Who Swallowed a Fly"? "She swallowed the cat to catch the bird, she swallowed the bird to catch the spider, she swallowed the spider to catch the fly." You could say that this song describes a food chain—the food and energy links in a community. Many food chains are as simple as the food chain in that old song.

Most organisms are part of many different food chains. Rabbits, for example, may be food for lynx, wolves, mink, raccoons, and wolverines. And the same grasses that rabbits eat are also food for deer, grasshoppers, and muskrats. The red fox eats mice, insects, and grass. So it makes more sense to speak of the network of food chains in an ecosystem as a food web. A food web consists of many food chains and may involve more than 100 species of living things.

Even though food webs vary greatly, they exist as an orderly system to transfer the energy of the sun throughout the biosphere. Even when it is cloudy and gloomy for a long stretch, God uses the sun to sustain all life on the earth. He structured creation so the earth receives energy from the sun. The sun provides plants the energy they need for photosynthesis. About 30% of the sun's energy is reflected back into space. Most of the incoming solar energy is absorbed by the atmosphere or by the surface of the earth itself, which is mostly covered by water. Living things use only a very small amount of the sun's energy. Through photosynthesis, this energy is converted into what fuels the growth of billions of metric tons of plants and animals

## TRY THIS

**Tropic Level Search**
Search an ecosystem near your school, and identify the organisms you find by their trophic level. Woodlands and meadows will provide the best results. Make five lists identifying producers, herbivores, carnivores, scavengers, and decomposers. (Even though you cannot see bacteria, you can see evidence of decomposing organic matter.) Use a field guide and a hand lens to find and identify different organisms.

Just because an animal is large does not mean it is a carnivore. Elephants eat only plants, which means they are herbivores and primary consumers.

## FYI

**Short Chained**

Food chains usually have three or four links. However, they have a limit of five links. Food chains are short because of the way that energy is transferred from one trophic nourishment level to the next. With each tropic level, more energy is lost and less is stored. About 10% of energy stored in plants is turned into herbivore biomass, and about 10% of herbivore biomass is turned into the organic material of primary carnivores, and so on. But this energy is not really lost as it travels up the food chain. Each organism in the food chain uses some of this energy in its own life processes and activities (such as metabolism, movement, and staying warm). Food chains, even though they are short, are part of the goodness of creation!

every year. The total mass of all the living organisms in an ecosystem at one time is called the **biomass**. Ecologists measure the biomass of an ecosystem to evaluate its general health. By knowing the biomass, scientists can determine how many species an ecosystem can support.

A food chain groups organisms based on what food they eat. A **trophic level** is a position in a food chain. Producers, which are organisms that use photosynthesis to make their own food, make up the first trophic level. Plants, algae, protists, and some bacteria are producers. Producers are the foundation of the food chain.

The next three trophic levels contain consumers—organisms that eat other organisms. Herbivores are animals that eat plants. Herbivores vary from ecosystem to ecosystem. Caterpillars, rabbits, deer, sea urchins, and elephants are all herbivores. Herbivores are the energy bridge between plants and the rest of animal life on Earth. Carnivores are animals that eat other animals. Like herbivores, carnivores are a diverse group. Spiders eat flies, hawks eat mice, and lions eat zebras. Omnivores are animals that eat both plants and animals. Rats,

**Trophic Levels**

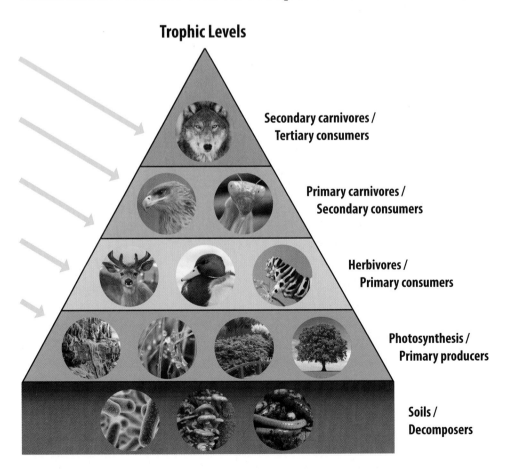

Secondary carnivores /
Tertiary consumers

Primary carnivores /
Secondary consumers

Herbivores /
Primary consumers

Photosynthesis /
Primary producers

Soils /
Decomposers

raccoons, and bears are examples of omnivores. Humans are also omnivores.

Detritivores are a fifth, and unique, trophic level. **Detritivores** are organisms that feed on dead matter. Detritivores include scavengers such as vultures, hyenas, snails, and worms that feed on the bodies of dead animals—and decomposers—organisms that get their energy by breaking down the remains of dead organisms or animal wastes. Decomposers recycle nutrients back into the soil for plants to use. Many bacteria and fungi are decomposers.

The owl is an example of a secondary consumer because it is eating a mouse—a primary consumer. However, if the owl were to eat a shrew, it would be an example of a tertiary consumer because the shrew is a secondary consumer.

These different trophic levels are related to the two main groups in food webs: producers and consumers. The foundation of food webs are the producers, which transmit the sun's energy to consumers. The consumers in a food web can serve one of the following different roles:

- **Primary consumers** are herbivores. These vary from ecosystem to ecosystem. In the ocean, shrimplike creatures called *krill* are the primary consumers. Insects, small rodents, and deer are the primary consumers in many forest ecosystems. Minnows are some of the primary consumers in a pond.
- **Secondary consumers**, or primary carnivores, are animals that eat herbivores. The largest animal on Earth, the blue whale,

Sharks are carnivores that eat other carnivores, making them tertiary consumers. Sharks are also top carnivores.

feeds on krill and is considered a primary carnivore. Both a weasel that eats rabbits and a robin that eats caterpillars are primary carnivores. This level also includes omnivores.
- **Tertiary consumers**, or secondary carnivores, are animals that eat primary carnivores. Ospreys are secondary carnivores when they eat trout, which are primary carnivores that eat aquatic insects. But when an osprey eats a goldfish, which eats algae, it is a primary carnivore.

Top carnivores, or top predators, are animals that have no predators as adults. Top carnivores are at the end of a food chain. As healthy adults, top carnivores hunt and eat many other animals. In aquatic ecosystems, top carnivores are usually secondary carnivores (tertiary consumers)—killer whales and great white sharks eat sea lions, which eat fish. On land,

Part of a clown fish's niche is its symbiotic relationship with a sea anemone. The sea anemone protects the fish, and the clown fish increases the rate of the anemone's growth.

## FYI

**Biogeochemical Cycles**
A biogeochemical cycle is the cycling of a chemical element through the living and nonliving parts of an ecosystem. Hydrogen, oxygen, carbon, and nitrogen make up the majority of all organic matter. Cycles allow these elements in the ecosystem to be reused without running out.

## TRY THIS

**Joining the Food Chain**
Make a list of all of the foods that you ate yesterday. Determine each food's place in the food chain. List the producers and the consumers.

top carnivores are often primary carnivores (secondary consumers)—wolves eat deer, moose, and small rodents; lions often eat zebras.

In some ecosystems, the top carnivores have disappeared or are in danger, mostly because they have been poached (illegally hunted) or their habitats have been destroyed. Many people believe these animals need to be saved—not only because they are a beautiful part of God's creation but also because top carnivores are important to ecosystems. They keep other populations in check. For example, when wolves are displaced in an ecosystem, the deer population may grow quickly. The increased number of deer eat so many plants that eventually some plant species may disappear from the area. The diminished plant supply may mean that some deer starve or leave the area to look for food. The decrease in plant and animal diversity cuts the strands of food webs and harms the ecosystem.

The tiniest members in food webs are bacteria. Some bacteria break down decaying matter into nutrients and other elements that many organisms need. If this dead matter were not decomposed, many organisms would starve. Some bacteria play an important role at the beginning of food chains. For example, cyanobacteria can photosynthesize, and nitrogen-fixing bacteria provide important nitrogen compounds for plants. Without these nitrogen compounds, plants could not carry out photosynthesis. So even the tiniest members of creation are important!

Every organism has a unique role in its environment. That role is called *a niche*. An organism's niche includes everything that the organism does and needs—for example, its habitat, its food, the organisms with which it interacts, and the physical conditions that it needs to thrive. A niche is like a person's address, job, and hobbies.

## LESSON REVIEW
1. Distinguish between a food chain and a food web.
2. What are the trophic levels in an ecosystem?
3. Name a producer in your area.
4. Describe a food web from an ecosystem in your neighborhood or town. Identify the organisms that are at each trophic level.
5. Why do you think God placed so many more producers on Earth than herbivores and carnivores?

Think of a simple community—perhaps a small pond. Even a pond community is home to many different species. Carp and tadpoles swim in the water. Frogs splash at the water's edge, catching flies or other insects. Cattails ring its shore. A turtle skims through the algae on its way across the pond. Mosquito larvae float on the water. God created the dozens of species in even the simplest communities to interact in many ways.

Sometimes the species do not directly affect one another. Such interactions are called *neutral interactions*. For example, a heron has little effect on a cattail that grows at the edge of a pond. At other times, however, species directly affect each other. **Commensalism** is a relationship between two organisms in which one organism benefits and the other is not affected. For example, trees provide birds with nesting sites and protection from predators, but the birds might not help the trees. Another example of commensalism is the remora fish—a small black fish that attaches itself to whales and large fish, such as sharks and manta rays. The remoras eat the leftover pieces of the large marine animal's food, but the large marine animal gets nothing in return from the remoras.

Mutualism is the relationship between two organisms that live and work together for the benefit of both. For example, certain fungi that live on tree roots provide trees with water and

## OBJECTIVES

- Identify several community interactions among different species.
- Compare the difference between predators and parasites.

## VOCABULARY

- **commensalism** a relationship between two organisms in which one benefits and the other is not affected

A whale shark with remoras attached

People provide food and shelter for dogs, and dogs supply protection and companionship for their owners.

minerals. In return, the fungi receive sugar and nitrogen compounds from the tree roots. Forests depend on these relationships. The algae that live inside coral are another example of mutualism. The food made by the algae provides energy for the coral to build the reef. The algae receive protection and vital nutrients from the coral for their own growth and development. Mutualism also exists between cattle egrets and cows. The egrets dine off the insects on the cow's fur, and the cows get their fur cleaned.

Not all species interact so agreeably. Sometimes they compete for food, water, space, or other needs. Intraspecific competition is competition within a species. Since members of the same species have exactly the same needs, this competition is much fiercer than interspecific competition, which is competition between two different species. For example, turtles and frogs may compete for the same food, but their competition is not as fierce as two deer competing for the same plant. The more similar two organisms are, the fiercer their competition is.

Other community interactions include predator-prey and parasite-host interactions. A predator is an organism that eats other living organisms but does not live on or in its prey (an organism eaten by a predator). Lions that kill and eat zebra are predators. So is a spider devouring a fly. A parasite is an organism that depends on another living organism and lives in, on, or near that organism. The organism on which a parasite depends is its host. The parasite may sometimes kill the host, but it usually does not—if it did, it would have to find a new host. Parasites include tapeworms living inside human intestines and lice or ticks living in the fur of mammals.

Many parasites, such as tapeworms, feed off their host, but other parasites depend on the host for things other than food. Sometimes they depend on another species' social behavior. Cow birds, for example, never build their own nests. They just remove an egg from another bird's nest and replace it with one of their own eggs. Some birds hatch the foreign egg and raise the chick as one of their own brood.

Most organisms have parasites. You may find the idea of parasites disgusting, but they do the important job of keeping

In this predator-prey interaction, can you tell which animal is the predator and which is the prey?

their host populations under control. Parasites are not usually a problem for people because of modern medicine and technology.

Some organisms live in a symbiotic relationship. Symbiosis is a relationship in which one organism lives on, near, or inside another organism, and at least one of the organisms benefits. For example, some fish, such as groupers, line up in front of rocks so certain shrimp or small fish can eat small parasites off their scales and from between their gills. This relationship feeds the cleaning shrimp and reduces parasitism for the fish. These fish do not eat the cleaning shrimp, and the shrimp are not afraid of the fish.

## LESSON REVIEW

**1.** What is commensalism?
**2.** Name three types of community interactions among different species and give an example of each.
**3.** How do the different interactions among organisms show God's wisdom?
**4.** What is the difference between a parasite and a predator?

### TRY THIS

**Human Interactions**
Think about the many interactions in the natural world: predator-prey, parasitism (parasites), mutualism, commensalism, interspecific competition, and intraspecific competition. How do human beings interact with each other in the same ways, either as individuals or as groups? Write down one comparison for each type of interaction.

Parasites are not just animals. Some plants, such as mistletoe, are parasites on trees and shrubs, stealing their food and water.

### OBJECTIVES

- Identify examples of physical features that help organisms survive.
- Discuss several defenses that prey use to avoid or startle predators.
- Explain how various animals reduce competition through different behaviors or physical developments.

### VOCABULARY

- **adapt** to adjust when survival is threatened
- **moment-of-truth defense** a behavior used by prey to startle a predator, giving the prey a chance to escape

Think about all the different species of plants and animals that live in your region. Why do these organisms all survive? Why do the herbivores not kill off all the producers? Why do the carnivores not kill off all the herbivores? Since each species is important, God equipped each one with ways to survive in the environment He designed. Overall, this achieves the balance that God intended for His world. Many species have specific responses to help them deal with predators or competition. Various species have an amazing array of body forms, colors, and patterns to help them survive in their communities.

The way an organism responds to changes in its environment is also part of its design. For example, when the weather turns cold in the fall, bears hibernate, but wolves in the same area grow a thicker coat of fur. Both animals respond to the change in the weather in the way God intended for them to. God also designed organisms to **adapt** when their survival is threatened. During a severe drought, herbivores must find a different local source of food because the plants they normally eat wither and die. Or the animals may move to another area with plenty of plants to eat. Both alterations in behavior are examples of how animals can adjust to changes in their environments. Some organisms are

better at adapting than others. However, certain behaviors, such as the ability to swim or fly, cannot be changed.

God gave organisms features that help them thrive in their environments. These features are especially interesting in the predator-prey relationship. The interactions between predator and prey are balanced by these features. Some predators have great speed. For example, the cheetah can reach speeds of up to 120 kph so it can chase down prey. Most vertebrate predators have eyes in the front of their heads to give them better depth perception as they pounce on prey. Predators also have built-in weapons. Birds of prey have talons and hooked beaks; cats have claws; wolves have strong jaws; sharks have razor-edged teeth; and moray eels have a keen sense of smell for following fish underwater. Some predators use camouflage so prey will not see them; the snow leopard's color helps it blend in with snow, and the tiger's stripes mimic shadows in the grass.

The black heron uses its wings to shade shallow water from the hot sun. It then snatches the minnows that come to enjoy the cooler water.

Predators also behave in ways that give them an advantage while hunting. Polar bears crash through thick ice in search of sleeping seals. Lions and chimpanzees hunt in groups to increase their chances of catching prey. Some predators use disguises. The scorpion fish looks like a rock on the ocean floor as it waits for unsuspecting prey. Other predators try to eliminate their competition for food. The main predators of cheetah cubs are lions seeking to cut down the number of competitors.

You may think that prey have no chance against such well-equipped predators. But God created prey with built-in defenses to help them escape predators. Rabbits have eyes on the sides of their heads for a greater range of vision to see predators, and they can also run very fast. Turtles and snails have hard shells that shield them from attackers. Camouflage in form, pattern, and color helps different kinds of prey blend into their environment: walking sticks, insects that look like dead sticks, blend in with trees; and chameleons change color to match their surroundings.

This owl blends in perfectly with the tree making it well camouflaged.

are slowed, hibernating animals wake up in response to danger. The biological clocks of other animals trigger migration. Many bird species travel great distances between winter and summer nesting sites. The Arctic tern, for example, breeds in the Arctic during the short summer. Within three months the hatchlings must be ready to migrate thousands of kilometers to Antarctica. In the summer, caribou and reindeer migrate from coniferous forests to the tundra to feed.

Some organisms have lunar rhythms, which are related to the cycles of the moon. Sea creatures, for example, respond to the moon to migrate or spawn. Corals in the Caribbean wait for a four-day period during a new moon in the spring to release their eggs into the sea. Thousands of horseshoe crabs respond to the spring tides brought on by the moon and scuttle onto beaches to lay millions of tiny green eggs.

## LESSON REVIEW

**1.** How did God design creatures to survive?
**2.** Name three physical features that help predators catch prey.
**3.** Name three physical features that help prey escape predators.
**4.** Name three defensive behaviors prey use to escape predators.
**5.** Explain how various animals reduce competition.

Terrestrial frogs usually hibernate on land. Frogs that are good diggers burrow deep underground below the frost line. Frog species that are not adept at digging find openings in logs or rocks to hibernate. Some dig down as far as they can in the leaf litter. These frogs are not as well protected from the cold temperatures and may freeze.

Natural disasters such as earthquakes, floods, hurricanes, or forest fires can devastate entire towns. People put their lives on hold while roads and bridges are repaired, and they try to save what they can from their ruined houses. Volunteers come in to help clean up the damage and to help residents rebuild and repair. If nobody cleaned up the mess, the ruins would not just disappear. They would remain for years—maybe even decades!

Ecosystems change constantly. Populations gain and lose individual organisms. Trees fall down and rot. Some animals migrate, and others hibernate for the winter. All the while God maintains the ecosystem so it adjusts smoothly to these routine but constant changes. Even when a natural disaster sweeps through an area, the affected ecosystem immediately begins to renew. Although many people see natural events like forest fires as disasters, they are not necessarily disastrous for the ecosystem. Fires, for example, can clear away dead wood and old trees, making room for new ones. The burning returns nutrients from the dead plants to the soil. In fact, some pine trees only release the seeds from their cones when they are burned.

**Ecological succession** is the process of gradual change in an ecosystem. There are two kinds of ecological succession: primary succession and secondary succession. **Primary succession** is the series of changes, including the establishment of plants, that occurs in a newly formed, barren habitat. One of the best examples of primary succession is found in the formation of a sand dune. During a low tide, wind dries the sand on shore and blows it beyond the point the water reaches at high tide. Beach

**OBJECTIVES**

- Explain the two processes of ecological succession.
- Describe human effects on ecosystems.

**VOCABULARY**

- **ecological succession** the process of gradual change in an ecosystem
- **primary succession** the series of changes occurring in a newly formed, barren habitat
- **secondary succession** the series of changes occurring in an area where the existing ecosystem has been disturbed

**TRY THIS**

**Succession**
Block off a square area of the school yard or your own yard. Arrange for this area not to be mowed, fertilized, weeded, or otherwise disturbed. What changes do you observe in eight weeks' time? Is there an increase in insects, other animals, or plants?

Plants begin to grow after a forest fire devastates the area.

Grasses help establish new sand dunes as they develop after a storm.

grass soon takes root, weaving a system of underground roots and stems that begins to hold the sand together. Couch grass traps more sand, which forms small meter-high sand dunes. The increased sand allows other plant species to colonize the dunes. As organic matter is added to the soil, more plants and fungus grow to provide habitats for animals.

**Secondary succession** is the series of changes that occur in an area where the existing ecosystem has been disturbed. For example, secondary succession takes place when plants recolonize farmland. New communities do not immediately include nearly as many species of plants and animals as the original communities. However, after only a few weeks, weeds sprout in the abandoned fields from seeds that have been dormant in the soil for many years. These plants also grow quickly. Eventually shrubs and tree saplings replace the weeds to fill the space. Later larger species of trees, such as oak, maple, and beech, crowd out the smaller trees. Secondary succession is common in areas damaged by fires or floods.

Until recently, many scientists thought serious damage to an ecosystem was permanent. However, researchers have found that most polluted or damaged ecosystems do recover over time. Ocean systems recover the fastest because the organisms there reproduce and grow more quickly than those in other ecosystems. Forest ecosystems are the slowest to recover because trees take a long time to grow. To compare, an ocean floor ecosystem recovers in less than 10 years whereas a forest ecosystem recovers in about 42 years.

Most ecosystems recover faster from natural disasters such as fires or hurricanes than they do from damage caused by people. Human-caused damage can be more widespread or complete than what is caused by a fire or storm. Very few damaged ecosystems will fail to recover over time. God designed all life to flourish and to overcome such changes in the environment.

Deforestation destroys ecosystems.

Restoration ecology is a separate field of study for ecologists. In some cases, a damaged area is allowed to regenerate naturally. This approach was used after devastating fires in Yellowstone National Park in 1988. The heat of the fire activated the seeds of the pine trees. Plants and trees grew quickly and filled in the blackened landscape. However, a fire in California's Cuyamaca Rancho State Park was so hot that seed cones were destroyed. The restoration effort there required human effort to replant thousands of acres of trees. Scientists are discovering how ecosystems function by putting them back together again, but careful planning helps people avoid polluting or damaging ecosystems in the first place.

## TRY THIS

**Regaining Balance**
Design an experiment that compares the recovery of an ecosystem from a natural disaster with the recovery of an ecosystem from disaster caused by humans.

## LESSON REVIEW

**1.** What is the difference between primary and secondary succession?
**2.** Name some things that disturb an ecosystem.
**3.** Why can human-made disturbances be more disastrous than natural disturbances?
**4.** What do you think is more common in your area—primary succession or secondary succession? Why?

 **BIBLE CONNECTION**

Most ecological changes are small, but some give people the opportunity to see how God designed communities in His world to regenerate over time. He even created the ability for communities that have been completely wiped out to come back! One example is the Indonesian island of Krakatau. On August 27, 1883, the volcano on the island erupted with a force of 200 megatons of TNT. The eruption killed everything on the island and generated tsunamis that swept into the neighboring islands of Sumatra and Java.

However, life on Krakatau regenerated. By August 1884, spiders had colonized the island by ballooning across the ocean on threads of silk. Shoots of grass cropped up. The wind brought bacteria, fungal spores, small seeds, and insects. Frogs, snakes, and small rodents rafted to the island on branches and logs. Other organisms traveled to Krakatau on or in birds. Bacteria, protozoa, spores, roundworms, mites, and lice traveled on the birds' feathers. Tree and shrub seeds passed through their digestive systems and grew on the island. Larger herbivores and carnivores, such as snakes and lizards, swam the long ocean distance to the island.

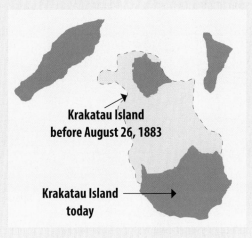

**Krakatau Island before August 26, 1883**

**Krakatau Island today**

There are now over 400 species of vascular plants, thousands of species of arthropods, and several species of mollusks, birds, reptiles, and bats. Today Krakatau is once again a tropical rain forest. The island is much smaller. It is not exactly the same as it was before the eruption, but biologists predict that within 100 years the island will once again have the variety of living things that it had before the volcano erupted.

# Introduction to Biomes

Have you ever wondered about the extreme climates on Earth? For example, the temperature in the deep freeze of the polar regions is –80°C; Old Faithful geyser in Yellowstone National Park reaches 200°C. Arica, Chile, averages about 0.06 cm of rain a year, but every day more than 2.5 cm of rain falls on parts of Hawaii. Twelve hours of sunlight a day brightens the equator's tropical forests, but at the earth's poles, residents wait six months to see one sunrise and one sunset each year. How do all of these extremes fit together on one planet?

To categorize the wide variety of environments on Earth, scientists group similar environments into divisions called *biomes*. A **biome** is a large area characterized by the prevailing climate and the types of organisms that thrive there. As you might expect, not all scientists divide Earth into the same number of biomes. Some scientists feel that there are two biomes: land and ocean. Others feel there are over a dozen! The six major land biomes are tundra, coniferous forests, deciduous forests, rain forests, grasslands, and deserts. A large biome not included with the land biomes is the ocean. Most biomes have undergone change over the years, especially as people have moved into

Sunrise in Antarctica

Old Faithful geyser at Yellowstone National Park in Wyoming

**FYI**

**Biome Barriers**

Most biomes overlap each other, but sometimes biomes are sharply divided. Mountain ranges often act as natural barriers or walls between biomes. For example, cool, moisture-filled air from the Pacific Ocean creates heavy rainfall on the western side of the Sierra Nevada mountains in California, which produces dense forests and vegetation. As the air moves over mountains and down the other side, it becomes drier and warms up, which creates a dry area called *a rain shadow* on the eastern side of the mountains. The desert area of Death Valley is in the rain shadow of the Sierra Nevada mountains.

Elevation is another natural barrier between biomes or ecosystems. For example, grasses and shrubs thrive in the warm conditions at the base of the Sierra Nevada mountains. Higher on the mountains, deciduous forests flourish in the cooler temperatures and extra moisture. Above the belt of mixed trees only conifer trees grow because they can tolerate the colder temperatures. Grasses and sedges thrive above the tree line (timberline), where conditions are too cold and dry for trees to grow. No vegetation grows on the highest mountain peaks, which are covered with ice and snow.

previously unpopulated areas. You may have seen a small section of prairie, but that is nothing like the prairie of even two hundred years ago, with tall grasses that stretched for hundreds of kilometers. The forests that you see now give only a glimpse of the glory that God gave them before the Fall.

The types of plants and animals that make up a biome depend on a region's climate, or the usual weather conditions in an area. Climate includes temperature, humidity, wind speed, cloud cover, and precipitation. These factors of climate are determined by vegetation, variations in sunlight, Earth's rotation, Earth's revolution around the sun, how far a region is from an ocean, and the elevation (height) of the land.

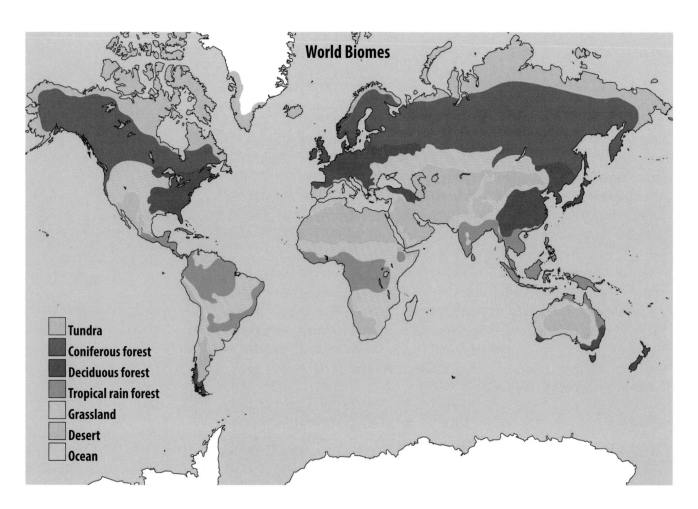

**World Biomes**

Tundra
Coniferous forest
Deciduous forest
Tropical rain forest
Grassland
Desert
Ocean

One difference between an ecosystem and a biome is size. A puddle can be an ecosystem, but biomes are not small. They usually spread out over large land masses. Therefore, biomes are also known as *major life zones*. God created several methods that scatter organisms throughout a biome. Animals migrate to different areas. Some spiders are carried to a new area by the wind. Fruit seeds are transported by birds, and burrs catch on animals' fur. Organisms can even raft between islands on debris or vegetation. Occasionally, violent storms produce waterspouts that can lift fish, lizards, and frogs out of the water and drop them into other areas! These methods help the ecosystems that make up a biome maintain a balance of organisms.

## LESSON REVIEW

**1.** What is a biome?

**2.** What are the seven major biomes?

**3.** What climate factors affect the type of organisms that live in a biome?

**4.** Are there any natural barriers in the biome in which you live? If so, what are they?

If you were to start at the top of Earth and travel to visit each biome, your first stop would be the tundra, the biome that circles the Arctic Ocean around the North Pole. It is located between the polar ice cap and the huge coniferous forests in North America, Europe, and Asia. Bring heavy winter clothes because the tundra is often very cold! The word *tundra* comes from the Finnish word *tuntura*, which means "treeless plain." A tundra is often described as *a cold desert*.

The tundra's harsh environment can be attributed to many limiting factors. A **limiting factor** is a condition that limits the number of species or the size of a population in an area. One such factor is the temperature—tundra temperatures range from 3°C in the summer to −34°C in the winter. Precipitation is also a limiting factor. Usually less than 25 cm of precipitation fall on the tundra each year. **Permafrost**, a layer of permanently frozen soil, is another limiting factor in the tundra. Water cannot drain through permafrost, and permafrost does not allow roots to grow deep into the soil. Because of this, most of the tundra's soil is waterlogged. Although bacteria and fungi live in the soil, they do not grow well in the cold, so tundra soil has very few nutrients.

Even so, the tundra is not dead—far from it! Well-drained areas serve as homes to dwarf willows, birches, low-growing grasses, and herbs. Wind-shielded bogs form depressions in the soil that provide habitats for sedges, rushes, and mosses. The alternate

Arctic willow

Winter in the tundra

Summer in the tundra

freezing and thawing of the soil heaves large boulders to the tundra's surface, providing a growing place for beautifully colored lichens. During winter, the snowpack insulates various mosses and herbs, and in the summer, it melts into liquid water that the plants use to grow. But bits of soil and ice blow across the flat terrain, acting like sandpaper on the plants that do take root, so they do not grow very tall.

Another reason that plants do not grow well in the tundra is because of the short (three-month) growing season. Summer is a beautiful time for the tundra. During these months, sunlight falls on the tundra 24 hours a day. This extended period of sunshine results in an amazing display of life. Plants and animals grow quickly during this long stretch of sun.

Although few insects live in the tundra, the marshes and ponds that form in the summer invite swarms of mosquitoes and black flies. The insects and plant life draw birds to the tundra to breed. Some birds, such as the willow grouse, migrate only a short distance, but the Arctic tern makes an incredible journey back from the Antarctic on the opposite side of the world.

Reindeer grazing in the Cairngorm mountains

Falcons, merlins, snow geese, and other birds also migrate to feed and breed in the tundra. Large herds of caribou and other mammals migrate to the tundra to take advantage of the new plant life. Grizzly bears and gray wolves follow the caribou to hunt them on the open tundra.

Arctic fox in summer

Young tundra animals mature quickly. Some must be ready to leave before winter. Several species of geese fly in for a couple of months to breed before migrating south for the winter. Reindeer give birth to their calves in the early summer, and the calves grow quickly to return to the southern forests with the adult herds before winter.

Some tundra animals are year-round residents, such as polar bears. Insects live through the winter in egg stages that resist the cold. The Arctic fox buries frozen food so that it can live through the winter. The feathers or fur of several tundra animal species—stoat, Arctic fox, and Arctic hare—change from brown to white for camouflage. Small mammals breed all winter and tunnel through the snow to get food. Most animals do not hibernate on the tundra because they are not able to tunnel through the frozen soil and because they do not have enough time to store fat during the short growing season. The ground squirrel is an exception. It hibernates for seven to eight months!

Arctic fox in winter

Another common small tundra mammal is the lemming. Because lemmings reproduce quickly, their populations can overgraze an area, so they migrate every four or five years to find food. According to a famous myth, lemmings migrating in large masses jump from cliffs to drown in order to keep their species from becoming overpopulated. That is untrue. The truth, though, is that lemmings are good swimmers and they often cross rivers to find better feeding grounds.

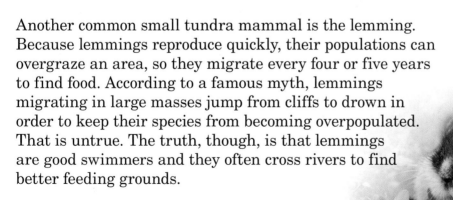
Norwegian lemming

## LESSON REVIEW
**1.** Where is the tundra?
**2.** What is the climate of the tundra?
**3.** What are the limiting factors of the tundra?
**4.** Why do so few plants grow on the tundra?
**5.** What organisms are native to the tundra?

A conifer's branches are designed to withstand heavy snow without breaking.

South of the tundra is the northernmost forest biome—the coniferous forest. In North America, these forests stretch from Alaska across Canada to Prince Edward Island. In Europe and Asia, they stretch from Scandinavia and all the way across Russia. The main feature of a coniferous forest is the **conifer**—a tree that has needlelike leaves and produces its seeds in cones. Coniferous forest biomes are generally cold, snowy areas that have frost at least six months of the year. These biomes receive little rain—about 30–90 cm a year. The cool temperatures keep water from evaporating, though, so this biome is not short on water. Because cold temperatures slow down the decomposition process of bacteria and fungi, the soil of coniferous forests is generally poor and acidic.

Conifers (such as spruces, pines, firs, and hemlocks) are designed especially for cold climates. Their thick, waxy, needlelike leaves protect them from the cold, keep them from drying out, and photosynthesize throughout the year (if the temperature is above freezing) because conifers do not drop their leaves. A conifer's shape and flexible branches allow it to shed heavy snow without drastically damaging its branches. These massive trees store plenty of water and nutrients to use during dry seasons. They help block the cold northern winds to shelter the inside of the forests from even the fiercest blizzards, which protects wildlife from the bitter cold temperatures.

The coniferous forest in Yoho National Park, Canada

# FYI

**Temperate Coniferous Forests**

The coniferous forest biome is also referred to as *the taiga* or *the boreal forest*. This large biome extends from the southern edge of the tundra to the northern portion of the deciduous forests. However, temperate coniferous forests grow in the lower latitudes of North America, Europe, and Asia, in the high elevations of mountains. Temperate coniferous forests receive up to 200 cm of precipitation annually as compared to the 30–90 cm that the northern boreal forests receive.

A temperate coniferous forest by Lake Misurina in Sexten Dolomites, Italy

Coniferous forests have fewer plant species and a shorter growing season than all other biomes except the tundra. The thick canopy of needles does not allow much light to reach the forest floor, and the layer of dry needles on the ground keeps most seeds from germinating. But mosses, ferns, and a few shrubs flourish on the dark forest floor, and mushrooms provide a rich habitat for many invertebrates. When the forest canopy breaks open to allow sunlight through, willow trees, alders, wild raspberries, aspen, and birch trees grow.

Many large animals make their homes in coniferous forests. Moose and mule deer are large browsing herbivores. Smaller herbivores—such as snowshoe hares, porcupines, and red-backed voles—feed on small plants. Black bears and grizzlies, the large omnivores of the forest, eat leaves, buds, fruits, nuts, berries, and fish. Wolves, wolverines, and lynx are the coniferous forests' primary predators. They keep the populations of herbivores in balance. During the winter, most of these animals stay within the forest's protection, but in the summer many animals migrate north to the tundra.

The forest floor is a dense mat of pine needles.

In winter, one way humans get around the forest is by using snowshoes because the snow is so deep. The concept of the snowshoe was inspired by God's design for coniferous forest animals that do not hibernate. Reindeer and moose have splayed feet to distribute their weight in the snow. The snowshoe hare, lynx, and willow grouse have wide, padded feet that allow them to travel easily through snow. God designed these creatures to thrive even in deep snow in a harsh environment.

The coniferous forest is home to many insects and the birds that eat them. Most of the birds migrate before winter, but the red crossbills of Europe and Russia and the white-winged crossbills of North America stay all winter. Their strong beaks are shaped for feeding on the seeds of pine cones, which are available all year long.

## LESSON REVIEW

1. Where is the coniferous forest biome located?
2. Describe the coniferous forest biome.
3. How are conifers designed to thrive in coniferous forest biome conditions?
4. What are the limiting factors of the coniferous forest biome?
5. What animals live in the coniferous forest biome?

A red crossbill feeding on seeds from pine cones

Farther south of the coniferous forest is the deciduous forest. The word *deciduous* comes from Latin and means "to fall down." This term is appropriate in the name of this biome because the trees that grow here—**deciduous trees**—lose their leaves at the end of the growing season. Deciduous trees include ash, elm, oak, maple, beech, and hickory trees. In the autumn, the fruits (acorns, nuts, or pods) of these trees are plump, ripe, and ready to fall. This fruit provides a bounty for the dozens of animal species that live in these forests.

Deciduous forests grow in the Northern Hemisphere in the eastern United States, throughout Europe, in eastern China, and in Japan. Smaller pockets of these forests grow at the tip of South America and in parts of New Zealand. Many of the deciduous forests of such areas as Britain and the United States have reduced in size because of expanding cities and increased farmlands.

The deciduous forest biome is warmer than the tundra or the coniferous forest biomes. Deciduous forests grow in temperate climates, which are not extremely hot or cold and are not extremely wet or dry. You usually will not need a heavy coat there, but you may need an umbrella—this biome gets plenty of rain. The annual rainfall totals 76–154 cm. The rain falls

## OBJECTIVES

- Describe the basic characteristics of the deciduous forest.
- Identify the limiting factors in the deciduous forest.
- Categorize the types of organisms that inhabit the deciduous forest.

## VOCABULARY

- **deciduous tree** a tree that loses its leaves at the end of its growing season

A grove of aspen trees in the autumn

Deciduous trees provide a gorgeous array of colors in the autumn.

## TRY THIS

### Using a Berlese Funnel

Many insects hide in moist soil and leaf litter in deciduous forests. Collect several buckets of moist leaf litter. Fill a funnel with leaf litter and suspend the funnel over a large jar. Suspend a lightbulb that gets warm over the litter. As the debris warms and dries, leaf litter invertebrates will scurry deeper into the funnel and eventually fall into the jar. Use a field guide to invertebrates or insects to identify the organisms you discover. What organisms live in the deciduous forest? Why is leaf litter an important part of this biome?

equally throughout the year, meaning this biome does not have a dry season. Deciduous forests are climate creators. The millions of trees transpire a lot of water, which creates high humidity. The humidity contributes to the rainfall. The trees also serve to break strong winds into gentle breezes.

These areas enjoy four distinct seasons, including four to six months of summer. Deciduous trees need at least four months of summer to thrive, and they need more nutrients in the soil than conifers do. The deciduous forest biome is very productive in the summer, but the winter is cold enough to stop plant growth.

In the autumn, the days shorten and the trees show off a blaze of colors as their leaves change. As the autumn days become shorter and offer less light, the green chlorophyll in the leaves becomes inactive, allowing brilliant red, yellow, and orange pigments to show. You may wonder why the leaves fall. The broad leaves of deciduous trees lose more water through evaporation than conifer needles. These trees cannot draw up water from cold or frozen soil, so they shed their leaves to conserve water in the winter. Shedding their leaves is one way that God designed them to thrive in their environment. God also created the fallen leaves to make fertile soil.

As you walk through the deciduous forest, notice the layers of plant growth. The top layer is formed by a continuous canopy of tree crowns. This layer is dominated by tall species of trees such as oak and maple. Underneath the canopy is the understory of shorter tree species and young trees. Hickory, dogwood, and other trees that thrive in a sheltered environment make up this layer. Next is the shrub layer of woody shrubs such as rhododendrons and huckleberries. Beneath the shrubs is the herb layer of grasses and flowers that bloom in the spring before the trees' leaves shade the ground. Mosses, liverworts, ferns, fungi, and wildflowers grow on the ground layer, which is covered with leaf litter. Leaf litter is dead and decaying plant and animal material, especially the fallen leaves that cover the ground. It keeps the soil moist and protects it from temperature swings.

The deciduous forest biome has rich soil because warm temperatures, spring rains, fungi, and bacteria turn fallen leaves into a soft mat of decaying material. Once the material starts decomposing, a host of organisms, such as earthworms, snails, slugs, millipedes, and wood lice, break down the leaf litter into soil.

Larger herbivores living in the deciduous forest include squirrels, opossums, and deer. The forests also contain slugs, snails, fly maggots, and beetle larvae that feed on the forest fungi. Thousands of tiny invertebrate predators—such as wolf spiders, harvestmen (daddy longlegs), and beetles—eat the smaller litter feeders. Small vertebrates, such as salamanders, mice, shrews, chipmunks, voles, and moles, eat the tiny predatory invertebrates. Dozens of bird species eat not only insects but also seeds, nuts, and leaf litter. Raccoons and badgers are larger omnivorous scavengers that eat anything remotely edible. Fox, lynx, and bobcat—and in Europe, badgers, polecats, and wild cats—are the biome's large predators.

God designed the animals that live in deciduous forests to thrive throughout the changing seasons. Hedgehogs and dormice hibernate through the winter. Some animals—such as squirrels, bears, and badgers—enter a temporary state of hibernation interrupted by wakeful periods. Deciduous forest animals also use and store food carefully. These qualities sustain the animals through the winter.

## LESSON REVIEW

**1.** What limiting factors are present in the deciduous forest?
**2.** Name and describe the layers of the deciduous forest.
**3.** Describe the deciduous forest.
**4.** What types of animals live in the deciduous forest?

Badgers have a reputation of being viciously fierce animals. They are known to fight off intruders much larger than themselves, including bears and wolves.

A deciduous forest consists of several layers, ranging from leaf litter to tall trees.

Amazon River and rain forest

The third forest biome is the rain forest. Three main regions of rain forests are South America's Amazon basin, Central Africa's Zaire basin, and the forests in southeast Asia and northern Australia. These vast stretches of tropical trees and plants encircle the equator.

You need only warm-weather clothes in the rain forest. Tropical climate temperatures do not change much from season to season, so the temperature in a rain forest is generally 20°C–34°C. You will need an umbrella! Rain forests get more rainfall than any other biome—usually 200–1,000 cm each year. Some Hawaiian rain forests receive more than 1,000 cm of rain a year—that is more than 2.5 cm a day!

The heavy rainfall is partly responsible for the poor soil of rain forests because it washes away the nutrients from the leaf litter. Many plants and fungi live up in the trees, so most of the nutrients in the biome are found not in the soil but in that layer. When tropical forests are cleared to raise crops, farmers have only two or three growing seasons before the soil becomes drained of nutrients. Fertilizers are then necessary for farming to continue.

Rain forests create high humidity because trees and other plants return moisture to the air through transpiration. More than half of the rain that falls on the forest results from transpiration from all the vegetation. Imagine how cutting down the trees affects the climate!

Rain forests play an important role in regulating Earth's water cycle. Like giant sponges, rain forests soak up water so the land does not flood. Some of this water evaporates and, along with the water that the trees transpire, forms clouds. So rain forests help provide rainwater to regions hundreds of kilometers away. When large areas of rain forests are removed, the water cycle is affected and the climate in the area changes.

The most breathtaking feature of rain forests is their great variety of life. The number and variety of organisms in a particular region or biome is called **biodiversity**. Even though rain forests cover less than 7% of the dry land on Earth, 50% of all known organisms live in rain forests.

Tens of thousands of different plant species grow in rain forests. This vegetation is arranged into three different layers.

Trees standing 35–42 m tall form the emergent canopy—the upper roof—of the forest. Flooded by year-round sunlight, these mammoth tropical trees are supported by huge flared roots called *buttressing roots* that are taller than a person and stretch across the forest floor for 9 m or more. The canopy is the next layer. Many plants and animals live in the canopy. The third layer, the understory, consists of sapling trees and other undergrowth. Some plants, called *epiphytes*, grow on other plants, which they use for support but not for nourishment. Epiphytes live in the branches of other rain forest plants and offer homes to countless animals. Little light reaches the rain forest floor, so few undergrowth shrubs grow there, except near riverbanks or where a large tree has fallen.

The rain forest is home to many animals, especially insects. Some entomologists—scientists who study insects—believe that over 30 million species of insects live in rain forests. Entomologists found 43 species of ants on one tree! Butterflies and moths as large as small birds flutter about the canopy. Goliath beetles grow almost as large as a person's hand, and army ant columns can stretch for almost 1 km. Tropical insects show all forms of camouflage. Some look like twigs or leaves, and you might mistake others for thorns, seeds, or flowers.

Rain forests provide homes for many other animals as well. The harpy eagle, one of the world's largest eagles, often soars over the rain forest treetops in search of prey. Colorful toucans and hornbills build their nests high in rain forest trees. The main canopy swarms with monkeys, lemurs, snakes, frogs, and

## TRY THIS

**Use the Rain Forest Wisely**
Research where the wood in your furniture and the paper you use come from, along with your food and other favorite products. If they come from the rain forest, determine if they are sustainable or nonsustainable uses of rain forest products. How can you do your part to conserve this part of creation?

Lantern fly

Emergent canopy in Bwindi Impenetrable National Park, an African rain forest

Agouti

insects. The paradise tree snake flattens its body to glide from tree to tree, and the flying frog has large webbed feet that act like parachutes to help it glide through the air. Some frogs lay their eggs in water pools formed in flowers and on the leaves of epiphytes. Tadpoles hatch in these tiny pools and go through metamorphosis without ever leaving the canopy. Many of these frogs, such as the poison dart frog, are brightly colored to advertise their deadly poison. Parrots and other exotic birds live in the canopy too.

Top predators live in the rain forest's understory, the lowest level. Margay and ocelots, which are small tropical cats, hunt for small herbivores. Jaguars hunt peccaries (a type of pig) and rodents such as agouti and paca. The golden cat is the primary carnivore in western Africa. In southeast Asia, the clouded leopard feeds on reptiles, rodents, birds, and ground-dwelling tapirs. The African rain forests are home to the largest animals, including gorillas, okapis, elephants, and tapirs. The largest lizard, the Komodo dragon, rules several small tropical forested islands.

Rain forests—the home for plants, insects, birds, monkeys, frogs, and many other organisms—are being cut down. The cutting or burning of a forest to clear it is called *deforestation*.

Margay

People destroy rain forests for many reasons. For example, native people slash and burn them for ranch or agricultural space. Many people use products that come from the rain forests. Some of these products are nonsustainable—once they are gone, they cannot be replaced. For example, tropical wood is often used for paper, cabinets, doors, furniture, insulation, windowsills, boats, and flooring. Even when trees are planted to replace those that were cut, the area does not return to its previous level of biodiversity because the canopy layer has been eliminated.

Okapi

The rain forest provides other products that are sustainable, which means they can be taken from the rain forest without destroying any part of it. These products include many spices (such as cinnamon, pepper, and vanilla), fruits (such as bananas, coconuts, guavas, pineapples, and oranges), other foods (such as cocoa, coffee, nuts, tea, and ginger), and products such as palm oil and rubber. About one quarter of the medicines people take are made from plants found in the rain forest. These medicines include pain relievers, cancer drugs, and antiviral drugs for diseases like HIV. One small rain forest flower, the Madagascar periwinkle, changed the outlook for children with acute lymphocytic leukemia. The disease was considered fatal until a treatment was developed using Madagascar periwinkle. Now over 80% of children with this form of cancer survive. So far, only about 1% of rain forest plants have been tested to see if they could be used as medicine.

Deforestation of the Amazon rain forest

The small Madagascar periwinkle is one of the many rain forest plants that has been used to make medicines for cancer, high blood pressure, and HIV.

Humans have a responsibility for caring for God's creation. One way people can be responsible is to find out where the products that they use come from. By choosing to buy sustainable products, people help protect the rain forest.

## LESSON REVIEW
**1.** Describe several characteristics of rain forests.
**2.** How do rain forests affect Earth's water cycle?
**3.** What are the limiting factors of the rain forest?
**4.** What animals live in the rain forest?
**5.** Why do you think it is important that the rain forest be protected?

Having explored the tundra and the three forest biomes, the next biome is the grasslands. Grasslands are found throughout the world. Grasslands once covered 25%–40% of Earth's land, stretching across North and South America, Europe, Asia, and Africa. Today, most of these grasslands, which include the **prairies** (North American grasslands), the **steppes** (Eurasian grasslands), **pampas** (South American grasslands), the **savannas** (African grasslands that are scattered with trees and shrubs), and the rangelands of Australia have been cleared for agriculture. However, some grassland ecosystems still flourish.

Most grasslands have flat, rolling hills. About 25–75 cm of rain fall on grasslands each year—just enough moisture to keep deserts from developing. Summers are hot and winters are cold. Even though droughts strike often and make agriculture a challenge, you only have to look around to realize that grasslands are not short on plants. Grassland plants have extensive root systems that hold the soil in place. If drought or overgrazing removes this vegetation, the soil loses these anchors—the winds sweep up the soil of this mostly treeless biome, hurling it through the air in violent dust storms. In the 1930s, these conditions produced the great Dust Bowl of North America's midwestern prairies. During this time, the soil erosion was so bad that entire houses were buried by blowing dust.

Grasses and soft-stemmed plants thrive in the grasslands, covering the land with richly colored carpets. Goldenrod and other flowers splash the North American prairies with deep yellow and purple, and emerald mosses and blue irises decorate

African savanna

the Russian steppes. The savannas of Africa also display a wide variety of spring colors. All of these grasslands turn brown during the dry season, so lightning often causes fires. These fires reduce tree and shrub growth, but they add minerals back into the soil, encouraging new plant growth. The parachute-shaped seeds of many grassland plants are dispersed by the wind. Other seeds have hooks, harpoons, anchors, or spikes to snag on the fur or feathers of passing animals that carry them along to new locations.

A dust storm approaching Stratford, Texas, in 1935

As you explore the grassland, many animals catch your attention. Insects thrive in the herbaceous layer, which is the layer of knee-high plants. Many species of grasshoppers and beetles eat these soft plants, and aphids and leafhoppers suck out the juices from the stems. Ants and earthworms live in the soil. In African savannas, termites build giant dirt mounds. The termite mounds are rock hard. Inside these colonies, the termites build nurseries for their larvae, air conditioning systems, and even fungi gardens, where they grow food! When the termites abandon the mounds, a wide variety of animals, such as snakes, porcupines, and anteaters, move in.

Lightning strikes on dry land can cause fires, but lightning also supplies the nitrogen plants need to the soil.

Grassland birds are usually small. They eat insects, seeds, or a combination of both. Many larger birds, such as partridges, quail, and guinea fowl, use their camouflage to hide from predators in the thick vegetation. Birds of prey, such as owls, also live in the grasslands.

Wildflowers on the prairie

Termite mound in Ethiopia, Africa

A wide variety of mammals make their homes in the grassland biomes, especially on the savannas of Africa. God gave grassland animals many features that help them thrive in this environment. Grazing animals have flat molar teeth for grinding grass and hooves for running fast over hard ground. With so many large mammals living on the grasslands of Africa, you might expect the plants to be overgrazed, but each animal eats a different part of the plants. Giraffes, elephants, and rhinos feed on trees and vegetation at their own height. Zebras eat the crowns of the grasses, wildebeest and topi feed on the middle layer, and gazelles eat the shorter grasses and fallen fruit.

The prairies and steppes have a smaller variety of herding grazers than the African savanna. Huge herds of bison and elk once roamed the prairies, but their numbers decreased steadily as people moved across North America in the 1800s. American prairie dogs, another common grassland mammal, build large underground villages that cover many acres. The burrowing activity helps aerate the grassland soil in much the same way as earthworm activity does. Other grassland mammals include the grazing pronghorn of North American prairies, the pampas deer and guanaco of the pampas, and the roe deer and Bactrian camel of the steppes.

Additional grassland animals include jumping mice, jerboas, and jackrabbits, which spot predators by leaping above the height of the grass. Several small predators hunt these herbivores. Ferrets and badgers hunt the North American prairies, the Corsac fox hunts the steppes of Europe, and the bat-eared fox prowls Africa's savanna. On the savannas, lions hunt zebras and wildebeest, cheetahs and Cape hunting dogs eat smaller gazelles and impalas, and foxes and eagles hunt the young herd animals. The scavengers of the savanna—such as jackals, hyenas, and

 **HISTORY**

### History: Bring Them Back

Native Americans hunted bison for hundreds of years. After the Spanish introduced the horse to the Native Americans and other European settlers came with their guns, the bison had little chance against hunters. The US government encouraged the hunting of buffalo during the great Westward Migration of the 19th century. But by 1880, the bison herds had been reduced to the point of near extinction. The European buffalo (wisent) experienced a similar fate and was placed on the Red List of Threatened Species in 1996. Conservation and breeding programs have brought these animals back from the brink of extinction.

North American pronghorn and bison

vultures—clean up carcasses left by these predators. On the North American prairies, large carnivores include wolves, coyotes, and foxes.

Grasslands throughout the world have been converted to farmland, and human activities have changed the ecosystems. Because they can be a threat to livestock, farmers and ranchers sometimes kill wolves, foxes, and other large predators. This action allows for overpopulation of burrowing animals and other prey such as deer. Grazing cattle can also strip the land of vegetation, which often causes the thin topsoil to blow away, creating a desertlike region. In fact, many deserts were once grasslands. The food and livestock produced by farming is important, but the rich grasslands are part of God's creation and are important as well.

## LESSON REVIEW
**1.** Where are grasslands found?
**2.** What is the grassland climate like?
**3.** What are the limiting factors in the grassland biome?
**4.** What animals live in the grasslands?

North of the African savanna is the Sahara Desert. The Sahara is almost as big as the United States, and it is growing, eating up the grasslands that you just read about.

Deserts receive 25 cm of rain or less a year. Some receive none; the Namib Desert in southwestern Africa receives its moisture in the form of fog. Depending on the moisture they receive, deserts can be sandy wastelands or dotted with small bushes.

The Sahara Desert is the world's largest desert. It stretches across Africa from the Atlantic coast in Africa to the Arabian peninsula. Other deserts lie across western North America, South America, western Asia, and central Australia. Deserts often form in the rain shadows on one side of a mountain range. For example, the Rocky Mountains helped form the deserts of the American southwest, and the mountains ringing the interior of Australia helped form the Outback desert region. Daytime temperatures in deserts can rise to more than 50°C because the sand reflects the sunlight. The sand gets hotter than the air. Even so, deserts are often cold at night. Without moisture, the night air does not hold the heat.

About one-fifth of the world's land is desert. Some scientists believe the desert land is increasing because livestock are allowed to overgraze and farmers use improper agricultural methods. Each year, over 300,000 hectares of desert are formed through **desertification**—the making of new deserts through a change in climate or destructive land use.

Even though human-made deserts do not support a wide variety of plants and animals, natural deserts are anything but lifeless. Most desert plant life is buried in the ground for much of the year. When it rains, seed plants quickly germinate and flowers bloom, decorating the desert with a beautiful tapestry. Extremely dry deserts do not have plant life except in wadis (dry stream beds).

After the rains, flowers spring up, painting the wadis with bursts of vibrant yellows and purples, pale pinks, reds, magentas, and cornflower blues.

Desert blooms

Butterflies, moths, bees, birds, and bats feast on the nectar and pollinate the blooming flowers. Some flowers and pollinators have mutualistic relationships. For example, the yucca moth cannot be fed without the spiky-leafed yucca, and the yucca cannot be pollinated without the yucca moth. Once the rains are gone and the flowers have seeded, the blooms die, and the seeds wait beneath the ground for the next rain.

When you think of deserts, you might think of cactus plants. You will not see any cacti in African deserts, but you will find lots of them in North American deserts. God created the cactus with a thick, round shape that slows down evaporation. Its shape presents the least amount of surface area to the heat. Instead of leaves, cacti have spikes, which discourage thirsty predators. They also have shallow roots to soak up any moisture that falls to the ground. Some cacti expand like giant balloons to retain more water. The giant saguaro cactus grows up to 15 m tall and can live in the harsh, desert climate for up to 200 years. In a single rainfall, its shallow roots—along with small root hairs

The saguaro cactus is found exclusively in the Sonoran Desert of Arizona and Mexico.

The giant hairy scorpion, found in the Mohave Desert, is the largest scorpion in the United States, but it is the least often seen. Although it is only mildly venomous to humans, its large size allows it to subdue large prey such as lizards. It prefers resting in burrows, often deep underground where the moisture is higher.

that draw moisture—may soak up as much as 760 L of water, enough to last a year.

There is a big difference in activity between the day and night in the desert. During the day, you see very few animals. Only lizards searching for insects and a few other heat-resistant animals brave the daytime heat. The horned toad in North American deserts does not even move if other animals threaten it during the day—it just stays still and squirts blood out of the corner of its eye to scare predators away. At night, though, the desert bursts into action because most desert animals respond better to the cooler night temperatures. Scorpions, spiders, centipedes, beetles, and wood lice crawl out from the soil and from under the rocks where they have spent the day. In North American deserts, the ferocious grasshopper mouse and pygmy mouse also come out at night. Elephant shrews, which have long noses that look like elephant trunks, and golden moles both enjoy the night temperatures in African deserts. Marsupial mice thrive at night in Australian deserts. Many desert rodents—such as gerbils, jerboas, and kangaroo rats—burrow during the day to avoid the heat and scurry around the rough terrain at night, trying to avoid the many snakes that live in the desert. African sand vipers are common in the Sahara. Rattlesnakes are common in North American deserts, along with Gila monsters, which are predatory lizards.

God gave each of these animals special features and behaviors to help them live in the extreme desert conditions. For example, reptiles, like other ectotherms, have a wide variety of ways to regulate their body temperatures. Many can raise their bodies off the ground so air circulates across the undersides of their bodies. Other reptiles change their color through pigment changes in the cells of their skin—lighter skin reflects heat, while a darker color absorbs it. To cool themselves, reptiles also move to the shade or completely or partially bury themselves.

God created desert mammals to have larger exposed body surfaces than mammals that live in colder environments. The fennec fox of North African deserts, for example, has large ears to radiate more heat, while the Arctic fox has small ears to conserve heat. And kangaroo rats rest during the day and plug up their mouths so that the air inside of them remains cool and moist. Other desert mammals include ferrets, badgers, skunks, desert foxes, and wildcats, which are among the large desert predators. Large desert herbivores include domesticated animals, such as camels, sheep, goats, and donkeys, which graze along with wild animals, such as gazelles and ibex.

Most desert birds eat insects, which provide them with both food and moisture. Seed eaters are rare. Desert birds must fly great distances for additional water. The sandgrouse of North Africa will fly several hundred kilometers to search for an oasis with water. This bird stores water in her beak and absorbs it in her feathers to take it back to her brood.

God even designed a few amphibians to live in the desert. For example, a few desert species of frogs and toads store diluted urine in their bodies. The spade foot toad spends up to nine months in its burrow waiting for rain. The amazing Australian water-holding frog leaves its underground cave only when it rains to gorge on insects, mate, and spawn. The tadpoles of this Australian frog grow at a rapid rate, fill their bodies with water as the rains stop, and return to their caves to secrete transparent, watertight layers of a cellophane-like substance around their bodies. Then they are dormant until the next rain.

## BIBLE CONNECTION

Read Psalm 23. Can you see why the images in this psalm were so restful to David, who lived in the arid environment of a desert biome?

## LESSON REVIEW

**1.** What are the characteristics of a desert?
**2.** What are the limiting factors in deserts?
**3.** How did God equip organisms to survive the desert's harsh conditions?

During warm weather, the Gila monster feeds at night on small mammals, birds, and eggs. Fat is stored in the tail and abdomen for the winter. Gila monsters are sluggish, but they have a strong, tenacious bite.

## OBJECTIVES

- Summarize the basic characteristics of the ocean biome.
- Describe the three groups of ocean organisms.
- Identify and describe four ocean zones.

## VOCABULARY

- **benthos** the organisms that live on or in the ocean floor
- **bioluminescence** the emission of visible light by organisms
- **intertidal zone** the area between land and sea where the tides go in and out
- **nekton** the free-swimming ocean organisms
- **neritic zone** the first ocean zone away from the shoreline
- **oceanic zone** the zone that extends from the edge of the continental shelf and covers the entire open ocean
- **photic zone** the ocean zone where photosynthesis occurs

You have learned about every land biome. The seventh biome is the ocean biome. About 71% of Earth is covered by water and about 97% of all water is contained in the earth's oceans. Although the ocean is often considered the biosphere's largest ecosystem, it is actually many different ecosystems of diverse undersea life. As you learn about the world's oceans, you will notice that many habitats are woven together, connected by various currents. These currents work together to circulate oxygen, nutrients, plants, spores, animal larvae, and sea-going animals such as turtles and whales.

How deep is the ocean biome? On average the ocean is around 4 km deep. But the deepest part of the ocean is about 11 km deep. The farther down you go, the more pressure you feel. The deepest parts of the ocean are under tremendous pressure from all the water above. Light also changes with depth. Photosynthesis can happen only where plants receive enough light to make their own food. The surface layer of the ocean where photosynthesis can occur is called the **photic zone**. This zone reaches only about 200 m down. Because sunlight penetrates and warms this zone, the average worldwide ocean temperature is about 17°C. Below this zone, plants cannot grow. However, oxygen is available even in the deepest ocean water. Nutrients fall from the ocean surface to the ocean floor, so organisms can live even in the deepest ocean canyons.

Ocean life can be divided into three groups: plankton, nekton, and benthos. These categories are defined by where the organisms live and how they move. Plankton are microscopic organisms that live near the surface of the water in the photic zone. Phytoplankton are plant plankton that make their own food through photosynthesis. They are the beginning of ocean food chains. The next link in the food chain is zooplankton, which are microscopic animals. They eat phytoplankton. Most species of fish, including nearly all tropical fish, spend the first part of their lives as zooplankton.

Mandarin fish are considered some of the most colorful in the world. These small fish live in tropical warm waters of coral reefs and lagoons.

**Nekton** are free-swimming ocean organisms, such as fish, sea turtles, whales, dolphins, and squid. Nekton eat plankton or each other. Most nekton live in the area of water that covers the continental shelf. This part of the ocean is ideal for marine life because it receives plenty of sunlight, the water pressure is low, and the temperature stays relatively constant. Some nekton roam the open ocean beyond the continental shelf. Here the water is cooler and the pressure is greater than it is closer to shore. The organisms that live here are very spread out.

Sea pens are benthic cnidarians. Each sea pen is a colony of polyps. When stimulated, sea pens glow with a bright, luminous light.

**Benthos** are organisms that live on or in the ocean floor. Benthic organisms include seagrass, kelp, water lilies, crabs, oysters, sponges, coral, worms, and sea stars. Some benthos, such as sponges, sea anemones, and plants, stay in one location. Other benthos, such as clams, lobsters, and octopi, move from place to place.

Scientists divide the ocean into various zones in order to study and get a better understanding of it. The **intertidal zone** is a marine environment between land and sea where the tides go in and out. Tides vary greatly. The Mediterranean Sea has

When the tide retreats, water remains in depressions called *tide pools*. Organisms that live in tide pools are designed to survive the constant pounding of the surf.

Coral reefs are found in the neritic zone. They are home to thousands of species of plankton, nekton, and benthos.

practically no tides, but the tide in the Bay of Fundy in southeast Canada can rise as high as 16 m. When the tide is out, organisms in the intertidal zone are exposed to air, ultraviolet radiation, wide temperature swings, and land predators. Despite these conditions, a rich variety of organisms, such as plankton, nekton, and benthos, live in this zone, including lichens, barnacles, mussels, anemones, sea stars, and fish. Water depth in the intertidal zone is relatively shallow, which means the intertidal zone is also in the photic zone.

The next ocean zone is the **neritic zone**, which extends over the continental shelf and reaches a depth of 200 m. This zone is also in the photic zone because it receives plenty of sunlight. The neritic zone has low water pressure, so many different plants and animals flourish. Seaweed covers the floor of the neritic zone. Clams, snails, lobsters, many kinds of fish, and even some whales live here. Most fishing is done in this zone. The neritic zone also contains coral reefs and estuaries. Hundreds of animal species live in coral reefs, including many young fish that find shelter in the reefs before they venture into deeper waters. Estuaries, the places where ocean biomes meet freshwater, are small but important habitats. They offer food and nutrients and are important nursery areas where many fish grow before moving out into deeper water.

The **oceanic zone** extends from the edge of the continental shelf and covers the entire open ocean. Migratory animals

Certain species of jellyfish are bioluminescent.

 **FYI**

**Ocean Food Webs**
Ocean food webs depend on phytoplankton, microscopic aquatic plants that use photosynthesis. Diatoms, beautiful microscopic organisms with geometric shapes that look like gemstones, are common phytoplankton in temperate waters. Phytoplankton make up most of the sea's producers. Other ocean producers include marine grasses, mangroves, and red, brown, and green seaweed. Zooplankton, which are microscopic aquatic animals, are the next vital link in ocean food chains. They eat phytoplankton. Copepods, which are crustaceans, and krill are common zooplankton.

such as sea turtles, whales, and dolphins roam the open ocean, but this marine habitat has very few nutrients. It is often considered a biological desert. Although the oceanic zone takes up more space than rain forests, it does not produce even twice the biomass of rain forests. In the deeper parts of the oceanic zone, temperatures are cold. Where there is no sunlight, **bioluminescence** is an important feature. Certain undersea creatures have the ability to produce their own light. This characteristic allows them to attract a mate or prey or to escape a predator.

It is likely that many yet-to-be-discovered ocean creatures swim in the deep, dark waters of the sea. Various discoveries seem to support this prediction. A coelacanth, a fish thought to have been extinct, was netted in 1938 near South Africa. The megamouth shark was not even discovered until 1976. Who knows what other marvelous creatures God created in the sea that have yet to be discovered!

## TRY THIS

**Ocean Creature**
Choose one of the ocean zones, and write a want ad or résumé for a creature that lives in that zone. What must its physical characteristics be? How does it eat? How does it thrive in its home? What challenges must it overcome?

## LESSON REVIEW

**1.** What are the basic characteristics of the ocean biome?
**2.** Explain what plankton, nekton, and benthos are.
**3.** Describe the photic, intertidal, neritic, and oceanic zones.

### Ocean Zones

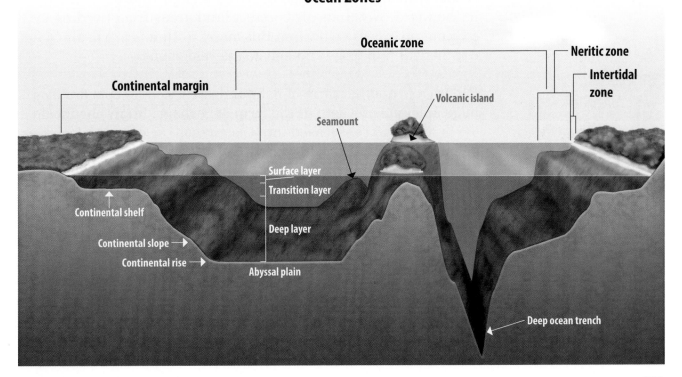

## OBJECTIVES

- Describe four ways in which habitats are being destroyed.
- Explain several reasons why species are endangered or extinct.
- Discuss ways to prevent the extinction of species.

## VOCABULARY

- **conservation** the protection and restoration of species and natural resources in an ecosystem
- **exotic species** a species that is not native but has been introduced into an ecosystem
- **native species** a species that is native to a particular ecosystem

God gave people the responsibility of caring for the living things in the world. God is the one who designed the biosphere, created a beautiful web of life from microorganisms, plants, animals, and fungi then supported them with the sun, air, water, and soil. Sometimes change to the biosphere cannot be avoided. But often people unnecessarily harm God's creation.

Every year, more land is cleared for agriculture, industry, timber, and grazing. Such clearing destroys the habitats of plants and animals. Most of these destroyed habitats are in forests. Deforestation, especially of tropical rain forests, is one of the most alarming forms of habitat destruction. It is estimated that over 32,000 hectares of rain forest are destroyed each day.

Because about one-fifth of the earth is covered with arid or semiarid terrain, desertification is a growing concern. Dry areas are expanding yearly in part because people overgraze domesticated animals and use improper agricultural methods. Human activity alone causes hundreds of thousands of hectares of desert to form every year.

Wetlands, which are transition zones between water and dry land, are also being destroyed. In the same way that kidneys act as filters for your body, wetlands filter and purify the earth's drinking water. They also offer a home to many animals, such as fish, shellfish, crustaceans, and waterfowl. Salt marshes, in fact, are among the most productive fish and wildlife habitats in the world. Wetlands also protect land areas from flooding by soaking up excess rainwater. But many wetlands are drained for commercial buildings and housing developments.

Pollution is another form of habitat destruction. People produce huge amounts of waste in the form of garbage. Many chemicals are released into the air, soil, or water by factories and people who do not properly dispose of wastes such as motor oil, paint, and batteries. Sewage, toxic metals, oil, fertilizers, and pesticides dumped into water harm ecosystems by depleting oxygen supplies. Some of these toxins build up in animal tissues, which can kill the animals. Even a little pollution can have a big impact on an ecosystem.

Habitat destruction is a factor in a larger issue: species extinction. Scientists agree that some organisms are becoming extinct much too quickly.

Cutting down the trees fundamentally changes habitats.

### What Can You Get from a Palm Tree?

How many different products can be made just from a simple palm tree? From the coconut tree alone, people make many products. People use material in the coconut husk to make ropes. They use the hard inner fruit layer for fuel and to make bottles. They extract coconut juice for a beverage. They also eat the coconut "meat" or use it as a source of cooking oil. They use the sap from the tree as a source of sugar and vinegar. The African oil palm is important chiefly for the palm oil obtained from the fruit coat of the coconut and for kernel oil from the seeds. These oils are used for cooking.

## TRY THIS

### Biodiversity

Imagine that you have been placed into a RAP (Rapid Assessment Program) team to determine the biodiversity of a particular ecosystem. Your job is to find and identify as many different species of organisms as you can in the given time period. Record the names of the animals, plants, algae, and fungi that you find in your assigned ecosystem.

They are concerned about the ripple effect such extinctions have on the biodiversity of the biosphere. But the extinction issue is complicated by the needs of people and the needs of the rest of creation.

Habitat destruction is the main human activity that causes species extinction. Remember the great biodiversity found in rain forests? Some species of animal, plant, invertebrate, or fungus often lives in a small patch of rain forest or even on just one tree. If that area is destroyed, such species may be lost completely.

Sometimes a species dies out from disease. For example, frog populations worldwide have been reduced by a fungal disease and some species of frogs are now extinct. All the exploration and resettling that took place in the past 600 years greatly increased the rate of organism extinctions, partly because of diseases brought from one part of the world to another. For example, from 1400 to 1700, many native North American species were killed by diseases that came from Europe with settlers.

Overhunting and poaching also contribute to species extinction. For example, the bison was nearly hunted to extinction in North America in the 1800s. Poaching is illegally killing an animal or removing a plant. Tigers, elephants, and many species of fish and birds are often poached. Cacti are frequently poached from the southwestern United States as are orchids from South America.

Littering is one way people pollute the world.

### Park Rangers

God gave people a beautiful world in which to live. Some areas have been designated as national, state, provincial, or regional parks to protect the area while allowing people to enjoy nature. The park rangers who work in these areas have many different duties.

Outdoor education is a major activity at some parks. Rangers lead visitors in activities that teach them about the ecosystems of the area. Some park rangers help control forest fires or gather scientific information for park displays and interpretive centers. Rangers also oversee backcountry trails and camping. Sometimes they even take part in search-and-rescue operations. Some parks have cultural or historical significance, and the rangers in those parks may be involved in crafts, folk art, or lifestyle demonstrations from that period of history. Rangers must also enforce laws and park rules. They investigate accidents and violations of park rules and laws that relate to fish and wildlife.

Park rangers need to understand how to responsibly manage natural resources such as forests, groundwater, lakeshores, seashores, and wildlife. Many park rangers attend a college or a university to study subjects such as earth science, natural resource management, forestry, history, archaeology, anthropology, law enforcement, or park and recreation management. Many rangers enjoy the outdoors and choose this type of work for that reason, although office work comes along with the other responsibilities too. Park rangers have the privilege of introducing others to the beauty of God's creation and teaching others about good stewardship.

The passenger pigeon, once probably the most numerous bird in the world, became extinct in 1914.

Hunters killed them for meat, and habitat destruction further reduced their population. The last passenger pigeon in the world died at the Cincinnati Zoo in 1914.

Some species, especially tropical birds, are in danger of extinction because they are popular pets. If too many birds are taken out of their natural ecosystem, the entire species will be in danger.

Introducing a new species to an ecosystem can also cause extinction. **Native species** are species that are naturally found in a particular ecosystem. **Exotic species** are species that are not native to an ecosystem but have been introduced to the area by humans. Sometimes an ecosystem is thrown out of balance when exotic species are introduced that threaten the survival of native species Exotic species often spread diseases to native species. Exotic species can grow out of control because they have no natural predators in their new ecosystem. For example, island ecosystems have been harmed by exotic species such as goats, dogs, snakes, and cats that compete with native iguanas and turtles for food.

The zebra mussel is a small freshwater mollusk native to the Caspian and Black Sea region. In 1988, this mussel was discovered in the United States in the Great Lakes region. The mussels were probably carried to North America in the ballast water of ships that crossed the Atlantic. Now they have spread throughout the waterways of the eastern United States and Canada. Zebra mussels consume large amounts of the microscopic plants and animals that form the base of the food chain for native aquatic species. By doing this, the mussels throw off the ecological balance that God created in those waterways. Zebra mussels often kill native species of mussels, and they also clog the pipes of water treatment plants.

Even though international trade on ivory was banned in 1990, the African elephant is still being poached.

Exotic species of plants cause problems as well. Purple loosestrife is a European plant that was brought to North America by immigrants who enjoyed its beautiful flowers. Ships also carried these seeds in their ballast waters, which were emptied into North American waters. Purple loosestrife is now a problem in pastures and wetlands. It crowds out other plants, which harms the species that use these plants for food and shelter.

Human activity is responsible for much of the damage to creation. However, people can also be part of the solution to habitat destruction and species extinction. Many people have gotten involved in **conservation**, the protection and restoration of species and natural resources in an ecosystem. The problems of habitat destruction and extinction may seem too big to solve, but many solutions are underway. For example, the Endangered Species Act was passed in the United States in 1973. Species in the United States that are at risk are now listed as either endangered or threatened. Endangered species are in danger of becoming extinct if they are not protected. Threatened species

Purple loosestrife

Zebra mussels smothering native Wisconsin mussels

will probably become endangered soon if they are not protected. The law protects endangered or threatened organisms from being harassed, hunted, or harmed. In 2002, Canada passed the Species at Risk Act, which requires recovery plans for all endangered and threatened species. Other nations have passed similar legislation as well.

In 1975, many countries joined together to write the Convention on International Trade in Endangered Species (CITES). By signing this treaty, countries agree not to allow endangered species to be bought or sold for profit. As of 2016, over 180 countries had signed this treaty.

Although many people care about big animals such as tigers that are in danger of becoming extinct, fewer people are as concerned about endangered plants or fungi. But these organisms have a role in the world too. Some environmental organizations use what are called *flagship species* to get donations that help save all endangered organisms. A flagship species is a well-known endangered animal that an environmental organization uses to attract donors. The use of these animals attracts people's attention so that they will support efforts that help not only the well-known animals, but also the plants and fungi. Chimpanzees, orangutans, giant pandas, and tree kangaroos are a few endangered animals that have been used as flagship species.

To save entire ecosystems, many countries have created national parks and wildlife preserves, which offer safe habitats for plants and animals while allowing people to enjoy the beauty of God's natural world. Sections of ocean have also been set aside in an effort to reduce water pollution and to protect coral reefs. The Great Barrier Reef Park in Australia, established in 1975, is a reef management success story. The reef is divided into sections for scientific research, tourism, and commercial fishing; the park authorities monitor and regulate use of the reef. Traditional cultures in the Pacific and southeast Asia have models for reef conservation that have worked for centuries. They limit visitors or close off areas and do not allow the use of some equipment.

One of the greatest global conservation challenges is rain forest preservation. Most rain forests are in nations with large, growing populations of poor people. Any preservation plans must consider these people because many of them rely on rain forest destruction for employment. People are working to balance the needs of people with preservation of the rain forest, however. In 2003, Brazil began the Amazon Region Protected Areas program. This conservation program seeks to preserve over 60 million hectares of the Amazon rain forest by strictly protecting some areas and allowing limited use in others. The program has improved the country's water quality and has provided sustainable jobs for people in the rain forest area. It has also improved the region's species biodiversity.

The red colobus of Africa has been used as a flagship species. Animals selected for this purpose must be on the endangered species list.

You may be thinking, "I don't run the government—I'm not even old enough to vote! I'm not a scientist. And my parents won't let me move to a rain forest to talk people out of chopping it down. I'm just one person. How can I make a difference?" But all people have a role to play in protecting God's creation.

Three key individual solutions can be summarized in the words *reduce*, *reuse*, and *recycle*. Reducing simply means cutting down on the amount of products that you use. Reducing saves not only raw materials, but also the energy that goes into making the products. It also reduces the amount of trash produced when people throw these items away. One simple way a family can practice reducing is to arrange to stop receiving junk mail. Nearly 100 million trees are cut down every year just to make the paper that goes into junk mail. Stopping junk mail helps reduce the problems of deforestation and habitat destruction.

Great Barrier Reef

Reusing things also makes a difference. Many items are designed to be used only once and then tossed. Other things are designed to last only a little while so that new ones have to be purchased. Think about how quickly some clothes, appliances, toys, and other items wear out or break. Most people throw things away that could be reused and then go buy new ones. For example, people who have paper or plastic grocery bags at home can reuse them for another trip to the grocery store. Washing disposable plastic dinnerware and reusing it also helps. What other things can you reuse?

Recycling—the reprocessing of products such as paper, wood, and aluminum in order to conserve natural resources—is another way to help care for the earth. Recycling paper, for example, reduces the need for cutting down trees. Glass, cardboard, batteries, toner cartridges, and metals can also be recycled. Fruit and vegetable scraps can be recycled by making a backyard compost pile. After all, God created His world to recycle organic matter, nutrients, and water, so His children should follow His example. Your actions can and do change the world.

## LESSON REVIEW

**1.** Give five reasons why so many organisms are in danger of extinction.
**2.** What can you do to help prevent extinctions?
**3.** What are four ways in which habitats are being destroyed?

Composting plant waste

428

## A

**abdomen** the hind body segment of an insect or other arthropod *4.2.1*

**abiotic factor** a nonliving component of an ecosystem *7.1.3*

**acid rain** a type of rainfall made acidic from pollution in the atmosphere *7.1.6*

**adapt** to adjust when survival is threatened *7.2.3*

**addiction** a physical or psychological dependence on a substance or behavior *5.4.6*

**adrenal gland** a gland that produces adrenaline *5.4.1*

**AIDS** (Acquired Immunodeficiency Syndrome) a condition that destroys the body's immune system *2.1.3*

**air bladder** a grapelike bulb that keeps an organism afloat *2.4.3*

**allele** a different form of a single gene *6.1.3*

**allergen** an antigen that triggers an allergic reaction *5.4.5*

**alveoli** a cluster of microscopic air sacs deep in the lungs *5.2.1*

**angiosperm** a vascular plant that produces seed in flowers *3.1.1*

**annual** a plant that completes its life cycle in one growing season *3.1.1*

**antibiotic** a drug used to kill harmful bacteria *2.2.4*

**antibody** a protein that attaches to a specific pathogen *5.1.4*

**antigen** a substance that causes the immune system to produce antibodies *5.1.4*

**arachnid** an arthropod with two body segments and eight legs *4.2.4*

**archaea** the prokaryotic unicellular organisms belonging to domain Archaea *1.3.3*

**arthropod** an invertebrate that has a segmented body, jointed appendages, and an exoskeleton *4.2.1*

**asexual reproduction** a form of reproduction in which offspring arise from a single parent *1.1.2*

**atrium** an upper chamber of the heart *5.1.2*

**autotroph** an organism that makes its own food *2.2.2*

**axon** a long extension of a neuron that transmits messages *5.3.3*

## B

**bacteria** the prokaryotic unicellular organisms belonging to domain Bacteria *1.3.3*

**bacteriophage** a virus that infects bacteria *2.1.2*

**benthos** the organisms that live on or in the ocean floor *7.3.8*

**biennial** a plant that completes its life cycle in two growing seasons *3.1.1*

**binary fission** the cell division process in which one cell splits into two identical cells, each having a complete set of DNA *1.2.6*

**binomial nomenclature** the two-part scientific naming system *1.3.1*

**biodiversity** the number and variety of organisms in a particular region or biome *7.3.5*

**biological clock** an inner mechanism that regulates an organism's biological rhythms *5.4.3*

**bioluminescence** the emission of visible light by organisms *7.3.8*

**biomass** the total mass of all the organisms in a given area *7.2.1*

**biome** a large region that is characterized by its climate and organisms *7.3.1*

**biosphere** the part of Earth and its atmosphere in which living organisms exist *7.1.3*

**biotic factor** a living component of an ecosystem *7.1.3*

**bivalve** a mollusk with two shells *4.1.5*

**blade** a leaflike structure of algae *2.4.3*

**botanist** a scientist who studies plants *1.3.4*

**bronchi** the branches of the trachea that extend into the lungs *5.2.1*

**budding** a type of asexual reproduction in which a small outgrowth of the parent develops into an independent organism *2.2.2*

**cambium** the layer of a woody stem that produces new vascular tissue *3.1.5*

**carrying capacity** the maximum number of organisms in a population that an environment can support over a long period of time *7.1.2*

**cell theory** a scientific explanation of the properties of cells *1.2.1*

**cellular respiration** the breaking down of food molecules by cells into usable energy *1.2.4*

**Celsius scale** a temperature scale in which 0° represents the freezing point of water and 100° represents the boiling point of water *F*

**centipede** a flat-bodied arthropod with one pair of legs on each body segment *4.2.3*

**central nervous system** the brain and spinal cord *5.3.3*

**centromere** the point at which sister chromatids are attached to each other *6.1.5*

**cephalopod** an ocean mollusk with a head, tentacles, and beaklike jaws *4.1.5*

**cephalothorax** the body segment of crustaceans and arachnids that consists of the head and thorax fused together *4.2.1*

**chlorophyll** the green pigment in plants that captures light energy for photosynthesis *3.2.1*

**chromatid** one of the two identical strands of a replicated chromosome *6.1.5*

**chromatin** the mixture of DNA and protein that makes up a chromosome *6.1.5*

**chromosome** a structure that contains genetic information and directs cell growth *1.2.2*

**cilia** the tiny, hairlike projections that help ciliates move *2.4.4*

**circadian rhythm** a regular rhythm of growth and activity that occurs in approximately 24-hour cycles *5.4.3*

**clone** a group of genetically identical cells that usually contains identical recombinant DNA molecules *6.2.2*

**codominant allele** an allele that is neither recessive nor dominant *6.1.7*

**collar cells** the cells that line the inner chamber of a sponge *4.1.2*

**colon** the large intestine *5.1.3*

**commensalism** a relationship between two organisms in which one benefits and the other is not affected *7.2.2*

**community** the populations of different species living in the same place *7.1.2*

**cone** a light-sensitive cell in the eye used for color vision and fine detail *5.3.4*

**conifer** a tree that has needlelike leaves and produces its seeds in cones *7.3.3*

**conjugation** the process of sexual reproduction in which two unicellular organisms exchange genetic information *2.4.4*

**conservation** the protection and restoration of species and natural resources in an ecosystem *7.3.9*

**constrictor** a snake that squeezes its prey *4.3.3*

**control** the sample in an experiment in which the variables are kept at a base level *F*

**cranium** the skull *5.3.1*

**crustacean** an arthropod that has five or more pairs of legs and two pairs of antennae *4.2.2*

**deciduous tree** a tree that loses its leaves at the end of its growing season *7.3.4*

**decomposer** an organism that breaks down dead plant and animal material to return it to the soil *2.2.2*

**dendrite** the short-branched extension of a neuron that receives messages from other cells *5.3.3*

**dependence** the state of being either physically or psychologically dependent on a substance or behavior *5.4.6*

**dermis** the thick layer of living skin below the epidermis *5.2.3*

**desertification** the making of new deserts through a change in climate or destructive land use *7.3.7*

**detritivore** an organism that feeds on dead matter *7.2.1*

**dicot** a plant that produces two seed leaves *3.1.5*

**diffusion** the movement of particles from an area of higher concentration to an area of lower concentration *1.2.5*

**DNA** the molecule within a cell that carries the genetic information of an organism *1.2.3*

**echolocation** a method of locating objects by interpreting reflected sound waves *4.3.8*

**ecological succession** the process of gradual change in an ecosystem *7.2.4*

**ecology** the science of how organisms interact with each other and their environment *7.1.3*

**ecosystem** a system in which organisms and their environment interact *7.1.3*

**ectotherm** an organism that uses the environment to regulate body temperature *4.3.1*

**electron microscope** a microscope that uses a beam of electrons to produce magnified images *2.1.1*

**element** a pure substance that cannot be broken down into simpler substances by chemical or physical means *5.1.1*

**embryo** an organism in its earliest stages of development *3.2.2*

**endoplasmic reticulum** an organelle that functions as the cell's transportation system *1.2.2*

**endoskeleton** an internal skeleton *4.1.6*

**endospore** a protective capsule that some bacteria form *2.2.2*

**endotherm** an organism that internally regulates its body temperature *4.3.1*

**entomologist** a scientist who studies insects *4.2.5*

**epidermis** the outer layer of cells covering an organism *3.1.5*

**epiphyte** a plant that grows on another plant, which it uses for support but not for nutrients *3.1.4*

**esophagus** the long tube that connects the throat to the stomach *5.1.3*

**eukaryote** an organism composed of one or more cells containing a visible nucleus *1.3.4*

**exoskeleton** an external skeleton *4.2.1*

**exotic species** a species that is not native but has been introduced into an ecosystem *7.3.9*

**extremophile** a microorganism that lives in extreme conditions *2.2.6*

**fallopian tube** the tube that connects each ovary with the uterus *5.4.2*

**flagellum** a thin, whiplike structure that helps an organism move through liquid *2.2.1*

**fragmentation** the separation of a parent plant into parts that develop into whole new plants *3.1.3*

**fronds** a large, divided leaf usually found on ferns *3.1.4*

**fungi** the nongreen organisms that reproduce from spores and absorb food *1.3.2*

**gamete** a male or female reproductive cell *6.1.2*

**gametophyte** the phase of a plant that produces sex cells (gametes) *3.1.1*

**gastropod** a mollusk with a single shell or no shell and a muscular foot *4.1.5*

**gene** a small segment of DNA that carries hereditary information *1.2.3*

**genetic code** the chemical instructions that translate genetic information *6.1.4*

**genetic engineering** the process used by scientists to transfer genes, or parts of DNA, from one organism to another *6.2.1*

**genetically modified organism** (GMO) a living thing whose DNA has been altered by the addition or deletion of genes *6.2.3*

**genetics** the study of how traits are passed from parents to offspring, from one generation to the next *6.1.1*

**genome** the total genetic material of an organism *6.1.7*

**genotype** the combination of an organism's dominant and recessive alleles for a trait *6.1.3*

**gland** an organ that secretes hormones and other chemicals *5.4.1*

**Golgi apparatus** an organelle that receives, packages, and disperses materials *1.2.2*

**gram** the standard unit for mass in the metric system *F*

**gravitropism** the growth of a plant in response to gravity *3.2.4*

**gymnosperm** a vascular cone-bearing plant with seeds that are not enclosed *3.1.1*

**habitat** the place where an organism lives and where all of its needs are met *7.1.1*

**heredity** the passing of traits from parent to offspring *6.1.1*

**herpetology** the study of reptiles and amphibians *4.3.3*

**heterotroph** an organism that obtains food from an outside source *2.2.2*

**heterozygous** having different alleles from each parent for a particular gene *6.1.3*

**HIV** (Human Immunodeficiency Virus) the virus that causes AIDS *2.1.3*

**holdfast** a rootlike structure an organism uses to anchor itself *2.4.3*

**homeostasis** the ability to maintain a stable internal environment *1.1.2*

**homozygous** having the same alleles from both parents for a particular gene *6.1.3*

**humus** partially or totally decayed organic matter *3.1.2*

**hybrid** the offspring produced as a result of cross-fertilization *6.1.3*

**hyphae** the threadlike filaments in fungi that produce enzymes *2.3.1*

**hypothesis** a prediction of what you think will happen and which can be tested to see if it is true *F*

**inbreeding** the mating of closely related organisms *6.2.1*

**inference** an educated guess based on observation *F*

**influenza** a contagious viral infection that causes muscle aches, inflammation of the respiratory system, fever, and chills *2.1.3*

**intertidal zone** the area between land and sea where the tides go in and out *7.3.8*

**karyotype** a tool used by geneticists to analyze the chromosomal characteristics of an individual cell *6.1.6*

**keel** the breastbone of a bird *4.3.7*

**lichen** an organism formed by a symbiotic relationship between a fungus and an alga or a cyanobacterium *1.3.4*

**life science** the study of living things *1.1.1*

**ligament** the connective tissue that holds bones together at moveable joints *5.3.1*

**limiting factor** a condition that limits the number of species or the size of a population in an area *7.3.2*

**lipid** an organic compound that is a fat or an oil *1.1.4*

**liter** the standard unit for the volume of liquid in the metric system *F*

**lysosome** an organelle that breaks down food particles and old cell organelles *1.2.2*

**malaria** a disease caused by a sporozoan that is characterized by periodic attacks of chills and high fevers *2.4.5*

**mammary gland** a structure in female mammals that produces and secretes milk *4.3.8*

**mandible** a mouth part *4.2.2*

**mantle** a growth organ that lines a mollusk's inner shell *4.1.5*

**marrow** the porous inner core of bones *5.3.1*

**marsupial** a mammal whose young are born and then develop further in the mother's pouch *4.3.8*

**maxilliped** the appendage that helps a crustacean eat *4.2.2*

**medusa** the bowl-shaped body plan of a cnidarian *4.1.3*

**meiosis** the cell division process that forms sex cells that contain half the usual number of chromosomes *1.2.6*

**melanin** the pigment that determines skin and hair color *5.2.3*

**metabolism** the sum of chemical activities taking place inside an organism *1.1.2*

**metamorphosis** a series of physical changes that certain animals undergo *4.2.5*

**meter** the standard unit for length in the metric system *F*

**metric system** a universal system of measurement based on the number 10 used by scientists around the world *F*

**migrate** to travel from one place to another in response to seasons or environmental conditions *4.3.7*

**millipede** a round-bodied arthropod with two pairs of legs on each body segment *4.2.3*

**mitochondrion** an organelle that produces energy for the cell *1.2.2*

**mitosis** the cell division process that forms new cells with an identical copy of the parent's chromosomes *1.2.6*

**mold** a type of fungus that grows on food or in damp places *2.3.2*

**molecule** a particle consisting of two or more atoms chemically bonded together *5.1.1*

**moment-of-truth defense** a behavior used by prey to startle a predator, giving the prey a chance to escape *7.2.3*

**monocot** a plant that produces one seed leaf *3.1.5*

**mushroom** a fungus with a cap on top of a stalk *2.3.2*

**mutagen** a physical or chemical agent that causes mutations to occur *6.1.4*

**mutation** a change in the sequence of one or more nucleotides in a DNA molecule *6.1.4*

**mutualism** the relationship between two organisms that live and work together for the benefit of both *2.3.3*

**mycelium** the large mass of hyphae that forms the growing structure of fungi *2.3.1*

**native species** a species that is native to a particular ecosystem *7.3.9*

**nekton** the free-swimming ocean organisms *7.3.8*

**nematocyst** the special stinging cell of cnidarians *4.1.3*

**nephron** a tiny filtering device in the kidney *5.2.2*

**neritic zone** the first ocean zone away from the shoreline *7.3.8*

**nitrogen-fixing bacteria** a type of bacteria that turns nitrogen into nitrogen compounds *2.2.3*

**nonvascular** a type of plant that does not have vessels to transport materials *3.1.1*

**nucleic acid** an organic compound that contains genetic information *1.1.4*

**nucleotide** the basic structural unit of DNA *1.2.3*

**nymph** a stage of incomplete metamorphosis that resembles the adult insect *4.2.5*

**observation** something noticed through the senses *F*

**oceanic zone** the zone that extends from the edge of the continental shelf and covers the entire open ocean *7.3.8*

**operculum** the covering that protects the gills in bony fish *4.3.5*

**organic compound** a compound that contains carbon *1.1.4*

**osmosis** the diffusion of water through a membrane *1.2.5*

**ovary** the rounded base of a pistil, or the gland in humans or animals where eggs develop *3.2.2, 5.4.1*

**ovulation** the release of an egg from the ovary *5.4.2*

**ovule** the structure that contains the egg cell of a seed plant *3.2.2*

**pampas** the South American grassland *7.3.6*

**pasteurization** the process of heating liquids to kill harmful bacteria *2.2.5*

**pathogen** a disease-causing organism *2.2.5*

**pelvis** the hip bones and lower part of the backbone *5.3.1*

**perennial** a flowering plant that lives for more than two years *3.1.1*

**peripheral nervous system** the nerves and sensory receptors *5.3.3*

**peristalsis** the wavelike muscle contractions that push food through the digestive system *5.1.3*

**permafrost** a layer of permanently frozen soil *7.3.2*

**phenotype** the observable traits of an organism *6.1.3*

**phloem** the vascular plant tissue that transports food from the leaves to the rest of the plant *3.1.5*

**photic zone** the ocean zone where photosynthesis occurs *7.3.8*

**photosynthesis** the process by which plants use light energy to make food *3.2.1*

**phototropism** the growth of a plant in response to light *3.2.4*

**pistil** the female reproductive structure of a flower *3.2.2*

**pith** the very center of a woody stem *3.1.5*

**pituitary gland** a gland in the brain that controls other glands *5.4.1*

**placenta** a tissue in the womb of certain mammals that nourishes the developing young *4.3.8*

**plankton** microscopic organisms that live near the surface of the water *2.4.2*

**plasma** the thin, yellow fluid part of the blood *5.1.2*

**platelet** a cell fragment that helps in clotting blood *5.1.2*

**pollutant** a substance or condition in the air, soil, or water that is harmful to living things *7.1.6*

**polyp** the vase-shaped body plan of a cnidarian *4.1.3*

**population** a group of organisms of one species that live in the same place at the same time *7.1.2*

**population density** the number of individual organisms of one species in a community *7.1.2*

**prairie** the North American grassland *7.3.6*

**preferred habitat** the place where an organism has an advantage over other organisms *7.1.1*

**preservative** a chemical added to food to slow the growth of bacteria and mold *2.2.5*

**primary consumer** an herbivore *7.2.1*

**primary succession** the series of changes occurring in a newly formed, barren habitat *7.2.4*

**prokaryote** a unicellular microorganism that lacks a distinct nucleus *1.3.3*

**protist** a eukaryotic organism that cannot be classified as a fungi, a plant, or an animal *1.3.4*

**protozoa** the animallike protists *2.4.4*

**pseudopod** a flowing extension of cytoplasm that amoebas use to move and obtain food *2.4.4*

**rabies** a potentially fatal viral disease transmitted by the bite of an infected animal *2.1.3*

**radula** a hard, strong band used to scrape food or to bore into objects *4.1.5*

**regeneration** the regrowth of a lost body part *4.1.2*

**replication** the process in which DNA molecules make exact duplicates *1.2.3*

**retrovirus** a virus that carries its genetic materials in the form of RNA *2.1.2*

**rhizome** a horizontal underground stem *3.1.4*

**ribosome** an organelle that produces proteins rhizome *1.2.2*

**RNA** a nucleic acid made of ribose rather than deoxyribose *6.1.4*

**rod** a light-sensitive cell in the eye used in noncolor vision and low-light situations *5.3.4*

**saprophyte** an organism that eats dead or decaying materials *2.4.6*

**savanna** the African grassland scattered with trees and shrubs *7.3.6*

**scientific law** a generalization based on observations that describe the ways an object behaves under specific conditions *F*

**scientific method** the series of steps that scientists follow when they investigate problems or try to answer questions *F*

**sebaceous gland** a gland in the dermis that produces sebum *5.2.3*

**secondary consumer** an animal that eats herbivores *7.2.1*

**secondary succession** the series of changes occurring in an area where the existing ecosystem has been disturbed *7.2.4*

**sediment** the solid material that settles to the bottom of a liquid *7.1.6*

**selective breeding** the intentional crossing of plants or animals that have desirable traits to produce offspring that have those traits *6.2.1*

**semen** a mixture of sperm cells and secretions *5.4.2*

**sepal** the leaflike structure that covers and protects the immature flower *3.2.2*

**sex-linked trait** a trait that is carried on the X chromosome *6.1.7*

**sexual reproduction** a form of reproduction in which offspring arise from two parents *1.1.2*

**somatic cell** a cell that has the full number of chromosomes for that organism *6.1.5*

**species** a group of organisms that can mate with one another and produce fertile offspring *1.3.2*

**spicule** the needlelike skeletal materials inside a sponge *4.1.2*

**spleen** a lymph organ that acts as a large blood filter *5.1.4*

**spore** a small reproductive cell that can develop into an adult without fusing with another cell *2.3.1*

**sporophyte** the spore-producing phase of a plant *3.1.1*

**sporozoan** a heterotrophic protist that cannot move on its own *2.4.5*

**stamen** the male reproductive structure of the flower *3.2.2*

**steppe** the European and Asian grassland *7.3.6*

**stigma** the tip of the pistil *3.2.2*

**stomata** the openings in a leaf through which gases pass *3.1.5*

**striated muscle** a type of muscle marked by light and dark stripes or bands *5.3.2*

**style** the long, slender part of the pistil *3.2.2*

**sweat gland** a gland in the dermis that produces sweat *5.2.3*

**swim bladder** a balloonlike organ that bony fish have that allows them to ascend or descend in water *4.3.5*

**symbiosis** a relationship in which one organism lives on, near, or inside another organism and at least one organism benefits *2.2.3*

**synapse** a gap between the axon of one neuron and the dendrites of another neuron *5.3.3*

**taproot** a large root that grows downward and has smaller roots branching off from it *3.1.5*

**taxonomy** the scientific classification of organisms *1.3.1*

**technology** the application of science *1.1.1*

**tendon** the connective tissue that attaches muscles to bones *5.3.1*

**tertiary consumer** an animal that eats primary carnivores *7.2.1*

**theory** an explanation of the scientific laws *F*

**thigmotropism** the growth of a plant in response to touch *3.2.4*

**thorax** the center body segment of an insect or other arthropod *4.2.1*

**trachea** the tube that carries air to and from the lungs *5.2.1*

**transpiration** the loss of water vapor by plants through stomata *3.2.1*

**trophic level** a group of organisms that share the same position in a food chain *7.2.1*

**tropism** a change in the growth of a plant in response to a stimulus *3.2.4*

**tube feet** the bulblike suction structures lining the arms of an echinoderm *4.1.6*

**tympanic membrane** the eardrum *5.3.4*

**unicellular** one-celled *2.4.1*

**ureter** the tube through which urine flows from the kidney to the bladder *5.2.2*

**urethra** the tube that carries urine from the bladder to the outside of the body *5.2.2*

**uterus** the muscular organ in which a fetus develops *5.4.2*

**vacuole** an organelle that stores materials *1.2.2*

**vagina** the tube that leads from the uterus to the outside of a female's body *5.4.2*

**variable** a changeable factor that could influence an experiment's outcome *F*

**vascular** a type of plant that has vessels to transport materials *3.1.1*

436

**vector** a carrier used to transfer genetic material from a donor organism to a target cell *6.2.1*

**ventricle** a lower chamber of the heart *5.1.2*

**viral infection** the penetration of a virus or its nucleic acid into a host cell *2.1.2*

**virus** a tiny particle that contains nucleic acid encased in protein *2.1.1*

**water hardness** the amount of magnesium and calcium in water *7.1.6*

**wetland** an area of land where the water level is near or above the ground surface for most of the year *7.1.5*

**withdrawal** the physical and psychological symptoms experienced when an addictive substance or behavior is discontinued *5.4.6*

**xylem** the vascular tissue that transports water and minerals from the roots throughout a plant *3.1.5*

**yeast** a unicellular fungus *2.3.2*

**zoologist** a scientist who studies animals *1.3.4*

food chain 7.2.1

forensic scientist 1.2.3

fragmentation 3.1.3

Franklin, Rosalind 1.2.3

fronds 3.1.4

fungi 1.3.2
structure of 2.3.1
types of 2.3.2

 G

gamete 6.1.2

gametophyte 3.1.1

gastroenterologist 5.1.1

gastropod 4.1.5

gene 1.2.3

general surgeon 5.1.1

genetic code 6.1.4

genetic counseling 6.1.7

genetic engineering 6.2.1

genetically modified
organism (GMO)
6.2.3

genetics 6.1.1
human 6.1.6, 6.1.7
plants 6.1.3

genome 6.1.7

genotype 6.1.3

gerontologist 5.1.1

gland 5.4.1

GMO 6.2.3

Golgi apparatus 1.2.2

Goodall, Jane 4.3.8

Gram, Hans
Christian 2.2.1

grassland 7.3.6

gravitropism 3.2.4

Great Chain of
Being 1.3.1

gymnosperm 3.1.1

gynecologist 5.1.1

 H

habitat 7.1.1
destruction of

heredity 6.1.2, 6.1.7, 6.2.4

Harvey, William 5.1.2

heredity 6.1.1

herpetology 4.3.3

heterotroph 2.2.2

heterozygous 6.1.3

HIV (Human
Immunodeficiency
Virus)
2.1.3

holdfast 2.4.3

homeostasis 1.1.2

homozygous 6.1.3

Hooke, Robert 1.2.1

hornwort 3.1.3

horsetail 3.1.4

horticulturalist 3.2.4

human body
development of 5.4.2
organization of 4.3.8, 5.1.1

human genome 6.2.4

humus 3.1.2

hybrid 6.1.3

hyphae 2.3.1

 I

immune system 5.1.4

inbreeding 6.2.1

influenza 2.1.3

insect 4.2.5

intertidal zone 7.3.8

invertebrate
cnidarian 4.1.3
echinoderm 4.1.6
mollusk 4.1.5
sponge 4.1.2
worm 4.1.4

 J

Jansen, Hans and
Zacharius 1.2.1

 K

karyotype 6.1.6

keel 4.3.7

 L

landscape architect 3.2.4

lichen 1.3.4

life science 1.1.1

ligament 5.3.1

limiting factor 7.3.2

Linnaeus, Carolus 1.3.1

lipid 1.1.4

Lister, Joseph 5.4.5

liverwort 3.1.3

living things 1.1.3, 1.1.4
characteristics of 1.1.2,
7.1.1
chemistry of 1.1.4
needs of 1.1.3, 7.1.1

lysosome 1.2.2

 M

malaria 2.4.5

mammal 4.3.8

mammary gland 4.3.8

mandible 4.2.2

mantle 4.1.5

marrow 5.3.1

marsupial 4.3.8

maxilliped 4.2.2

medusa 4.1.3

meiosis 1.2.6

melanin 5.2.3

metabolism 1.1.2

metamorphosis 4.2.5

methanogen 2.2.6

microbiologist 2.2.1

migrate 4.3.7

millipede 4.2.3

mitochondrion 1.2.2

mitosis 1.2.6

mold 2.3.2

molecule 5.1.1

mollusk 4.1.5, 7.3.8

moment-of-truth
    defense 7.2.3

monocot 3.1.5

moss 3.1.4, 7.3.1, 7.3.2

muscular system 5.3.2

mushroom 2.3.2

mutagen 6.1.4

mutation 6.1.4

mutualism 2.3.3

mycelium 2.3.1

native species 7.3.9

nekton 7.3.8

nematocyst 4.1.3

nephrologist 5.2.2

nephron 5.2.2

neritic zone 7.3.8

nervous system 5.3.3

neurologist 5.1.1

nitrogen-fixing
    bacteria 2.2.3

nonvascular 3.1.1

nucleic acid 1.1.4

nucleotide 1.2.3

nutrition 5.4.4

nymph 4.2.5

ocean 7.3.8

oceanic zone 7.3.8

operculum 4.3.5

ophthalmologist 5.1.1

organic compound 1.1.4

organism 6.2.3, 7.2.1

orthopedic surgeon 5.1.1

osmosis 1.2.5

ovary 3.2.2

ovulation 5.4.2

ovule 3.2.2

P

pampas 7.3.6

park ranger 7.3.3

Pasteur, Louis 1.2.1

pasteurization 2.2.5

pathogen 2.2.5

pediatrician 5.1.1

pelvis 5.3.1

Penfield, Wilder 5.3.3

perennial 3.1.1

peripheral nervous
    system 5.3.3

peristalsis 5.1.3

permafrost 7.3.2

phenotype 6.1.3

phloem 3.1.5

photic zone 7.3.8

photosynthesis 3.2.1

phototropism 3.2.4

pistil 3.2.2

pith 3.1.5

pituitary gland 5.4.1

placenta 4.3.8

plankton 2.4.2

plant physiologist 3.2.4

plants
    behavior of 3.2.4
    life processes of 3.2.1
    needs of 3.1.2
    types of 3.1.1

plasma 5.1.2

plastic surgeon 5.1.1

platelet 5.1.2

pollutant 7.1.6

polyp 4.1.3

population 7.1.2

population density 7.1.2

prairie 7.3.6

preferred habitat 7.1.1

preservative 2.2.5

primary consumer 7.2.1

primary succession 7.2.4

prokaryote 1.3.3

**protist** 1.3.4, 2.4.1
  animallike 2.4.4
  funguslike 2.4.6
  plantlike 2.4.2

**protozoa** 2.4.4

**pseudopod** 2.4.4

### R

**rabies** 2.1.3

**radula** 4.1.5

**regeneration** 4.1.2

**replication** 1.2.3

**reproduction** 1.2.6, 2.1.2,
  3.2.2, 3.2.3, 5.4.2, 6.1.2

**reproductive system** 5.4.2

**reptile** 4.3.3

**respiratory system** 5.2.1

**retrovirus** 2.1.2

**rhizome** 3.1.4

**ribosome** 1.2.2

**RNA (ribonucleic
  acid)** 6.1.4

**rod** 5.3.4

### S

**Salk, Jonas** 2.1.3

**saprophyte** 2.4.6

**savanna** 7.3.6

**Schleiden, Matthias** 1.2.1

**Schulz, Heide** 2.2.1

**Schwann, Theodor** 1.2.1

**sebaceous gland** 5.2.3

**secondary consumer** 7.2.1

**secondary
  succession** 7.2.4

**sediment** 7.1.6

**seed plants** 3.1.5

**selective breeding** 6.2.1

**semen** 5.4.2

**senses** 5.3.4

**sepal** 3.2.2

**sex-linked trait** 6.1.7

**sexual reproduction** 1.1.2

**skeletal system** 5.3.1

**skin** 5.2.3

**somatic cell** 6.1.5

**species** 1.3.2

**spicule** 4.1.2

**spleen** 5.1.4

**sponge** 4.1.2

**spontaneous
  generation** 1.1.2

**spore** 2.3.1

**sporophyte** 3.1.1

**sporozoan** 2.4.5

**sports medicine
  specialist** 5.1.1

**stamen** 3.2.2

**steppe** 7.3.6

**Stevens, Nettie** 6.1.5

**stigma** 3.2.2

**stimulus** 5.3.3

**stomata** 3.1.5

**striated muscle** 5.3.2

**style** 3.2.2

**succession** 7.2.4

**sweat gland** 5.2.3

**swim bladder** 4.3.5

**symbiosis** 2.2.3

**synapse** 5.3.3

### T

**taproot** 3.1.5

**taxonomy** 1.3.1

**technology** 1.1.1

**tendon** 5.3.1

**tertiary consumer** 7.2.1

**thigmotropism** 3.2.4

**thorax** 4.2.1

**trachea** 5.2.1

**trait** 6.1.1

**transpiration** 3.2.1

**transport** 1.2.5

**trophic level** 7.2.1

**tropism** 3.2.4

**tube feet** 4.1.6

**tundra** 7.3.2

**tympanic membrane** 5.3.4

### U

**unicellular** 2.4.1

**ureter** 5.2.2

**urethra** 5.2.2

**urinary system** 5.2.2

**urologist** 5.1.1

**uterus** 5.4.2

### V

**vacuole** 1.2.2

**vagina** 5.4.2

**Van Leeuwenhoek,
  Antonie** 1.2.1

**vascular** 3.1.1

**vector** 6.2.1

**ventricle** 5.1.2

**vertebrate**
 ectotherm 4.3.2
 endotherm 4.3.6

**veterinarian** 4.3.1

**viral infection** 2.1.2

**Virchow, Rudolf** 1.2.1

**virus**
 reproduction of 2.1.2
 structure of 2.1.1

**water hardness** 7.1.6

**Watson, James** 1.2.3

**wetland** 7.1.5

**Whittaker, Robert
 H.** 1.3.1

**withdrawal** 5.4.6

**worm** 4.1.4

**xylem** 3.1.5

**yeast** 2.3.2

**zoologist** 1.3.4